INTERNET THEORY, TECHNOLOGY AND APPLICATIONS

BROADBAND

DEPLOYMENT, ACCESS AND THE DIGITAL DIVIDE

INTERNET THEORY, TECHNOLOGY AND APPLICATIONS

Additional books and e-books in this series can be found on Nova's website under the Series tab.

INTERNET THEORY, TECHNOLOGY AND APPLICATIONS

BROADBAND

DEPLOYMENT, ACCESS AND THE DIGITAL DIVIDE

ORLANDO HARRISON
EDITOR

snova
New York

Library of Congress Cataloging-in-Publication Data

ISBN: 978-1-53616-409-1

Published by Nova Science Publishers, Inc. † New York

Contents

PREFACE

Broadband—whether delivered via fiber, cable modem, mobile or fixed wireless, copper wire, or satellite—is increasingly the technology underlying telecommunications services such as voice, video, and data. Chapter 1 focuses on the gaps specifically related to broadband availability and adoption. How broadband is defined and characterized in statute and in regulation can have a significant impact on federal broadband policies and how federal resources are allocated to promote broadband deployment in unserved and underserved areas as discussed in chapter 2.

The move to place restrictions on the owners of the networks that comprise and provide access to the internet, to ensure equal access and nondiscriminatory treatment, is referred to as "net neutrality." While there is no single accepted definition of net neutrality most agree that any such definition should include the general principles that owners of the networks that comprise and provide access to the internet should not control how consumers lawfully use that network; and should not be able to discriminate against content provider access to that network as reported in chapters 3 and 4.

The "digital divide" is a term that has been used to characterize a gap between "information haves and have-nots," or in other words, between those Americans who use or have access to telecommunications and information technologies and those who do not. Chapter 5 focuses on the one important subset of the digital divide debate which concerns high-speed internet access and advanced telecommunications services, also known as broadband.

While there are many examples of rural communities with state-of-the-art telecommunications facilities, recent surveys and studies have indicated that, in general, rural areas tend to lag behind urban and suburban areas in broadband deployment. The Rural Utilities Service (RUS) at the U.S. Department of Agriculture (USDA) houses three ongoing assistance programs exclusively created and dedicated to financing broadband deployment: the Rural Broadband Access Loan and Loan Guarantee Program, the Community Connect Grant Program, and the ReConnect Program. Chapter 6 discusses each of these programs.

Tribal lands are generally in remote and rugged areas and broadband access can help residents develop online businesses, access telemedicine services, and use online educational tools. However, residents of tribal lands have lower levels of broadband access than residents of non-tribal lands. Chapters 7 through 11 report on the status of broadband on tribal lands.

Chapter 1 - Broadband—whether delivered via fiber, cable modem, mobile or fixed wireless, copper wire, or satellite—is increasingly the technology underlying telecommunications services such as voice, video, and data. Since the initial deployment of high-speed internet in the late 1990s, broadband technologies have been deployed throughout the United States primarily by the private sector. These providers include telephone, cable, wireless, and satellite companies as well as other entities that provide broadband services to residential, business, and institutional customers. The Federal Communications Commission's (FCC's) 2010 National Broadband Plan identified broadband as a basic infrastructure necessary for improving economic growth, job creation, civic engagement, global competitiveness, and a better quality of life. Broadband enables or enhances applications such as e-commerce, telemedicine, distance education, telework, entertainment, public safety, and energy conservation. Increasingly viewing broadband as a basic infrastructure, Congress and successive Administrations have focused on addressing gaps specifically related to broadband availability and adoption. Broadband availability refers to whether or not broadband service is offered, while broadband adoption refers to the extent to which American households actually subscribe to and use broadband.

Chapter 2 - Broadband—whether delivered via fiber, cable modem, copper wire, satellite, or mobile wireless—is increasingly the technology underlying telecommunications services such as voice, video, and data. Since the initial deployment of high-speed internet in the late 1990s, broadband technologies have been deployed throughout the United States primarily by the private sector. These providers include telephone, cable, wireless, and satellite companies as well as other entities that provide commercial telecommunications services to residential, business, and institutional customers. How broadband is defined and characterized in statute and in regulation can have a significant impact on federal broadband policies and how federal resources are allocated to promote broadband deployment in unserved and underserved areas. One way broadband can be defined is by setting a minimum threshold speed for what constitutes "broadband service." Section 706 of the Telecommunications Act of 1996 requires the Federal Communications Commission (FCC) to regularly initiate an inquiry concerning the availability of broadband to all Americans and to determine whether broadband is "being deployed to all Americans in a reasonable and timely fashion." If the determination is negative, the act directs the FCC to "take immediate action to accelerate deployment of such capability by removing barriers to infrastructure investment and by promoting competition in the telecommunications market." Starting in 1999, there have been 11 Section 706 reports, each providing a snapshot and assessment of broadband deployment. As part of this assessment, and to help determine whether broadband is being deployed in "a reasonable and timely fashion," the FCC has set a minimum broadband speed that essentially serves as the benchmark the FCC uses to determine what it considers broadband service for the purposes of its Section 706 determination. In 2015 the FCC, citing changing broadband usage patterns and multiple devices using broadband within single households, raised its minimum fixed broadband benchmark speed from 4 Mbps (download)/1 Mbps (upload) to 25 Mbps/3 Mbps. On August 8, 2017, the FCC adopted and released its *Thirteenth Section 706 Notice of Inquiry* (*NOI*). One proposal under consideration was establishing a lower benchmark speed specifically for mobile broadband. Stakeholders who supported an FCC determination that broadband is *not* being deployed in a reasonable and timely fashion generally opposed lowering the broadband benchmark by considering the presence of either fixed (at 25 Mbps/3 Mbps) or mobile (at 10 Mbps/1 Mbps) broadband as an indication that an area has adequate broadband service. On the other hand, stakeholders who supported an FCC determination that broadband *is* being deployed in a reasonable and timely fashion generally supported changing the FCC's broadband benchmark methodology to include the presence of either fixed or mobile broadband as an indication that an area is receiving adequate broadband service. On February 2, 2018, the FCC adopted and released its latest 706 report, the *2018 Broadband Deployment Report.* While the FCC concluded that advanced telecommunications capability is being deployed to all Americans in a reasonable and timely fashion, it concluded that adoption of a single mobile benchmark is currently unworkable and that mobile broadband service is not a full substitute for fixed service at this time. As broadband technology advances, commercially available download and upload speeds will likely increase, and the level at which broadband benchmark threshold speeds should be set is likely to remain controversial. Accordingly, the FCC's annual Section 706 determination is likely to be an issue in Congress as long as it is seen by stakeholders as providing a justification for current or future FCC regulatory or deregulatory policies.

Chapter 3 - The move to place restrictions on the owners of the networks that comprise and provide access to the internet, to ensure equal access and nondiscriminatory treatment, is referred to as "net neutrality." While there is no single accepted definition of net neutrality most agree that any such definition should include the general principles that owners of the networks that comprise and provide access to the internet should not control how consumers lawfully use that network; and should not be able to discriminate against content provider access to that network. Determining the appropriate framework to ensure an open internet is central to the debate over broadband access, and is an issue that the Federal Communications Commission (FCC) has been grappling with for decades. Some policymakers contend that more proscriptive regulations, such as those contained in the FCC's 2015 Open Internet Order (2015 Order), are necessary to protect the marketplace from potential abuses which could threaten the net neutrality concept. Others contend that existing laws and the current, less restrictive approach, contained in the FCC's 2017 Restoring Internet Freedom Order (2017 Order), provide a more suitable framework. There is also a growing consensus that Congress should amend the 1934 Communications Act, as amended, (Communications Act) to address the debate.

Chapter 4 - As congressional policymakers continue to debate telecommunications reform, a major discussion point revolves around what approach should be taken to ensure unfettered access to the internet. The move to place restrictions on the owners of the networks that compose and provide access to the internet, to ensure equal access and nondiscriminatory treatment, is referred to as "net neutrality." There is no single accepted definition of "net neutrality," but most agree that any such definition should include the general principles that owners of the networks that compose and provide access to the internet should not control how consumers lawfully use that network, and they should not be able to discriminate against content provider access to that network. The Federal Communications Commission (FCC) in its February 26, 2015, open meeting voted 3-2, along party lines, to adopt open internet rules and released these rules on March 12, 2015. One of the most controversial aspects of the rules was the decision to reclassify broadband internet access service (BIAS) as telecommunications service under Title II, thereby subjecting internet service providers to a more stringent regulatory framework. With limited exceptions, the rules went into effect June 12, 2015. Various parties challenged the legality of the FCC's 2015 Open Internet Order, but the U.S. Court of Appeals for the D.C. Circuit, in a June 14, 2016, ruling, voted (2-1) to uphold the legality of all aspects of the 2015 FCC Order. A petition for full U.S. Appeals Court review was denied and a subsequent petition for U.S. Supreme Court review was declined. The FCC on December 14, 2017, adopted (3-2) an Order that largely reverses the 2015 regulatory framework. The 2017 Order, among other things, reverses the 2015 classification of BIAS as a telecommunications service under Title II of the Communications Act, shifts much of the oversight from the FCC to the Federal Trade Commission and the Department of Justice, and provides for a less regulatory approach. This action has once again opened up the debate over what the appropriate framework is to ensure an open internet. Reaction to the 2017 Order has been mixed. Some see the 2015 FCC rules as regulatory overreach and welcome a more "light-touch" approach, which they feel will stimulate broadband investment, deployment, and innovation. Others support the 2015 regulations and feel that their reversal will result in a concentration of power to the detriment of content, services, and applications providers, as well as consumers, and refute the claim that these regulations have had a negative impact on broadband investment, expansion, or innovation. The 2017 Order was published in the *Federal Register* on February 22, 2018, and went into effect on June 11, 2018. *Federal Register* publication triggered timelines for both court challenges and Congressional Review Act (CRA) consideration. Petitions for review have been consolidated in the U.S. Court of Appeals, D.C. Circuit. CRA resolutions (S.J.Res. 52, H.J.Res. 129) to overturn the 2017 Order were introduced in the 115th Congress. S.J.Res. 52 passed (52-47) the Senate, but H.J.Res. 129 was not considered in the House. Additional bills to provide a regulatory framework to outline FCC authority over broadband internet access services were introduced, but not acted on, in the 115th Congress. Debate over what the appropriate regulatory framework should be for broadband access has continued in the 116th Congress. Two bills (H.R. 1644, S. 682) add a new title to the Communications Act that overturns the 2017 Order and restores the 2015 Order. An amended version of H.R. 1644 passed (232-190) the House on April 10, 2019. Additional bills (H.R. 1006, H.R. 1096, H.R. 1101, and H.R. 1860) that address the net neutrality debate have also been introduced.

Chapter 5 - The "digital divide" is a term that has been used to characterize a gap between "information haves and have-nots," or in other words, between those Americans who use or have access to telecommunications and information technologies and those who do not. One important subset of the digital divide debate concerns high-speed internet access and advanced telecommunications services, also known as broadband. Broadband is provided by a series of technologies (e.g., cable, telephone wire, fiber, satellite, mobile and fixed wireless) that give users the ability to send and receive data at volumes and speeds necessary to support a number of applications including voice communications, entertainment, telemedicine, distance education, telework, ecommerce, civic engagement, public safety, and energy conservation. Broadband technologies are currently being deployed primarily by the private sector throughout the United States. While the numbers of new broadband subscribers continue to grow, studies and data suggest that the rate of broadband deployment in urban/suburban and high-income areas is outpacing deployment in rural and low-income areas. Some policymakers, believing that disparities in broadband access across American society could have adverse economic and social consequences on those left behind, assert that the federal government should play a more active role to address the "digital divide" in broadband access. With the conclusion of the grant and loan awards established by the American Recovery and Reinvestment Act of 2009 (P.L. 111-5), there remain two primary ongoing federal vehicles which direct federal money to fund broadband infrastructure: the broadband and telecommunications programs at the Rural Utilities Service (RUS) of the U.S. Department of Agriculture and the Universal Service Fund (USF) programs under the Federal Communications Commission (FCC). RUS broadband programs were reauthorized and modified by the 2018 farm bill. The USF High Cost Fund is undergoing a major transition to the Connect America Fund, which is targeted to the deployment, adoption, and utilization of both fixed and mobile broadband. Meanwhile, the Consolidated Appropriations Act, 2018 (P.L. 115-141) appropriated $600 million to RUS to conduct a new broadband loan and grant pilot program (called the ReConnect Program), and appropriated $7.5 million to the National Telecommunications and Information Administration to update the national broadband availability map in coordination with the FCC. Additionally, P.L. 115-141 contained provisions seeking to facilitate deployment of broadband infrastructure on federal property, as well as making more spectrum available for wireless broadband. To the extent that the 116th Congress may consider various options for further encouraging broadband deployment and adoption, a key issue is how to strike a balance between providing federal assistance for unserved and underserved areas where the private sector may not be providing acceptable levels of broadband service, while at the same time minimizing any deleterious effects that government intervention in the marketplace may have on competition and private sector investment.

Chapter 6 - Given the large potential impact broadband access may have on the economic development of rural America, concern has been raised over a "digital divide" between rural and urban or suburban areas with respect to broadband deployment. While there are many examples of rural communities with state-of-the-art telecommunications facilities, recent surveys and studies have indicated that, in general, rural areas tend to lag behind urban and suburban areas in broadband deployment. According to the Federal Communications Commission's *Communications Marketplace Report*, as of 2017, 24% of Americans in rural areas lacked coverage from fixed terrestrial 25 Mbps/3 Mbps broadband, as compared to only 1.5% of Americans in urban areas. The comparatively lower population density of rural areas is likely a major reason why broadband is less deployed than in more highly populated suburban and urban areas. Particularly for wireline broadband technologies—such as cable modem and fiber—the greater the geographical distances among customers, the larger the cost to serve those customers. The Rural Utilities Service (RUS) at the U.S. Department of Agriculture (USDA) houses three ongoing assistance programs exclusively created and dedicated to financing broadband deployment: the Rural Broadband Access Loan and Loan Guarantee Program, the Community Connect Grant Program, and the ReConnect Program. Additionally, the Telecommunications Infrastructure Loan and Loan Guarantee Program (previously the Telephone Loan Program) funds broadband deployment in rural areas. Distance Learning and Telemedicine (DLT) grants—while not principally supporting connectivity—fund equipment and software that operate via telecommunications to rural end-users of telemedicine and distance learning applications. The Consolidated Appropriations Act, 2019 (P.L. 116-6) provided $5.83 million to subsidize a rural broadband loan level of $29.851 million, $30 million to Community Connect broadband grants, $47 million for DLT grants, and $1.725 million in loan subsidies for a total loan level of $690 million for the Telecommunications Infrastructure Loan and Loan Guarantee Program. P.L. 116-6 also provided $550 million for the ReConnect Program, which is in addition to the $600 million provided in the 2018 Consolidated Appropriations Act. The Administration's

FY2020 budget proposal requested zero funding for Rural Broadband Access Loans, $200 million for the ReConnect Program, $1.933 million in budget authority to subsidize a loan level of $690 million for Telecommunications Infrastructure Loans and Loan Guarantees, $30 million for Community Connect Grants, and $43.6 million for Distance Learning and Telemedicine Grants. On December 20, 2018, the President signed the 2018 farm bill (P.L. 115-334, Agriculture Improvement Act of 2018). Regarding the RUS broadband programs, the act includes provisions authorizing a grant component in combination with the broadband loan program; increasing the annual authorization level from $25 million to $350 million; raising the proposed service area eligibility threshold of unserved households from 15% to 50% for broadband loans; authorizing grants, loans, and loan guarantees for middle mile infrastructure; directing improved federal agency broadband program coordination; and providing eligible applicants with technical assistance and training to prepare applications. In the 116th Congress, appropriations will determine the extent to which these programs will be funded.

Chapter 7 - Broadband furthers economic development, educational attainment, and public health and safety; however, residents of tribal lands have lower levels of broadband access relative to the U.S. population. Congress has prioritized identifying and targeting funds to unserved areas. FCC uses data from broadband providers to develop maps and reports depicting broadband availability in the United States, with specific information on tribal lands. GAO was asked to review FCC's efforts to collect broadband data for tribal lands. This chapter examines the extent to which: (1) FCC's approach to collecting broadband data accurately captures broadband access on tribal lands and (2) FCC obtains tribal input on the data. GAO interviewed stakeholders from 25 tribal governments or tribally owned providers, and visited nine tribal lands. The selected tribes varied geographically and in levels of broadband availability, among other characteristics. GAO also reviewed FCC's rulemakings on broadband data and interviewed other tribal stakeholders, FCC officials, and 13 non-tribal broadband providers selected to include a diversity of technologies. Provider and tribal interviews were based on non- generalizable samples.

Chapter 8 - In 2018, FCC estimated that 35 percent of Americans living on tribal lands lack broadband service compared to 8 percent of Americans overall. Various federal programs support increasing broadband deployment in unserved areas, including tribal lands. Tribes can form partnerships with private sector companies and others to deploy broadband infrastructure on tribal lands. GAO was asked to provide information on these partnerships. This chapter discusses (1) examples and outcomes of tribal partnership arrangements, (2) the amount of federal funding provided to tribal entities for broadband deployment, and (3) stakeholder-identified barriers that tribes face in obtaining federal funding and the extent to which federal agencies have addressed those barriers. GAO identified partnerships by reviewing federally funded broadband projects that included a partnership component; analyzed federal funding dedicated to broadband deployment; interviewed agency and tribal government officials, tribal associations, tribally owned broadband providers, and industry stakeholders; and assessed RUS's efforts to address the regulatory funding barriers tribes may face. The information presented is illustrative and is not generalizable to all tribes or all partnerships.

Chapter 9 - This is an edited, reformatted and augmented version of United States Government Accountability Office; Report to Congressional Requesters, Publication No. GAO-19-134T, dated October 2018.

Chapter 10 - In 2018, FCC estimated that 35 percent of Americans living on tribal lands lack broadband service compared to 8 percent of Americans overall. Broadband service can be delivered through wireless technologies using radio frequency spectrum. According to FCC, increasing tribal access to spectrum would help expand broadband service on tribal lands. GAO was asked to review spectrum use by tribal entities— tribal governments and tribally owned telecommunications providers. This chapter examines (1) tribal entities' ability to obtain and access spectrum to provide broadband services and the reported barriers that may exist, and (2) the extent to which FCC promotes and supports tribal efforts to obtain and access spectrum. GAO interviewed 16 tribal entities that were using wireless technologies. Selected entities varied geographically, among other characteristics. GAO analyzed FCC's license and auction data as of September 2018, reviewed FCC's rulemakings on spectrum for broadband services, and interviewed other tribal and industry stakeholders and FCC officials. The information presented is not generalizable to all tribes or industry participants.

Chapter 11 - Tribal areas and communities continue to lag behind other areas and segments of American society with respect to broadband and telecommunications services. High poverty rates and low income levels in tribal lands—along with the fact that many tribal communities are located in remote rural areas (often with rugged terrain)—are major factors that may explain why tribal areas have comparatively poor levels of broadband access, and why providers may lack an economic incentive to serve those areas. Until recently, data on tribal broadband deployment had been scarce. However, the Federal Communications Commission (FCC) and the Department of Commerce have begun to collect and compile data on tribal broadband deployment. The most recent data show that, as of December 31, 2017, approximately 32% of Americans living on tribal lands lacked access to broadband at speeds of at least 25 Mbps download/3 Mbps upload. This compares unfavorably to about 6% of all Americans lacking access to broadband at those speeds. Tribal areas that are the most lacking in broadband service are rural Alaskan villages and rural tribal lands in the lower 48 states. Because the presence of robust broadband and improved digital connectivity in tribal areas could play a significant role in revitalizing many tribal communities, the federal government continues to provide some financial assistance to tribal lands for broadband deployment. The Government Accountability Office, in its 2016 report, *Challenges to Assessing and Improving Telecommunications for Native Americans on Tribal Lands*, identified programs in two federal agencies that serve as the primary source of funding for deploying broadband infrastructure in tribal lands and communities. These federal agencies are the FCC and the Rural Utilities Service (RUS) in the U.S. Department of Agriculture. Tribal entities and projects in tribal areas are eligible for virtually all federal broadband programs. With a few exceptions, however, there are no carve-outs or dedicated funding streams specifically for tribal applicants or nontribal entities proposing to serve tribal lands. Thus, annual amounts of federal financial assistance vary, depending on the number and quality of tribal-related applications received, and the number of tribal-related broadband awards subsequently made by the funding agencies. Debate has centered on whether federal funding for tribal broadband is sufficient, and the extent to which portions of federal funds available for broadband should be specifically targeted for tribal broadband. In the 115th Congress, bills were introduced to direct federal funding specifically for tribal broadband. Notwithstanding whether federal broadband funding programs target tribal lands, whether or not tribal lands will receive additional funding for broadband in the 116th Congress will likely be determined by the ongoing trajectory of overall federal funding for broadband.

In: Broadband
Editor: Orlando Harrison
ISBN: 978-1-53616-409-1

Chapter 1

BROADBAND DEPLOYMENT: STATUS AND FEDERAL PROGRAMS*

Colby Leigh Rachfal

ABSTRACT

Broadband—whether delivered via fiber, cable modem, mobile or fixed wireless, copper wire, or satellite—is increasingly the technology underlying telecommunications services such as voice, video, and data. Since the initial deployment of high-speed internet in the late 1990s, broadband technologies have been deployed throughout the United States primarily by the private sector. These providers include telephone, cable, wireless, and satellite companies as well as other entities that provide broadband services to residential, business, and institutional customers.

The Federal Communications Commission's (FCC's) 2010 National Broadband Plan identified broadband as a basic infrastructure necessary for improving economic growth, job creation, civic engagement, global competitiveness, and a better quality of life. Broadband enables or enhances applications such as e-commerce, telemedicine, distance education, telework, entertainment, public safety, and energy conservation. Increasingly viewing broadband as a basic infrastructure, Congress and successive Administrations have focused on addressing gaps specifically related to broadband availability and adoption. Broadband availability refers to whether or not broadband service is offered, while broadband adoption refers to the extent to which American households actually subscribe to and use broadband.

---

* This is an edited, reformatted and augmented version of Congressional Research Service, Publication No. IF10441, dated June 6, 2019.

AVAILABILITY

The lack of adequate broadband is most pressing in rural America (especially tribal lands), where the costs of serving large geographical areas, coupled with low population densities, often reduce economic incentives for telecommunications providers to invest in and maintain broadband infrastructure and service. According to the latest FCC data, in Dec. 2017, 93.5% of Americans had fixed terrestrial broadband (e.g., fiber and cable modem connections) at minimum speeds of 25 Mbps (download speed)/3 Mbps (upload speed). Table 1 shows the broadband availability gap between urban, rural, and tribal lands in recent years. Table 2 shows percentages of Americans with access to multiple providers.

Section 706 of the Telecommunications Act of 1996 (P.L. 104-104) requires the FCC to regularly initiate an inquiry and release a report (commonly called the "706 report") assessing the status of broadband deployment to all Americans. In its *2019 Broadband Deployment Report*, the FCC concluded that broadband is being deployed to all Americans in a reasonable and timely fashion, asserting that FCC policies are continuing to remove barriers to infrastructure investment and promote competition in the telecommunications market.

Table 1. Percentage of Americans with Access to Fixed Terrestrial Broadband (25/3 Mbps)

	2014	2015	2016	2017
United States	89.4%	89.9%	91.9%	93.5%
Urban Areas	96.4%	96.7%	97.7%	98.3%
Rural Areas	60.4%	61.5%	67.8%	73.6%
Tribal Lands	57.2%	57.8%	63.1%	67.9%

Source: FCC, *2019 Broadband Deployment Report,* released May 29, 2019, p. 16, available at https://docs.fcc.gov/public/attachments/FCC19-44A1.pdf.

Table 2. Percentage of Americans With Multiple Options for Fixed Terrestrial Broadband (25/3 Mbps)

	No provider	1 provider	2 providers	3 or more
Nationwide	6.6%	29.4%	43.5%	20.5%
Urban	1.9%	25.8%	48.4%	24%
Rural	26.4%	44.4%	23.1%	6.1%
Tribal	32.1%	36.3%	20.8%	10.8%

Source: CRS, derived from FCC Fixed Broadband Deployment Map, data as of December 2017. (https://broadbandmap.fcc.gov/).

The FCC determined that the current speed benchmark of 25 Mbps/3 Mbps remains an appropriate measure by which to assess whether a fixed service provides advanced telecommunications capability. Regarding mobile broadband, the FCC found that adoption of a single mobile benchmark is still currently unworkable, given certain data limitations and the inherent variability of mobile services. As with its *2018 Report*, the FCC concluded that mobile broadband service is not a full substitute for fixed service at this time.

ADOPTION

The National Broadband Plan also identified broadband adoption as a problem, with a significant number of Americans having broadband available, but choosing not to subscribe. According to a November 2017 Census Bureau survey compiled and reported by the National Telecommunications and Information Administration (NTIA), 85.9% of American households use wired broadband at home.

Table 3. Percentage of Households with Home Internet Use Having Wired Broadband Service

Total U.S.	85.9%
Native American	75.0%
African American	80.7%
Hispanic	81.5%
White	87.3%
Unemployed	80.9%
Employed	87.1%
Disabled	80.7%
Not disabled	86.6%
<$25K family income	75.3%
$25K-$49K	82.0%
$50K-$74K	88.1%
$75K-$99K	90.4%
$100K or more	92.7%
No diploma	70.6%
High-school graduate	79.3%
Some college	86.4%
College graduate	92.1%
Rural	75.6%
Urban	87.5%

Source: NTIA, *Digital Nation Data Explorer*, November 2017 data. (https://www.ntia.doc.gov/data/ digital-nation-data-explorer).

Table 3 shows that populations continuing to lag behind in broadband adoption include people with low incomes, certain minority populations, the less-educated, the unemployed, the disabled, and households in rural areas. According to the Pew Research Center, in its broadband adoption survey *Home Broadband 2015*, the cost of monthly subscriptions is the leading reason people do not have broadband connections.

FEDERAL PROGRAMS

Federal programs exist that can provide support for broadband availability and adoption to unserved and underserved communities and populations. NTIA has published a *Guide to Federal Funding of Broadband Projects* and companion searchable database, which provides an overview of funding sources across the federal government. Major broadband funding streams are centered in two federal entities: the FCC and the Rural Utilities Service (RUS) at the U.S. Department of Agriculture. Programs at the FCC include the following:

- Connect America Fund (High-Cost Fund). Subsidizes the cost of operating and extending broadband infrastructure to serve consumers and small businesses in rural, high-cost areas.
- E-Rate (Schools and Libraries) Program. Provides discounts of up to 90% for broadband to and within elementary and secondary schools (public and private), and public libraries in rural and nonrural areas.
- Rural Health Care Program. Subsidizes broadband connectivity for public and nonprofit health care providers, with a focus on rural areas.
- Lifeline Program. Subsidizes eligible subscribers to cover the recurring monthly service charges associated with broadband subscribership; support is not given directly to the subscriber but to the service provider.
- Programs at the RUS include the following:
- Farm Bill Broadband Loans. Funds the costs of construction, improvement, or acquisition of facilities and equipment needed to provide service in eligible rural areas.
- ReConnect Program. Offers grants, loans, and loan/grant combinations for new or upgraded broadband service in areas where at least 90% of households do not have adequate broadband.
- Telecommunications Infrastructure Loans and Loan Guarantees. Funds the construction, maintenance, improvement, and expansion of telephone service and broadband in extremely rural areas with a population of 5,000 or fewer.

- Community Connect Grants. Funds broadband deployment in rural communities where it is not yet economically viable for private-sector providers to deliver service.
- Distance Learning and Telemedicine Grants. Funds end-user equipment and broadband facilities to help rural communities use telecommunications to link teachers and medical service providers in one area to students and patients in another.

The 2018 farm bill (P.L. 115-334) included provisions authorizing a grant component in combination with the broadband loan program; increasing the annual authorization level from $25 million to $350 million; and authorizing grants, loans, and loan guarantees for middle mile infrastructure.

The Consolidated Appropriations Act, 2018 (P.L. 115-141) established the $600 million ReConnect Program at RUS, appropriated $7.5 million to NTIA to update the national broadband availability map in coordination with the FCC, facilitated deployment of broadband infrastructure on federal property, and made more spectrum available for wireless broadband. The Consolidated Appropriations Act, 2019 (P.L. 116-6) provided an additional $550 million to RUS for the ReConnect Program and an additional $7.5 million to NTIA for mapping.

In: Broadband
Editor: Orlando Harrison
ISBN: 978-1-53616-409-1

Chapter 2

DEFINING BROADBAND: MINIMUM THRESHOLD SPEEDS AND BROADBAND POLICY*

Lennard G. Kruger

ABSTRACT

Broadband—whether delivered via fiber, cable modem, copper wire, satellite, or mobile wireless—is increasingly the technology underlying telecommunications services such as voice, video, and data. Since the initial deployment of high-speed internet in the late 1990s, broadband technologies have been deployed throughout the United States primarily by the private sector. These providers include telephone, cable, wireless, and satellite companies as well as other entities that provide commercial telecommunications services to residential, business, and institutional customers.

How broadband is defined and characterized in statute and in regulation can have a significant impact on federal broadband policies and how federal resources are allocated to promote broadband deployment in unserved and underserved areas. One way broadband can be defined is by setting a minimum threshold speed for what constitutes "broadband service." Section 706 of the Telecommunications Act of 1996 requires the Federal Communications Commission (FCC) to regularly initiate an inquiry concerning the availability of broadband to all Americans and to determine whether broadband is "being deployed to all Americans in a reasonable and timely fashion." If the determination is negative, the act directs the FCC to "take immediate action to accelerate deployment of such capability by removing barriers to infrastructure investment and by promoting competition in the telecommunications market."

Starting in 1999, there have been 11 Section 706 reports, each providing a snapshot and assessment of broadband deployment. As part of this assessment, and to help

* This is an edited, reformatted and augmented version of Congressional Research Service, Publication No. R45039, dated May 10, 2018.

determine whether broadband is being deployed in "a reasonable and timely fashion," the FCC has set a minimum broadband speed that essentially serves as the benchmark the FCC uses to determine what it considers broadband service for the purposes of its Section 706 determination. In 2015 the FCC, citing changing broadband usage patterns and multiple devices using broadband within single households, raised its minimum fixed broadband benchmark speed from 4 Mbps (download)/1 Mbps (upload) to 25 Mbps/3 Mbps.

On August 8, 2017, the FCC adopted and released its *Thirteenth Section 706 Notice of Inquiry* (*NOI*). One proposal under consideration was establishing a lower benchmark speed specifically for mobile broadband. Stakeholders who supported an FCC determination that broadband is *not* being deployed in a reasonable and timely fashion generally opposed lowering the broadband benchmark by considering the presence of either fixed (at 25 Mbps/3 Mbps) or mobile (at 10 Mbps/1 Mbps) broadband as an indication that an area has adequate broadband service. On the other hand, stakeholders who supported an FCC determination that broadband *is* being deployed in a reasonable and timely fashion generally supported changing the FCC's broadband benchmark methodology to include the presence of either fixed or mobile broadband as an indication that an area is receiving adequate broadband service.

On February 2, 2018, the FCC adopted and released its latest 706 report, the *2018 Broadband Deployment Report*. While the FCC concluded that advanced telecommunications capability is being deployed to all Americans in a reasonable and timely fashion, it concluded that adoption of a single mobile benchmark is currently unworkable and that mobile broadband service is not a full substitute for fixed service at this time.

As broadband technology advances, commercially available download and upload speeds will likely increase, and the level at which broadband benchmark threshold speeds should be set is likely to remain controversial. Accordingly, the FCC's annual Section 706 determination is likely to be an issue in Congress as long as it is seen by stakeholders as providing a justification for current or future FCC regulatory or deregulatory policies.

INTRODUCTION

Broadband—whether delivered via fiber, cable modem, copper wire, satellite, or mobile wireless—is increasingly the technology underlying telecommunications services such as voice, video, and data. Since the initial deployment of high-speed internet in the late 1990s, broadband technologies have been deployed throughout the United States primarily by the private sector. These providers include telephone, cable, wireless, and satellite companies as well as other entities that provide commercial telecommunications services to residential, business, and institutional customers.

The Federal Communications Commission's (FCC's) 2010 National Broadband Plan identified broadband as a basic infrastructure necessary for improving economic growth, job creation, civic engagement, global competitiveness, and a better quality of life. Broadband enables or enhances applications such as entertainment, telemedicine, distance education, telework, e-commerce, public safety, and energy conservation. Increasingly viewing broadband as a basic infrastructure, Congress and successive Administrations have focused on addressing gaps specifically related to broadband availability and adoption.[1] In general, rural areas tend to be less served by broadband than urban or suburban areas.[2] Tribal areas and populations stand out as being particularly underserved and unserved by broadband service, as compared to the nation as a whole.[3]

How broadband is defined and characterized in statute and in regulation can have a significant impact on federal broadband policies and how federal resources are allocated to promote broadband deployment in unserved and underserved areas. The issue of how the FCC defines broadband with respect to minimum threshold speeds has come to the forefront, as stakeholders recognize that minimum speed threshold levels can play a role in determining what steps the FCC may or may not take with respect to an array of broadband policies.

DEFINITIONS OF BROADBAND

The FCC's working definition of broadband is in Section 706 of the Telecommunications Act of 1996 (P.L. 104-104), which defines "advanced telecommunications capability" as

> high-speed, switched, broadband telecommunications capability that enables users to originate and receive high quality voice, data, graphics, and video telecommunications using any technology.[4]

A similar definition can be found in the statute that authorizes the Rural Broadband Access Loan and Loan Guarantee program in the Rural Utilities Service (RUS) at the U.S. Department of Agriculture (USDA):

> The term broadband service means any technology identified by the Secretary as having the capacity to transmit data to enable a subscriber to the service to originate and receive high-quality voice, data, graphics, and video.[5]

---

[1] Broadband availability refers to whether or not broadband service is offered, while broadband adoption refers to the extent to which American households actually subscribe to and use broadband.

[2] CRS In Focus IF10441, *Broadband Deployment: Status and Federal Programs*, by Lennard G. Kruger.

[3] See CRS Report R44416, *Tribal Broadband: Status of Deployment and Federal Funding Programs*, by Lennard G. Kruger.

[4] 47 U.S.C. 1302(d)(1). Another FCC definition of broadband internet access service can be found in the FCC's *Open Internet Order* (47 C.F.R 8.2(a)): "A mass-market retail service by wire or radio that provides the capability to transmit data to and receive data from all or substantially all Internet endpoints, including any capabilities that are incidental to and enable the operation of the communications service, but excluding dial-up Internet access service. This term also encompasses any service that the Commission finds to be providing a functional equivalent of the service described in the previous sentence, or that is used to evade the protections set forth in this part."

[5] 7 USC 950bb(b)(1).

Another way broadband can be defined is by setting a minimum threshold speed for what constitutes "broadband service." Broadband speeds are characterized with respect to both download speeds and upload speeds. The Government Accountability Office (GAO) defines download speed as the rate at which data are transferred to the consumer, and upload speed as the rate at which data are transferred from the consumer to the internet.[6] Both are measured by the number of bits of data transferred per second (typically expressed as megabits per second or Mbps).[7]

Table 1. Minimum download speeds for typical online activities

Activity	Minimum Download Speed (Mbps)
General Usage	
General Browsing and Email	1
Streaming Online Radio	Less than 0.5
VoIP Calls	Less than 0.5
Student	5-25
Telecommuting	5-25
File Downloading	10
Social Media	1
Watching Video	
Streaming Standard Definition Video	3-4
Streaming High Definition (HD) Video	5-8
Streaming Ultra HD 4K Video	25
Video Conferencing	
Standard Personal Video Call (e.g., Skype)	1
HD Personal Video Call (e.g., Skype)	1.5
HD Video Teleconferencing	6
Gaming	
Game Console Connecting to the Internet	3
Online Multiplayer	4

Source: FCC, *Broadband Speed Guide*, last updated/reviewed February 6, 2018, available at https://www.fcc.gov/ reports-research/guides/broadband-speed-guide.

Note: Speeds are based on running one activity at a time. Additional speed may enhance performance.

Typically, broadband delivered by fixed wired technologies (e.g., cable modem and fiber) provides higher speeds than mobile wireless and satellite broadband technologies. For example, according to the *akamai state of the internet report*, the average mobile speed in the United States was 10.7 Mbps versus an overall average (for all broadband technologies) of 18.7 Mbps in the first quarter of 2017.[8] Similarly, the FCC's *Measuring*

[6] Government Accountability Office, *Broadband Performance: Additional Actions Could Help FCC Evaluate Its Efforts to Inform Consumers*, GAO-15-363, April 2015, p. 4, available at http://www.gao.gov/ assets/670/669739.pdf.

[7] Other technical characteristics—such as latency or service reliability—could also be used to define and measure the adequacy of broadband service. According to GAO, latency is "a measure of the time that it takes for data to travel from a computer to a server and back again." See Ibid.

[8] Akamai Technologies, Inc., *akamai's [state of the internet] Q1 2017 report*, p. 24, 44, available at https://www.akamai.com/fr/fr/multimedia/documents/state-of-the-internet/q1-2017-state-of-the-internet-

Fixed Broadband Report—2016 shows that cable and fiber provide significantly higher median download speeds than satellite broadband.[9]

Download and upload speeds can determine what kinds of applications are possible for broadband users. For example, Table 1 shows minimum download speeds for typical online activities, Table 2 shows minimum download speed needs for light, moderate, and high household use, and Table 3 shows baseline download speed requirements for community institutions.

Table 2. Minimum download speed needs for light, moderate, and high household use

	Light Use (Basic functions: email, browsing, basic video, VoIP, internet radio)	Moderate Use (Basic functions plus one high-demand application: streaming HD video, multiparty video conferencing, online gaming	High Use (Basic functions plus more than one high-demand application running at the same time)
1 user on 1 device	Basic	Basic	Medium
2 users or devices at a time	Basic	Medium	Medium/Advanced
3 users or devices at a time	Medium	Medium	Advanced
4 users or devices at a time	Medium	Advanced	Advanced

Source: FCC, House Broadband Guide, last updated/reviewed February 6, 2018, available at https://www.fcc.gov/research-reports/guides/household-broadband-guide.

Notes: Basic Service = 3 to 8 Mbps; Medium Service = 12 to 25 Mbps; Advanced Service = more than 25 Mbps.

Table 3. Baseline download speed requirements for community institutions

Institution	Typical Application	Download Speeds
Hospital	sharing health records performing virtual consultations connecting first responders	1 Gbps+
Library	operating public computer centers mobile hotspot lending enabling maker spaces	100 Mbps-1 Gbps+
School	sharing educational material online testing accessing databases	100 Mbps-1 Gbps+
Small Business	managing inventory operating point-of-sale terminals coordinating shipping	50 Mbps+
Home	completing homework streaming video web browsing	25 Mbps+

Source: BroadbandUSA, National Telecommunications and Information Administration, What Speed Do You Need? available at https://www2.ntia.doc.gov/files/what_speed_061917.pdf.

Notes: Download speeds in this table are suggested by BroadbandUSA, which states that "download speed requirements vary based on the activity, location and number of users, and these needs will continue to change as technology advances."

connectivityreport.pdf. Akamai operates global Content Delivery Network services and publishes quarterly state of the internet reports based on data from its global cloud delivery platform.

[9] FCC, *Measuring Fixed Broadband Report—2016*, December 2016, available at https://www.fcc.gov/reports-research/ reports/measuring-broadband-america/measuring-fixed-broadband-report-2016#_Toc458762536.

MINIMUM SPEEDS IN FEDERAL BROADBAND PROGRAMS

Currently, there are two primary streams of federal funding to support broadband infrastructure deployment in rural areas: the FCC's Universal Service Fund programs and the portfolio of RUS broadband grant and loan programs at USDA (see Table 4). Each of these programs has set a minimum broadband speed threshold to determine which areas are or are not eligible for support. As Table 4 shows, minimum broadband speed thresholds are not uniform across programs—they differ based on the particular needs and goals of the individual program.

Because broadband technology is always advancing, the statutes that have established these programs have built in flexibility for the implementing agency to set minimum threshold broadband speeds. For example, the 2014 farm bill (P.L. 113-79), which authorizes the Rural Broadband Access Loan program, established a "minimum broadband acceptable level of broadband service for a rural area" of at least 4 Mbps downstream transmission capacity and 1 Mbps upstream transmission capacity.[10] However, the statute states that

> At least once every 2 years, the Secretary shall review, and may adjust through notice published in the Federal Register, the minimum acceptable level of broadband service ... to ensure that high quality, cost-effective broadband service is provided to rural areas overtime.[11]

The statute further adds that in adjusting the minimum acceptable level of broadband service, the Secretary may consider establishing different transmission rates for fixed broadband service and mobile broadband service, and that the Secretary shall not establish requirements for bandwidth or speed that have the effect of precluding the use of evolving technologies appropriate for rural areas.[12] Accordingly, the latest Notice of Solicitation of Applications (NOSA) for the Rural Broadband Access Loan program defined existing broadband service as a minimum of 25 Mbps (download)/3 Mbps (upload) for both fixed (wired) and mobile (wireless) broadband.[13]

---

[10] 7 U.S.C. 950bb(e)(1).

[11] 7 U.S.C. 950bb(e)(2).

[12] 7 U.S.C. 950bb(e)(3).

[13] Department of Agriculture, Rural Utilities Service, "Rural Broadband Access Loan and Loan Guarantees Program," Notice of Solicitation of Applications (NOSA), 83 *Federal Register* 13225-13226, March 28, 2018, available at https://www.gpo.gov/fdsys/pkg/FR-2018-03-28/pdf/2018-06175.pdf.

Table 4. Federal broadband programs and minimum broadband speeds

Program	Description	Minimum Broadband Speeds
Federal Communications Commission		
Connect America Program)	Program provides support for eligible telecommunications carriers to help offset the higher-thanaverage costs of providing telecommunications services in rural, insular, or other high cost areas.	Requires providers to offer a minimum speed of 10 Mbps (download)/1 Mbps (upload).[a]
Schools and Libraries (E-rate)	Provides discounts to providers offering telecommunications services, internet access, and internal connections to qualifying K12 schools and libraries.	100 Mbps per 1,000 users (short term goal); 1 Gbps per 1,000 users (long term goal).[b]
Rural Health Care	Provides support to qualifying rural public and nonprofit health care providers for telecommunications and broadband connectivity.	No minimum bandwidth requirement for supported services; however, certain types of support are limited to connections that provide actual speeds of 1.5 Mbps (symmetrical) or higher.[c]
Low Income Program (Lifeline)	Provides support to assist eligible households pay monthly service charges. Support is not given directly to the subscriber but to their designated telecommunications provider, who in turn charges the subscriber lower rates.	Minimum service standard for fixed broadband speed set at 15 Mbps (download)/2 Mbps (upload). Minimum is 4 Mbps/1 Mbps in cases where the provider does not offer any generally available residential fixed broadband package that meets the 15 Mbps/2 Mbps standard.[d]
USDA Rural Utilities Service		
Farm Bill Broadband Loans	Provides loans for the costs of construction, improvement, or acquisition of facilities and equipment needed to provide broadband service in rural areas.	Existing broadband service is defined as a minimum of 25 Mbps (download)/3 Mbps (upload) for both fixed (wired) and mobile (wireless) broadband. Funded projects must provide minimum speeds of 25 Mbps (download/3 Mbps (upload) for fixed and mobile broadband.[e]
Telecommunications Infrastructure Loans	Provides loans for the construction, maintenance, improvement and expansion of telephone service and broadband in extremely rural areas (less than 5,000 population).	All facilities financed must be capable of supporting broadband services, defined as "any technology identified by the Administrator as having the capacity to transmit data to enable a subscriber to originate and receive high-quality voice, data, graphics, and video."[f]
Community Connect Grants	Provides grants to fund broadband deployment into rural areas where it is not yet economically viable for private sector providers to deliver service.	Existing broadband service is defined as a minimum of 10 Mbps (download)/1 Mbps (upload) for both fixed and mobile broadband. Funded projects must provide minimum speeds of 25 Mbps (download)/3 Mbps (upload) for both fixed (wired) and mobile (wireless) broadband.[g]

Source: CRS, compiled from agency program regulations and documents.

[a]FCC, "In the Matter of Connect America Fund," Report and Order, FCC 14-190, WC Docket No. 10-90, adopted December 11, 2014, p. 6, available at https://apps.fcc.gov/edocs_public/attachmatch/ FCC-14- 190A1.pdf.

[b]FCC, "In the Matter of Modernizing the E-rate Program for Schools and Libraries," Report and Order and Further Notice of Proposed Rulemaking, FCC 14-99, WC Docket No. 13-184, adopted July 11, 2014, p. 16, available at https://transition.fcc.gov/Daily_Releases/Daily_Business/2014/db0723/FCC-14-99A1.pdf.

[c]FCC, "In the Matter of Rural Health Care Support Mechanism," Report and Order, FCC 12-150, WC Docket No. 02-60, adopted December 12, 2012, p. 52, available at https://apps.fcc.gov/edocs_ public/attachmatch/ FCC-12-150A1.pdf.

[d]FCC, Public Notice, "Wireline Competition Bureau Announced Updated Lifeline Minimum Service Standards and Indexed Budget Amount," June 26, 2017, available at https://apps.fcc.gov/edocs_ public/attachmatch/DA17-619A1.pdf.

[e]Department of Agriculture, Rural Utilities Service, "Rural Broadband Access Loan and Loan Guarantees Program," Notice of Solicitation of Applications (NOSA), 83 Federal Register, pp. 13225-13226, March 28, 2018, available at https://www.gpo.gov/fdsys/pkg/FR-2018-03-28/pdf/2018-06175.pdf.

[f]Department of Agriculture, Rural Utilities Service, Infrastructure Loan Application Guide, updated as of December 1, 2015, p.41, available at https://www.rd.usda.gov/programs-services/ telecommunicationsinfrastructure-loans-loan-guarantees.

[g]Department of Agriculture, Rural Utilities Service, "Announcement of Grant Application Deadlines and Funding Levels," Notice of Solicitation of Applications (NOSA), 83 Federal Register, pp. 11494-11499, March 15, 2018, available at https://www.gpo.gov/fdsys/pkg/FR-2018-03-15/pdf/2018-05200.pdf.

SECTION 706 AND THE FCC'S DEFINITION OF MINIMUM BROADBAND SPEED

Section 706 of the Telecommunications Act of 1996 requires the FCC to regularly initiate an inquiry concerning the availability of broadband to all Americans and to determine whether broadband is "being deployed to all Americans in a reasonable and timely fashion." If the determination is negative, the act directs the FCC to "take immediate action to accelerate deployment of such capability by removing barriers to infrastructure investment and by promoting competition in the telecommunications market."[1]

Starting in 1999, there have been 11 Section 706 reports, each providing a snapshot and assessment of broadband deployment. As part of this assessment, and to help determine whether broadband is being deployed in "a reasonable and timely fashion," the FCC has set a minimum broadband speed that essentially serves as the benchmark it uses to determine what constitutes broadband service for the purposes of its Section 706 determination. Table 5 shows each Section 706 report, the date adopted, the FCC determination of whether broadband was being deployed in a reasonable and timely fashion, the minimum broadband benchmark speed used in that FCC determination, and the vote of the FCC Commissioners. As Table 5 shows, the minimum speeds have risen, from 200 kbps in 1999, to 4 Mbps/1 Mbps in 2010, to 25 Mbps/3 Mbps in 2015.

As Table 5 also shows, there has seldom been agreement among FCC commissioners as to whether broadband is being deployed in a timely and reasonable manner. For example, during the George W. Bush Administration, the finding that broadband deployment was reasonable and timely was associated with the FCC's deregulatory policies, whereas during the Obama Administration, the finding that broadband deployment was not reasonable and timely has been associated with FCC policies favoring a greater level of government intervention in the marketplace.

An important factor in the Section 706 determination has proven to be how the FCC chooses to define broadband service in terms of minimum speed. In 2015, the FCC, citing changing broadband usage patterns and multiple devices using broadband within single households, raised its minimum broadband benchmark speed from 4 Mbps/1 Mbps to 25 Mbps/3 Mbps.

Critics asserted that the FCC excessively raised broadband benchmark speeds in order to produce a negative finding and to justify the FCC's pursuit of policies addressing such regulatory issues as Universal Service Fund reform and net neutrality.

---

[1] 47 U.S.C. 1302(b).

Table 5. Section 706 Reports, 1999-2018

	Date Adopted	Is Broadband Deployed to All Americans in Reasonable and Timely Fashion?	Broadband Definition (threshold speed)	Vote of FCC Commissioners
First Broadband Progress Report	1/28/1999	Yes	200 kbps in both directions	5-0
Second Broadband Progress Report	8/3/2000	Yes	200 kbps in at least one direction	4-0
Third Broadband Progress Report	2/6/2002	Yes	200 kbps in at least one direction	3-1
Fourth Broadband Progress Report	9/9/2004	Yes	200 kbps in at least one direction	3-2
Fifth Broadband Progress Report	3/19/2008	Yes	200 kbps in at least one direction	3-2
Sixth Broadband Progress Report	7/16/2010	No	4 Mbps/1 Mbps	3-2
Seventh Broadband Progress Report	5/20/2011	No	4 Mbps/1 Mbps	3-1
Eighth Broadband Progress Report	8/14/2012	No	4 Mbps/1 Mbps	3-2
2015 Broadband Progress Report	1/29/2015	No	25 Mbps/3 Mbps	3-2
2016 Broadband Progress Report	1/28/2016	No	25 Mbps/3 Mbps	4-1
2018 Broadband Deployment Report	2/2/2018	Yes	25 Mbps/3 Mbps	3-2

Source: FCC, Archive of Released Broadband Progress Notices of Inquiry, available at https://www.fcc.gov/general/archive-released-broadband-progress-notices-inquiry.

On February 2, 2018, the FCC adopted and released its latest 706 report, the *2018 Broadband Deployment Report.*[2] The FCC concluded that advanced telecommunications capability is being deployed to all Americans in a reasonable and timely fashion. This determination was based on evaluating progress—comparing deployment in the present year to deployment in previous years. According to the *Report*

> We find that analyzing progress to determine whether deployment is occurring in a reasonable and timely fashion is the approach that is most consistent with the language of section 706, as the analysis of such progress enables the Commission to determine whether advanced telecommunications capability "is being deployed" in the manner that section 706 requires. The use of the present progressive tense—"is being deployed"—as well as the language requiring an evaluation of whether that deployment is "reasonable and timely" indicates that Congress intended that the Commission evaluate the current

[2] Federal Communications Commission, *2018 Broadband Deployment Report*, In the Matter of Inquiry Concerning Deployment of Advanced Telecommunications Capability to All Americans in a Reasonable and Timely Fashion, GN Docket no. 17-199, FCC 18-10, adopted and released February 2, 2018, 89 pp., available at https://www.fcc.gov/ document/fcc-releases-2018-broadband-deployment-report.

state of deployment to all Americans, not a rigid requirement that each and every American be served *at this moment*.[3]

The FCC's latest 706 determination that broadband is being deployed in a reasonable and timely fashion is a departure from the FCC's previous five determinations that broadband was not being deployed in a reasonable and timely fashion. The latest 706 determination was approved by the three Republican FCC Commissioners, with the remaining two Democratic commissioners dissenting. According to FCC Commissioner Rosenworcel's dissent.

> This chapter concludes that in the United States the deployment of broadband to all Americans is reasonable and timely. This is ridiculous—and irresponsible. Today there are 24 million Americans without access to broadband. There are 19 million Americans in rural areas who lack the ability to access high-speed services at home. There are 12 million school-aged children who are falling into the Homework Gap because they do not have the broadband at home they need for nightly schoolwork. Ask any one of them if they think the deployment of the most essential digital age infrastructure is reasonable and timely and you will get a resounding "No." To call these numbers a testament to our national success is insulting and not credible.[4]

THIRTEENTH SECTION 706 REPORT NOTICE OF INQUIRY

On August 8, 2017, the FCC adopted and released its *Thirteenth Section 706 Notice of Inquiry* (*NOI*).[5] Comments and data received from the *NOI* were used to develop the FCC's latest Section 706 broadband progress report, the *2018 Broadband Deployment Report*.[6]

In the *NOI*, the FCC proposed to maintain the 25 Mbps/3 Mbps benchmark for fixed broadband, while at the same time soliciting comments on how the presence and nature of mobile broadband should factor into the determination of whether broadband is being deployed to all Americans in a timely and reasonable fashion. On the one hand, the FCC asked for comments on whether the presence of fixed *or* mobile broadband should indicate that an area is served:

> Given that Americans use both fixed and mobile broadband technologies, we seek comment on whether we should evaluate the deployment of fixed and mobile broadband

[3] *2018 Broadband Deployment Report*, p. 4.

[4] Ibid., p. 89.

[5] FCC, *Thirteenth Section 706 Report Notice of Inquiry*, GN Docket No. 17-199, FCC 17-109, August 8, 2017, available at https://transition.fcc.gov/Daily_Releases/Daily_Business/2017/db0808/FCC-17-109A1.pdf.

[6] In two past instances, a Section 706 NOI did not lead to a Section 706 report: the *Ninth NOI*, released on February 4, 2015, and the *Twelfth NOI* released on August 4, 2016.

> as separate and distinct ways to achieve advanced telecommunications capability. Taking into account the differences between the various services and the geographic, economic, and population diversity of our nation, we seek comment on focusing this Section 706 Inquiry on whether some form of advanced telecommunications capability, be it fixed or mobile, is being deployed to all Americans in a reasonable and timely fashion.[7]

Alternatively, the *NOI* sought comment on whether the FCC should "evaluate broadband deployment based on the presence of both fixed *and* mobile services," given that "mobile and fixed broadband have different technical characteristics and limitations, and broadband providers choose to market their fixed and mobile products in different ways."[8]

Within the context of seeking whether to assess fixed *or* mobile versus fixed *and* mobile, the FCC sought comment on the benchmarks they should use to define "advanced telecommunications capability." While proposing to maintain the current speed benchmark of 25 Mbps/3 Mbps for fixed broadband, the FCC sought comment on potential benchmarks and metrics for mobile broadband:

> The Commission has not previously set a mobile speed benchmark. Our consideration of whether and how to set a speed benchmark will be informed by assessing the mobile broadband services and speeds that are available to consumers today, as well as evidence regarding what services consumers are choosing today, and what might be available in the near future.... Should the Commission set a mobile speed benchmark, and if so, what it should be? We anticipate that any speed benchmark we set would be lower than the 25 Mbps/3 Mbps benchmark adopted for fixed broadband services, given differing capabilities of mobile broadband.... We seek comment on whether a mobile speed benchmark of 10 Mbps/1 Mbps is appropriate for mobile broadband services. Would a download speed benchmark higher or lower than 10 Mbps be appropriate for the purpose of assessing American consumers' access to advanced telecommunications capability?[9]

DEBATE OVER BROADBAND SERVICE BENCHMARKS

How and at what level the FCC sets broadband service benchmarks in its Section 706 reports has been a controversial issue in Congress. For example, in a January 2016 letter to FCC Chairman Tom Wheeler regarding the *2016 Broadband Progress Report*, a group of Republican Senators disputed the FCC's determination that broadband was not being deployed to all Americans in a reasonable and timely fashion. They asserted that the

---

[7] *Thirteenth Section 706 Report Notice of Inquiry*, p. 5.
[8] Ibid.
[9] Ibid., p. 7.

benchmark of 25 Mbps/3 Mbps was set at an arbitrarily high level and "fails to accurately capture what most Americans consider broadband." Specifically, the Senators argued that most commonly used applications require speeds considerably below the 25/3 level, that "the majority of Americans who can purchase 25 Mbps service choose not to," and that the use of the 25/3 benchmark "discourages broadband providers from offering speeds at or above the benchmark." The letter also made the point that the FCC's 25/3 benchmark is inconsistent with the Connect America Fund benchmark of 10/1 for rural broadband, and with the FCC's *Open Internet Order*'s definition of broadband, which includes any service above dial-up.[10]

Conversely, the August 2017 *NOI* led other Members of Congress to object that the FCC may be preparing to effectively lower the broadband benchmark speed by potentially deeming an area as served by broadband if it has either fixed broadband (which the FCC proposes to maintain at 25/3 Mbps) *or* mobile broadband (which the FCC proposes to set at a level of 10 Mbps/1 Mbps). Such a change in setting the broadband benchmark could, they asserted, lead to an unwarranted FCC determination that advanced telecommunications services *are* being deployed to all Americans in a timely and reasonable fashion. For example, an August 2017 letter from a group of Democratic Senators to the FCC noted that "the Commission appears ready to decide that mobile broadband could be a substitute, rather than a complement, to fixed broadband service, and that slower-speed mobile service substitutes as effectively." The letter argued that mobile broadband is not yet equivalent to fixed broadband services and that "such a striking change in policy would significantly and disproportionately disadvantage Americans in rural, tribal, and low income communities across the nation, whose livelihoods depend on a reliable and affordable broadband connection."[11] Similarly, a group of Democratic Senators and Representatives, in an October 2017 letter to the FCC, expressed concern "with any effort to weaken the FCC's current policy finding that every American should have access to broadband services with speeds of at least 25 Mbps download/3 Mbps upload," and that "[f]inding instead that only mobile service of 10 Mbps download/1 Mbps upload is sufficient would result in significantly slower and less reliable Internet access for millions of Americans."[12]

Much of the debate over the FCC's broadband benchmark played out in comments and letters filed in response to the FCC's *Thirteenth Section 706 NOI*. In general, those parties who support an FCC determination that broadband is *not* being deployed to all Americans in a timely and reasonable fashion were opposed to altering the current

---

[10] Letter from 6 Republican Senators to Chairman Tom Wheeler, Federal Communications Commission, January 21, 2016, available at https://assets.documentcloud.org/documents/2697501/Letter-to-Fcc-Chairman-Wheeler-Broadband.pdf.

[11] Letter from 12 Democratic Senators to the Federal Communications Commission, August 31, 2017, available at https://prodnet.www.neca.org/publicationsdocs/wwpdf/0901sen.pdf.

[12] Letter from 9 Democratic Senators and 36 Democratic House Members to the Federal Communications Commission, October 5, 2017, available at https://ecfsapi.fcc.gov/file/100571368034/10.5.2017% 20FCC%20 Chariman%20Pai%20letter.pdf.

broadband benchmark by considering the presence of fixed or mobile broadband as an area that is "served." They asserted that mobile broadband is not a satisfactory substitute or replacement for fixed broadband service. Among their arguments were that mobile broadband currently lacks the capacity, speed, and reliability of fixed broadband;[13] that mobile wireless data caps and overage charges make the cost of mobile data prohibitively high;[14] that few businesses rely solely on mobile broadband and are unlikely to locate in "mobile-only" communities;[15] and that low-income communities and communities of color are more smartphone-dependent than wealthier and white communities, with smartphones not providing full access to the benefits of the internet such as applying for a job, doing homework, or taking online classes or training.[16]

On the other hand, parties who favor an FCC determination that broadband *is* being deployed to all Americans in a timely and reasonable fashion supported the notion of altering the current broadband benchmark by considering the presence of fixed or mobile broadband as an area that is "served." Among their arguments were that LTE 4G mobile broadband deployment and usage is ubiquitous, with private sector investment continuing to grow for future deployments such as 5G;[17] that the FCC should consider whether *progress* in broadband deployment is reasonable and timely, as opposed to whether deployment already is complete and ubiquitous;[18] that consumers increasingly use mobile devices in fixed locations for many voice, data, graphics, and video applications, and that mobile broadband need not be a perfect or complete economic substitute for fixed broadband for the FCC to determine that both meet the definition of "advanced telecommunications capability";[19] and that many consumers in fact do view mobile and

---

[13] Comments of the California Public Utilities Commission before the FCC, In the Matter of Inquiry Concerning the Deployment of Advanced Telecommunications Capability to All Americans in a Reasonable and Timely Fashion, GN Docket No. 17-199, October 5, 2017, p. 4-5, available at https://ecfsapi.fcc.gov/file/1005131709819/ GN%20Docket%2017-199%20CPUC%20Comments%20to%20the%20FCC.pdf.

[14] Comments of the Open Technology Institute at New America before the FCC, In the Matter of Inquiry Concerning the Deployment of Advanced Telecommunications Capability to All Americans in a Reasonable and Timely Fashion, GN Docket No. 17-199, September 21, 2017, p. 8-11, available at https://ecfsapi.fcc.gov/file/10921256530521/OTI%20FCC%20Section%20706%20Comments.pdf.

[15] Comments of NTCA—The Rural Broadband Association before the FCC, In the Matter of Inquiry Concerning the Deployment of Advanced Telecommunications Capability to All Americans in a Reasonable and Timely Fashion, GN Docket No. 17-199, p. i-iii, available at https://ecfsapi.fcc.gov/file/10921173109916/ 09.21.17%20NTCA%20Comments% 20on%20Section%20706%20NOI%2C%20GN%2017-199.pdf.

[16] Leadership Conference on Civil and Human Rights, Letter to the FCC, November 3, 2017, p. 2-3, available at https://ecfsapi.fcc.gov/file/1103708927319/Leadership%20Conference %20706%20Letter%2011-03-17.pdf.

[17] Comments of Verizon before the FCC, In the Matter of Inquiry Concerning the Deployment of Advanced Telecommunications Capability to All Americans in a Reasonable and Timely Fashion, GN Docket No. 17-199, September 21, 2017, p. 2-8, available at https://ecfsapi.fcc.gov/file/10921261086205/ 2017%2009%2021%20Verizon%20706%20comments_final.pdf.

[18] Comments of CTIA before the FCC, In the Matter of Inquiry Concerning the Deployment of Advanced Telecommunications Capability to All Americans in a Reasonable and Timely Fashion, GN Docket No. 17-199, September 21, 2017, p. 1-2, 4, available at https://ecfsapi.fcc.gov/file/109212607904338/ 170921%20CTIA%20Section%20706%20Comments.pdf.

[19] Comments of USTelcom Association before the FCC, In the Matter of Inquiry Concerning the Deployment of Advanced Telecommunications Capability to All Americans in a Reasonable and Timely Fashion, GN Docket

fixed broadband as interchangeable substitutes and that the two services provide a competitive constraint on each other and should be considered to be in the same market.[20]

Ultimately, the *2018 Broadband Deployment Report* concluded that adoption of a single mobile benchmark is currently unworkable, given available data and the inherent variability of actual mobile speeds, and that mobile broadband service is not a full substitute for fixed service at this time:

> ... we disagree with those that argue that mobile services are currently full substitutes for fixed service. Both fixed and mobile services can enable access to "information, entertainment, [and] employment options," but there are salient differences between the two technologies. Beyond the most obvious distinction that mobile services permit their users mobility, there are clear variations in consumer preferences and demands for fixed and mobile services. Each clearly provides capabilities that satisfy the statutory definition of advanced telecommunications capability, and are important services that provide different functionalities, tailored to serve different consumer needs.[21]

Concluding Observations

During the FCC's consideration of its latest 706 Report, three issues for Congress were how and at what level broadband benchmarks should be set, whether the FCC would determine that broadband is being deployed in a reasonable and timely fashion, and how that determination and those benchmarks would impact current and future broadband policies and programs intended to improve broadband availability and adoption throughout the nation. As broadband technology advances, commercially available download and upload speeds will likely increase, and the level at which broadband benchmark threshold speeds should be set is likely to remain controversial. Accordingly, the FCC's annual Section 706 determination is likely to remain contentious as long as it is seen by stakeholders as providing a justification for current or future FCC regulatory or deregulatory policies.

The debate over the *Thirteenth NOI* followed this paradigm. Stakeholders who supported an FCC determination that broadband is not being deployed in a reasonable and timely fashion generally opposed lowering the broadband benchmark by considering the presence of either fixed (at 25 Mbps/3 Mbps) or mobile (at 10 Mbps/1 Mbps)

No. 17 199, September 21, 2017, p. 5-6, available at https://ecfsapi.fcc.gov/file/1092177736005/ 2016%2013th%20BB%20Progress%20NOI%20Comments.pdf.

[20] Comments of the Free State Foundation before the FCC, In the Matter of Inquiry Concerning the Deployment of Advanced Telecommunications Capability to All Americans in a Reasonable and Timely Fashion, GN Docket No. 17 199, September 21, 2017, p. 3-4, available at https://ecfsapi.fcc.gov/file/1092136091140/ FSF%20 Comments%20in%20Section%20706%20Inquiry%20-%20Final%20-%20092117.pdf.

[21] *2018 Broadband Deployment Report*, pp. 7-8.

broadband as an indication that an area has adequate broadband service.[22] On the other hand, stakeholders who supported an FCC determination that broadband is being deployed in a reasonable and timely fashion generally supported changing the FCC's broadband benchmark methodology to include the presence of either fixed or mobile broadband as an indication that an area is receiving adequate broadband service.[23]

Ultimately, the levels at which broadband speed benchmarks are set can have an impact on broadband speeds, particularly those offered in unserved and underserved rural areas served by providers utilizing federal subsidies or assistance such as the Connect America Fund or the loan and grant programs in the RUS. Higher broadband download and upload speeds can translate into more robust and varied applications, but are generally more costly to deploy. Lower broadband speeds, because they are less costly to deploy, may be built out to more rural communities, but may offer less capacity and economic benefits than the typically faster broadband connections available in urban and suburban areas. Given the finite amount of federal funding available, policymakers face a tradeoff between the quality of rural broadband connections versus the quantity of rural areas that will be considered "served."

---

[22] For example, those stakeholders typically supported policies such as Open Internet and Title II classification, removing state barriers to community broadband, and a more robust Lifeline program to support improved digital inclusion programs.

[23] For example, those stakeholders typically supported policies such as removing local and government barriers to wireless and wireline siting; identifying and allocating more licensed and unlicensed spectrum for commercial broadband services; removing Title II requirements; and removing requirements that incumbent providers maintain and lease their legacy networks that employ copper and outdated technology.

In: Broadband
Editor: Orlando Harrison

ISBN: 978-1-53616-409-1

Chapter 3

ACCESS TO BROADBAND NETWORKS: NET NEUTRALITY*

Angele A. Gilroy

INTRODUCTION

The move to place restrictions on the owners of the networks that comprise and provide access to the internet, to ensure equal access and nondiscriminatory treatment, is referred to as "net neutrality." While there is no single accepted definition of net neutrality most agree that any such definition should include the general principles that owners of the networks that comprise and provide access to the internet should not control how consumers lawfully use that network; and should not be able to discriminate against content provider access to that network.

Determining the appropriate framework to ensure an open internet is central to the debate over broadband access, and is an issue that the Federal Communications Commission (FCC) has been grappling with for decades. Some policymakers contend that more proscriptive regulations, such as those contained in the FCC's 2015 Open Internet Order (2015 Order), are necessary to protect the marketplace from potential abuses which could threaten the net neutrality concept. Others contend that existing laws and the current, less restrictive approach, contained in the FCC's 2017 Restoring Internet Freedom Order (2017 Order), provide a more suitable framework.

---

* This is an edited, reformatted and augmented version of Congressional Research Service Publication No. IF10955, Updated April 10, 2019.

There is also a growing consensus that Congress should amend the 1934 Communications Act, as amended, (Communications Act) to address the debate.

THE FCC 2015 OPEN INTERNET ORDER

The FCC in February 2015, adopted the Open Internet Order to establish a regulatory framework to address access to broadband internet access service (BIAS), that is, the retail broadband service Americans buy from cable, phone, satellite, and wireless providers. The 2015 Order which applies to both mobile as well as fixed BIAS included provisions that reclassified BIAS as a telecommunications service under Title II of the Communications Act; placed a ban on blocking, throttling, and paid prioritization; created a general conduct standard; enhanced transparency requirements; permitted providers to engage in "reasonable network management" and provide "specialized services" (e.g., heart or energy consumption monitors); and in general did not apply the rules to internet interconnection.

THE FCC 2017 RESTORING INTERNET FREEDOM ORDER

In December 2017 the FCC vacated the 2015 Order and adopted a new framework (2017 Order) that reclassifies BIAS and embraces a less proscriptive approach. The 2017 Order, which went into effect on June 11, 2018, among other things: restores the classification of BIAS as an information service under Title I and Federal Trade Commission (FTC) authority; removes the no blocking, no throttling, and no paid prioritization rules; eliminates the general conduct standard; removes FCC authority over internet interconnection; expands the public transparency rules by requiring internet service providers to publicly disclose information about their network management practices including blocking, throttling, paid prioritization, and commercial terms of service; and preempts local and state laws that conflict with this framework.

SELECTED POLICY ISSUES

As Congress continues to debate what the appropriate framework should be for broadband access, some of the major policy issues focus on: the regulatory classification of BIAS; the role of the FCC versus the FTC; the impact of paid prioritization; and the role of the states.

Regulatory Classification—Title I versus Title II

The FCC derives its authority to establish its policies and regulations from the Communications Act. The Telecommunications Act of 1996 added the definitions of information service and telecommunications service to the Communications Act and applied the law to these services in different ways. Telecommunications service providers must be treated as common carriers under Title II of the act, which grants the FCC broad regulatory powers. Information service providers are defined under Title I, but may not be regulated as common carriers, and the FCC's authority over those services is more circumscribed. There is no provision in the current statute defining BIAS, but most agree that BIAS arguably could be interpreted as a telecommunications or an information service. This had led to one of the most contentious issues surrounding the net neutrality debate: Under which definition should BIAS fall?

The FCC chose, when adopting the 2015 Order, to classify BIAS as a telecommunications service placing it under the extensive regulatory framework, originally established to regulate monopoly voice telephone service, of Title II. On the other hand, the FCC, when it adopted the 2017 Order, reversed that classification and reclassified BIAS as an information service, effectively limiting the FCC's authority to regulate BIAS directly. The FCC's change in the classification of BIAS in less than a three-year period, has reopened the debate over whether congressional action is needed to provide clarity and stability. There is increasing agreement that legislation to outline regulatory authority and provide a stable framework for BIAS is desirable, but the specifics remain elusive.

Role of the FCC versus the FTC

The FCC, an independent regulatory agency, regulates interstate wire and radio communications as delineated in the Communications Act. The FCC generally promulgates and enforces rules/regulations using its public interest standard, to establish regulatory frameworks as applied to the communications sector. The FTC has broad authority to oversee the trade practices of entities across numerous market sectors under Section 5 of the Federal Trade Commission Act (FTC Act), which prohibits unfair methods of competition and unfair and deceptive trade practices. The FTC generally enforces Section 5 by examining particular factual circumstances and bringing enforcement actions on a case-by-case basis against practices the FTC believes violate Section 5. In contrast to the FCC, the FTC rarely issues prescriptive rules, unless specifically directed to do so by Congress.

However, certain entities and activities are exempted from Section 5's coverage. These exemptions include one for common carriers regulated under the Communications

Act, such as landline or wireless telephone services. The classification of BIAS, therefore, has a major impact not only on the regulatory role of the FCC but the FTC as well. The FTC has no authority over BIAS if it is classified as a telecommunications service, as it was in the 2015 Order, because telecommunications services are common carriers under the Communications Act. On the other hand, the FTC does have authority to enforce Section 5 against BIAS providers when BIAS is an information service, because information services cannot be regulated as common carrier services under the Communications Act.

In the 2017 Order, the FCC argued that classifying BIAS as an information service and returning authority to the FTC was the preferable course of action due to, among other considerations, the FTC's extensive experience overseeing data security and privacy practices across numerous sectors. The FTC supported the FCC's decision. Other stakeholders disagreed, arguing that the FTC's practice of case-by-case adjudication lacked certainty, and the establishment of a regulatory framework would better serve the public interest.

Network Management and Paid Prioritization

In the past most internet traffic was delivered on what is known as a "best efforts" basis, a quality standard that does not guarantee that the traffic will be delivered by a certain time or speed. Under best efforts some data packets arriving at congestion points will be dropped and held until a future date while others will be forwarded in real time. Earlier common applications (e.g., email) are not time sensitive and the use of best efforts will not degrade the user experience. Newer applications (e.g., telemedicine) are sensitive to interruption and latency making network management practices that affect how packets travel over the network of greater concern. Content providers offering congestion-sensitive services may be given the option to pay a fee, to network managers, to ensure quality of service by being given priority of transit at network congestion points. This practice is known as *paid prioritization*.

Whether a BIAS regulatory framework should contain provisions to address the practice of paid prioritization has become a major discussion point in the net neutrality debate. Some see paid prioritization as a reasonable management tool to ensure that time-sensitive applications get priority over nonsensitive applications. Such prioritization may be necessary, they state, given the growth in such applications and may be needed to spur innovation. Others feel that paid prioritization is anticompetitive and should be prohibited. Paid prioritization, they state, could have a negative impact on small, nascent companies that cannot afford to pay such fees, may be used to discriminate among content, and could thwart innovation. Whether paid prioritization is necessary to accommodate the growth of time-sensitive applications or is a practice that, absent regulation, will cause marketplace or consumer harm remains a contentious issue.

State Activity

The controversy over net neutrality is also playing out in the states as they take three approaches: issuing executive orders; promulgating laws; and/or filing legal challenges. Some states are using their financial power and procurement policies to regulate internet service provider (ISP) behavior by issuing executive orders that require ISPs that conduct business with the state to adhere to various net neutrality rules. Many ISPs, not wishing to forgo state procurement contracts, may adhere to executive order requirements. In other cases, states are promulgating laws establishing specific net neutrality regulations that would apply to all ISP activity within their states. Individual legislation varies, with certain states embracing all or some of the regulations contained in the 2015 Order, while others address issues that go beyond its scope.

How successful states will be in achieving their objectives is subject to question. The FCC's 2017 Order states that BIAS is an interstate service and contains provisions that preempt state or local requirements that are inconsistent with the Order leaving the authority of the states to enforce these actions unclear.

State attorneys general from 22 states and the District of Columbia are among the petitioners challenging the legality of the 2017 Order. Petitions have been consolidated in the U.S. Court of Appeals, D.C. Circuit (*Mozilla v. FCC, et al.*, No. 18-1051). Oral arguments were held February 1, 2019.

CONGRESSIONAL ACTIVITY—116TH CONGRESS

Debate over what the appropriate regulatory framework should be for broadband access has continued in the 116th Congress. Two bills (H.R. 1644, S. 682) add a new title to the Communications Act that overturns the 2017 Order and restores the 2015 Order and its subsequent regulations. The bills would, among other things, reclassify both mobile and fixed BIAS as a telecommunications service under Title II of the Communications Act. H.R. 1644, as amended, passed the House (232-190) on April 10, 2019.

Three bills (H.R. 1006, H.R. 1096, H.R. 1101) have been introduced to provide a regulatory framework to outline FCC authority over BIAS by amending Title I of the Communications Act, and one (H.R. 1860) to prohibit FCC regulation of rates charged for BIAS.

In: Broadband
Editor: Orlando Harrison
ISBN: 978-1-53616-409-1

Chapter 4

THE NET NEUTRALITY DEBATE: ACCESS TO BROADBAND NETWORKS*

Angele A. Gilroy

ABSTRACT

As congressional policymakers continue to debate telecommunications reform, a major discussion point revolves around what approach should be taken to ensure unfettered access to the internet. The move to place restrictions on the owners of the networks that compose and provide access to the internet, to ensure equal access and nondiscriminatory treatment, is referred to as "net neutrality." There is no single accepted definition of "net neutrality," but most agree that any such definition should include the general principles that owners of the networks that compose and provide access to the internet should not control how consumers lawfully use that network, and they should not be able to discriminate against content provider access to that network.

The Federal Communications Commission (FCC) in its February 26, 2015, open meeting voted 3-2, along party lines, to adopt open internet rules and released these rules on March 12, 2015. One of the most controversial aspects of the rules was the decision to reclassify broadband internet access service (BIAS) as telecommunications service under Title II, thereby subjecting internet service providers to a more stringent regulatory framework. With limited exceptions, the rules went into effect June 12, 2015. Various parties challenged the legality of the FCC's 2015 Open Internet Order, but the U.S. Court of Appeals for the D.C. Circuit, in a June 14, 2016, ruling, voted (2-1) to uphold the legality of all aspects of the 2015 FCC Order. A petition for full U.S. Appeals Court review was denied and a subsequent petition for U.S. Supreme Court review was declined.

The FCC on December 14, 2017, adopted (3-2) an Order that largely reverses the 2015 regulatory framework. The 2017 Order, among other things, reverses the 2015

* This is an edited, reformatted and augmented version of Congressional Research Service, Publication No. R40616, dated April 15, 2019.

classification of BIAS as a telecommunications service under Title II of the Communications Act, shifts much of the oversight from the FCC to the Federal Trade Commission and the Department of Justice, and provides for a less regulatory approach. This action has once again opened up the debate over what the appropriate framework is to ensure an open internet. Reaction to the 2017 Order has been mixed. Some see the 2015 FCC rules as regulatory overreach and welcome a more "light-touch" approach, which they feel will stimulate broadband investment, deployment, and innovation. Others support the 2015 regulations and feel that their reversal will result in a concentration of power to the detriment of content, services, and applications providers, as well as consumers, and refute the claim that these regulations have had a negative impact on broadband investment, expansion, or innovation. The 2017 Order was published in the *Federal Register* on February 22, 2018, and went into effect on June 11, 2018. *Federal Register* publication triggered timelines for both court challenges and Congressional Review Act (CRA) consideration. Petitions for review have been consolidated in the U.S. Court of Appeals, D.C. Circuit. CRA resolutions (S.J.Res. 52, H.J.Res. 129) to overturn the 2017 Order were introduced in the 115th Congress. S.J.Res. 52 passed (52-47) the Senate, but H.J.Res. 129 was not considered in the House. Additional bills to provide a regulatory framework to outline FCC authority over broadband internet access services were introduced, but not acted on, in the 115th Congress.

Debate over what the appropriate regulatory framework should be for broadband access has continued in the 116th Congress. Two bills (H.R. 1644, S. 682) add a new title to the Communications Act that overturns the 2017 Order and restores the 2015 Order. An amended version of H.R. 1644 passed (232-190) the House on April 10, 2019. Additional bills (H.R. 1006, H.R. 1096, H.R. 1101, and H.R. 1860) that address the net neutrality debate have also been introduced.

INTRODUCTION

As congressional policymakers continue to debate telecommunications reform, a major discussion point revolves around what approach should be taken to ensure unfettered access to the internet. The move to place restrictions on the owners of the networks that compose and provide access to the internet, to ensure equal access and nondiscriminatory treatment, is referred to as "net neutrality." There is no single accepted definition of "net neutrality." However, most agree that any such definition should include the general principles that owners of the networks that compose and provide access to the internet should not control how consumers lawfully use that network, and they should not be able to discriminate against content provider access to that network.

A major focus in the debate is concern over whether the regulatory framework as delineated in the Federal Communications Commission's (FCC's) 2015 Open Internet Order is the appropriate approach to ensure access to the internet for content, services, and applications providers, as well as consumers, or whether a less regulatory approach contained in the 2017 Order is more suitable. The issue of regulation of and access to broadband networks is currently being addressed in three venues

- at the FCC, where the commissioners on December 14, 2017, adopted (3-2) an Order that went into effect on June 11, 2018, that revokes the 2015 regulatory framework in favor of one that reverses the 2015 classification of broadband internet access services as a telecommunications service under Title II of the Communications Act, provides for a less regulatory approach, and shifts much of the oversight from the FCC to the Federal Trade Commission (FTC) and the Department of Justice (DOJ);
- in the courts, where consolidated petitions for review of the 2017 Order are under consideration in the U.S. Court of Appeals, D.C. Circuit; and a suit filed in the U.S. District Court of the Eastern District of California by the DOJ and various trade groups, challenging the legality of a California internet regulation law, is pending;
- in the 116th Congress, where debate over what the appropriate regulatory framework should be for broadband access continues.

Whether Congress will take broader action to amend existing law to provide guidance and more stability to FCC authority over broadband access remains to be seen.

FEDERAL COMMUNICATIONS COMMISSION ACTIVITY

The Information Services Designation and Title I

In 2005 two major actions dramatically changed the regulatory landscape as it applied to broadband services, further fueling the net neutrality debate. In both cases these actions led to the classification of broadband internet access services as Title I information services, thereby subjecting them to a less rigorous regulatory framework than those services classified as telecommunications services. In the first action, the U.S. Supreme Court, in a June 2005 decision (*National Cable & Telecommunications Association v. Brand X Internet Services*), upheld the Federal Communications Commission's (FCC's) 2002 ruling that the provision of cable modem service (i.e., cable television broadband internet) is an interstate information service and is therefore subject to the less stringent regulatory regime under Title I of the Communications Act of 1934.[1] In a second action, the FCC, in an August 5, 2005, decision, extended the same regulatory relief to telephone company internet access services (i.e., wireline broadband internet access, or DSL), thereby also defining such services as information services

---

[1] 47 U.S.C. 151 et seq.

subject to Title I regulation.[2] As a result, neither telephone companies nor cable companies, when providing broadband services, are required to adhere to the more stringent regulatory regime for telecommunications services found under Title II (common carrier) of the 1934 act.[3] However, classification as an information service does not free the service from regulation. The FCC continues to have regulatory authority over information services under its Title I, ancillary jurisdiction.[4] Similarly, classification under Title II does not mean that an entity will be subject to the full range of regulatory requirements, as the FCC is given the authority, under Section 10 of the Communications Act of 1934, to forbear from regulation.

The 2005 Internet Policy Statement

Simultaneous to the issuing of its August 2005 information services classification order, the FCC also adopted a policy statement (internet policy statement) outlining four principles to "encourage broadband deployment and preserve and promote the open and interconnected nature of [the] public Internet." The four principles are (1) consumers are entitled to access the lawful internet content of their choice; (2) consumers are entitled to run applications and services of their choice (subject to the needs of law enforcement); (3) consumers are entitled to connect their choice of legal devices that do not harm the network; and (4) consumers are entitled to competition among network providers, application and service providers, and content providers. Then-FCC Chairman Martin did not call for their codification. However, he stated that they would be incorporated into the policymaking activities of the commission.[5] For example, one of the agreed-upon conditions for the October 2005 approval of both the Verizon/MCI and the SBC/AT&T mergers was an agreement made by the involved parties to commit, for two years, "to conduct business in a way that comports with the commission's (2005) Internet policy statement."[6] In a further action, AT&T included in its concessions to gain FCC approval of its merger to BellSouth an agreement to adhere, for two years, to significant net neutrality requirements. Under terms of the merger agreement, which was approved on

---

[2] See http://hraunfoss.fcc.gov/edocs_public/attachmatch/DOC-260433A2.pdf for a copy of former FCC Chairman Martin's statement. For a summary of the final rule see Appropriate Framework for Broadband Access to the Internet over Wireline Facilities. *Federal Register*, Vol. 70, No. 199, October 17, 2005, p. 60222.

[3] For example, Title II regulations impose rigorous antidiscrimination, interconnection and access requirements. For a further discussion of Title I versus Title II regulatory authority see CRS Report R43971, *Net Neutrality: Selected Legal Issues Raised by the FCC's 2015 Open Internet Order*.

[4] Title I of the 1934 Communications Act gives the FCC such authority if assertion of jurisdiction is "reasonably ancillary to the effective performance of [its] various responsibilities." The FCC in its order cites consumer protection, network reliability, or national security obligations as examples of cases where such authority would apply (see paragraph 36 of the final rule summarized in the *Federal Register* cite in footnote 2, above).

[5] See https://apps.fcc.gov/edocs_public/attachmatch/FCC-05-151A1.pdf. August 5, 2005. *FCC Adopts Policy Statement on Broadband Internet Access.*

[6] See http://hraunfoss.FCC.gov/edocs_public/attachmatch/DOC-261936A1.pdf. It should be noted that applicants offered certain voluntary commitments, of which this was one.

December 29, 2006, AT&T not only agreed to uphold, for 30 months, the FCC's internet policy statement principles, but also committed, for two years (expired December 2008), to stringent requirements to "maintain a neutral network and neutral routing in its wireline broadband Internet access service."[7]

Then-FCC Chairman Genachowski announced, in a September 21, 2009, speech,[8] a proposal to consider the expansion and codification of the 2005 internet policy statement and suggested that this be accomplished through a notice of proposed rulemaking (NPR) process. Shortly thereafter, an NPR on preserving the open internet and broadband industry practices was adopted by the FCC in its October 22, 2009, meeting. (See "The FCC 2010 Open Internet Order," below).

The FCC August 2008 Comcast Decision

In perhaps one of its most significant actions relating to its internet policy statement to date, the FCC, on August 1, 2008, ruled that Comcast Corp., a provider of internet access over cable lines, violated the FCC's policy statement when it selectively blocked peer-to-peer connections in an attempt to manage its traffic.[9] This practice, the FCC concluded, "unduly interfered with Internet users' rights to access the lawful Internet content and to use the applications of their choice." Although no monetary penalties were imposed, Comcast was required to stop these practices by the end of 2008. Comcast complied with the order, and developed a new system to manage network congestion. Comcast no longer manages congestion by focusing on specific applications (such as peer-to-peer), nor by focusing on online activities, or protocols, but identifies individual users within congested neighborhoods that are using large amounts of bandwidth in real time and slows them down, by placing them in a lower priority category, for short periods.[10] This new system complies with the FCC internet principles in that it is application agnostic; that is, it does not discriminate against or favor one application over another but manages congestion based on the amount of a user's real-time bandwidth usage. As a result of an April 6, 2010, court ruling, the FCC's order was vacated. Comcast, however, stated that it will continue to comply with the internet principles issued in the FCC's August 2005 internet policy statement.[11] (See "Comcast v. FCC," below).

---

[7] See http://hraunfoss.fcc.gov/edocs_public/attachmatch/DOC-269275A1.pdf.

[8] "*Preserving a Free and Open Internet: A Platform for Innovation, Opportunity, and Prosperity*," prepared remarks of FCC Chairman Julius Genachowski, at the Brookings Institution, September 21, 2009. Available at http://hraunfoss.fcc.gov/edocs_public/ attachmatch/DOC-293568A1.pdf.

[9] See http://hraunfoss.fcc.gov/edocs_public/attachmatch/FCC-08-183A1.pdf.

[10] Comcast, *Frequently Asked Questions and Network Management.* Available at http://help. comcast.net/content/faq/Frequently-Asked-Questions-about-Network-Management.

[11] *Comcast Statement on U.S. Court of Appeals Decision on Comcast v. FCC.* Available at http://www.comcast.com/ About/PressRelease/PressReleaseDetail.ashx?PRID=984.

Comcast v. FCC

Despite compliance, however, Comcast filed an appeal[12] in the U.S. Court of Appeals for the District of Columbia, claiming that the FCC did not have the authority to enforce its internet policy statement, therefore making the order invalid. The FCC argued that while it did not have express statutory authority over such practices, it derived such authority based on its ancillary authority contained in Title I of the 1934 Communications Act. The court, in an April 6, 2010, decision, ruled (3-0) that the FCC did not have the authority to regulate an internet service provider's (in this case Comcast's) network management practices and vacated the FCC's order.[13] The court ruled that the exercise of ancillary authority must be linked to statutory authority and that the FCC did not in its arguments prove that connection; it cannot exercise ancillary authority based on policy alone. More specifically, the Court ruled that the FCC "failed to tie its assertion of ancillary authority over Comcast's Internet service to any ['statutorily mandated responsibility']."[14] Based on that conclusion the court granted the petition for review and vacated the order.

The impact of this decision on the FCC's ability to regulate broadband services and implement its broadband policy goals remains unclear. Regardless of the path that is taken, then-FCC Chairman Genachowski stated that the court decision "does not change our broadband policy goals, or the ultimate authority of the FCC to act to achieve those goals." He further stated that "[T]he court did not question the FCC's goals; it merely invalidated one, technical, legal mechanism for broadband policy chosen by prior Commissions."[15] Consistent with this statement, the FCC in a December 21, 2010, action adopted the Open Internet Order to establish rules to maintain network neutrality. (See "The FCC 2010 Open Internet Order," below).

The FCC 2010 Open Internet Order

The FCC adopted, on December 21, 2010, an Open Internet Order establishing rules to govern the network management practices of broadband internet access providers.[16] The order, which was passed by a 3-2 vote,[17] intended to maintain network neutrality by

---

[12] Comcast Corporation v. FCC, No. 08-129 (D.C. Cir. September 4, 2008).

[13] Comcast Corporation v. FCC decided April 6, 2010. Available at http://pacer.cadc.uscourts.gov/common/opinions/201004/08-1291-1238302.pdf.

[14] Comcast v. FCC decision, issued April 6, 2010, part V, p. 36.

[15] FCC Announces Broadband Action Agenda, released April 8, 2010. Available at http://hraunfoss.fcc.gov/edocs_public/attachmatch/DOC-297402A1.pdf.

[16] *In the Matter of Preserving the Open Internet, Broadband Industry Practices*. GN Docket No. 09-191; WC Docket No. 07-52, released December 23, 2010. Available at https://apps.fcc.gov/edocs_public/attachmatch/FCC-10-201A1_Rcd.pdf.

[17] The vote fell along party lines with Chairman Genachowski approving, Commissioner Clyburn approving in part and concurring in part, former Commissioner Copps concurring, and Commissioner McDowell and former Commissioner Baker dissenting.

establishing three rules covering transparency,[18] no blocking, and no unreasonable discrimination. More specifically

- fixed and mobile broadband internet service providers were required to publicly disclose accurate information regarding network management practices, performance, and commercial terms to consumers as well as content, application, service, and device providers;
- fixed and mobile broadband internet service providers were both subject, to varying degrees, to no blocking requirements. Fixed providers were prohibited from blocking lawful content, applications, services, or nonharmful devices, subject to reasonable network management. Mobile providers were prohibited from blocking consumers from accessing lawful websites, subject to reasonable network management, nor were they allowed to block applications that compete with the provider's voice or video telephony services, subject to reasonable network management; and
- fixed broadband internet service providers were subject to a "no unreasonable discrimination rule" that states that they shall not unreasonably discriminate in transmitting lawful network traffic over a consumer's broadband internet access service. Reasonable network management shall not constitute unreasonable discrimination.[19]

Additional provisions in the order included those which provided for ongoing monitoring of the mobile broadband sector and created an Open Internet Advisory Committee[20] to track and evaluate the effects of the rules and provide recommendations to the FCC regarding open internet policies and practices; while not banning paid prioritization, stated that it was unlikely to satisfy the "no unreasonable discrimination" rule; raised concerns about specialized services and while not "adopting policies specific to such services at this time," would closely monitor such services; called for review, and possible adjustment, of all rules in the order no later than two years from their effective date; and detailed a formal and informal complaint process. The order, however, did not prohibit tiered or usage-based pricing (see "Metered/Usage-Based Billing," below).

---

[18] The FCC, on June 30, 2011, released a public notice offering initial guidance regarding compliance with the transparency rule. *FCC Enforcement Bureau and Office of General Counsel Issue Advisory Guidance for Compliance with Open Internet Transparency Rule*. Available at http://hraunfoss.fcc.gov/edocs_public/attachmatch/DA-11- 1148A1.pdf.

[19] A network management practice is considered reasonable if "it is appropriate and tailored to achieving a legitimate network management purpose, taking into account the particular network architecture and technology of the broadband Internet access service." Cited examples include ensuring network security and integrity; providing parental controls; or reducing or mitigating the effects of congestion on the network.

[20] The FCC announced the creation of an Open Internet Advisory Committee April 21, 2011, *Federal Register*, Vol. 76, No. 77, April 21, 2011, p. 22395. The committee, which includes members from a broad range of industry, academic and community representatives held its first meeting in July 2012. For additional information see http://www.fcc.gov/ encyclopedia/open-internet-advisory-committee.

According to the order, the framework "does not prevent broadband providers from asking subscribers who use the network less to pay less, and subscribers who use the network more to pay more" since prohibiting such practices "would force lighter end users of the network to subsidize heavier end users" and "would also foreclose practices that may appropriately align incentives to encourage efficient use of networks."[21]

The authority to adopt the order abandoned the "third way approach" previously endorsed by then-Chairman Genachowski and other Democratic commissioners,[22] and treated broadband internet access service as an information service under Title I. The order relied on a number of provisions contained in the 1934 Communications Act, as amended, to support FCC authority. According to the order the authority to implement these rules lies in Section 706 of the 1996 Telecommunications Act, which directs the FCC to "encourage the deployment on a reasonable and timely basis" of "advanced telecommunications capability" to all Americans and to take action if it finds that such capability is not being deployed in a reasonable and timely fashion;[23] Title II of the Communications Act and its role in protecting competition and consumers of telecommunications services; Title III, which gives the FCC the authority to license spectrum, subject to terms that serve the public interest, used to provide fixed and mobile wireless services; and Title VI, which gives the FCC the duty to protect competition in video services.

The Order went into effect November 20, 2011, which was 60 days after its publication in the *Federal Register*.[24] Since the Order's publication multiple appeals were filed and subsequently consolidated for review in the U.S. Court of Appeals, D.C. Circuit.[25] Verizon Communications was the remaining challenger seeking review[26] claiming, among issues, that it was a violation of free speech and that the FCC had exceeded its authority in establishing the rules.[27] The court issued its ruling on January

---

[21] *In the Matter of Preserving the Open Internet, Broadband Industry Practices,* paragraph 72.

[22] This approach consists of pursuing a bifurcated, or separate, regulatory approach by applying the specific provisions of Title II to the transmission component of broadband access service and subjecting the information component to, at most, whatever ancillary jurisdiction may exist under Title I. See *The Third Way: A Narrowly Tailored Broadband Framework*, FCC Chairman Julius Genachowski, May 6, 2010. Available at http://hraunfoss.fcc.gov/edocs_public/ attachmatch/DOC-297944A1.pdf. Also see *A Third-Way Legal Framework for Addressing the Comcast Dilemma*, Austin Schlick, FCC General Counsel, May 6, 2010. Available at http://hraunfoss.fcc.gov/edocs_public/attachmatch/ DOC-297945A1.pdf.

[23] The FCC made such a finding, that is that "broadband is not being deployed to all Americans in a reasonable and timely fashion" in its *Sixth Broadband Deployment Report,* adopted on July 16, 2010. Available at https://apps.fcc.gov/ edocs_public/attachmatch/FCC-10-129A1.pdf.

[24] Preserving the Open Internet; Final Rule. *Federal Register*, Vol.76, No. 185, September 23, 2011, pp. 59192-59235.

[25] Order Granting Mot. Cons., DC/1:11-ca-01356 (J.P.M.L., October 6, 2011).

[26] Earlier appeals by both companies were filed but dismissed by the court. See *Verizon v. FCC*, Case No. 11-1014, (D.C. Cir. January 20, 2011); and *MetroPCS Communications et al. v. FCC*, Case No. 11-1016 (D.C. Cir. January 24, 2011). The U.S. Court of Appeals, on April 4, 2011, rejected both filings as premature, stating that the Order is a rulemaking and therefore must first be published in the *Federal Register* before it can be subject to judicial review. *Verizon v. FCC*, Order Granting Mot. Dismiss, Case No. 11-1014 (D.C. Cir. April 4, 2011).

[27] *Verizon Communications Inc. v. FCC,* Case No. 11-1355 (D.C. Cir. October 18, 2011).

14, 2014, and remanded the decision to the FCC for consideration. (See "The 2014 Open Internet Order Court Ruling and the FCC Response," below).

The 2014 Open Internet Order Court Ruling and the FCC Response

Verizon Communications Inc. v. Federal Communications Commission

On January 14, 2014, the U.S. Court of Appeals, D.C. Circuit, issued its ruling on the challenge to the FCC's Open Internet Order (*Verizon Communications Inc. v. Federal Communications Commission*, D.C. Cir., No. 11-1355).[28] The court upheld the FCC's authority to regulate broadband internet access providers, and upheld the disclosure requirements of the Open Internet Order, but struck down the specific antiblocking and nondiscrimination rules contained in the Order. (See "The FCC 2010 Open Internet Order," above).

Citing the decision by the FCC to classify broadband providers as information service providers (see "The Information Services Designation and Title I"), not common carriers, the court stated that the Communications Act expressly prohibits the FCC from regulating them as such. The court was of the opinion that the Order's nondiscrimination rules, applied to fixed broadband providers, and antiblocking rules, applied to both fixed and wireless broadband providers, were an impermissible common carrier regulation of an information service and could not be applied.

However, the court upheld the disclosure rules, and more importantly upheld the FCC's general authority to use Section 706 (advanced communications incentives) of the Telecommunications Act of 1996 (P.L. 104-104) to regulate broadband internet providers. Therefore the court concluded that the FCC does, within limitations, have statutory authority, under Section 706, to establish rules relating to broadband deployment and broadband providers' treatment of internet traffic. The court remanded the case to the FCC for further action.

The Federal Communications Commission Response

In response to the court remand, then FCC Chairman Wheeler issued, on February 19, 2014, a statement outlining the steps proposed "to ensure that the Internet remains a platform for innovation, economic growth, and free expression."[29] Chairman Wheeler proposed that the FCC establish new rules under its Section 706 authority that enforce and enhance the transparency rule that was upheld by the court; fulfill the "no blocking" goal; fulfill the goals of the nondiscrimination rule; leave open as an option the possible

---

[28] *Verizon v. FCC*, Case No. 11-1355 (D.C. Cir. January 14, 2014).

[29] *Statement of FCC Chairman Tom Wheeler on the FCC's Open Internet Rules*, released February 19, 2014. Available at http://transition.fcc.gov/Daily_Releases/Daily_Business/2014/db0219/DOC-325654A1.pdf.

reclassification of internet access service as telecommunications service subject to Title II authority; forgo judicial review of the appeals court decision; solicit public comment; hold internet service providers to their commitment to honor the safeguards articulated in the 2010 Open Internet Order; and seek opportunities to enhance competition in the internet access market.

In conjunction with this statement the FCC established a new docket (GN Docket No. 14-28) to seek input on how to address the remand of the FCC's Open Internet rules.[30] This docket was released February 19, 2014, to seek opinion on "what actions the commission should make, consistent with our authority under section 706 and all other available sources of Commission authority, in light of the court's decision." However, it should be noted that FCC Commissioners O'Rielly and Pai issued separate statements expressing their disagreement with then-Chairman Wheeler's proposal to establish new rules to regulate the internet.[31] Despite this opposition, the FCC, on a 3-2 vote, initiated a proceeding to establish rules to address the court's remand of its 2010 open internet order. (See "The FCC 2014 Open Internet Notice of Proposed Rulemaking," below).

The FCC 2014 Open Internet Notice of Proposed Rulemaking

On May 15, 2014, the FCC adopted, by a 3-2 party line vote, a Notice of Proposed Rulemaking (NPRM) seeking public comment on "how best to protect and promote an open Internet." The NPRM (GN Docket No.14-28) solicited comment on a broad range of issues to help establish a policy framework to ensure that the internet remains an open platform and retains the concepts adopted by the FCC in its 2010 Open Internet Order, of transparency, no blocking, and nondiscrimination.[32]

Following the guidance of the January 2014 D.C. Circuit Appeals Court decision, the NPRM tentatively concluded that the FCC should rely on Section 706 of the 1996 Telecommunications Act for its legal authority. However, the NPRM noted that the FCC "will seriously consider the use of Title II of the Communications Act as the basis for legal authority" and recognizes that Section 706 and Title II are both "viable solutions."[33] The NPRM also recognized the use of Title III for mobile services and sought comment,

---

[30] *New Docket Established to Address Open Internet Remand*. GN Docket No. 14-28, released February 19, 2014. Available at http://transition.fcc.gov/Daily_Releases/Daily_Business/2014/db0219/DA-14-211A1.pdf.

[31] *Statement of FCC Commissioner O'Rielly on FCC Chairman Wheeler's "Open Internet" Announcement*, released February 19, 2014. *Statement of Commissioner Ajit Pai on FCC Internet Regulation*, released February 19, 2014. Available at http://transition.fcc.gov/Daily_Releases/Daily_Business/2014/db0219/DOC-325660A1.pdf and http://transition.fcc.gov/Daily_Releases/Daily_Business/2014/db0219/DOC-325659A1.pdf, respectively.

[32] *In the Matter of Protecting and Promoting the Open Internet*. GN Docket No. 14-28, adopted and released May 15, 2014. Available at http://transition.fcc.gov/Daily_Releases/Daily_Business/2014/db0515/FCC-14-61A1.pdf.

[33] *In the Matter of Protecting and Promoting the Open Internet,* para. 4.

in general, on other sources of authority the FCC may utilize. The degree to which the FCC should use forbearance was also discussed.

The NPRM retained the definition and scope contained in the 2010 Open Internet Order which address the actions of broadband internet access service providers, and as defined did not, for example, cover the exchange of traffic between networks (e.g., peering), enterprise services (i.e., services offered to large organizations through individually negotiated offerings), data storage services, or specialized services. However, the NPRM did seek comment on whether the scope of services as defined in the 2010 Open Internet Order should be modified. The question of whether broadband provider service to edge providers, that is, their function as edge providers' carriers, should be addressed was also raised. Furthermore, the NPRM sought comment on whether it should revisit its different standard applied to mobile services regarding its no-blocking rule and its exclusion from the unreasonable discrimination rule, and whether technological and marketplaces changes are such that the FCC should consider if rules should be applied to satellite broadband internet access services.

The FCC tentatively concluded that the nonblocking rule established in the 2010 Open Internet Order should be upheld, but that "the revived no-blocking rule should be interpreted as requiring broadband providers to furnish edge providers with a minimum level of access to their end-user subscribers."[34] However, the NPRM proposed that the conduct of broadband providers permissible under the no-blocking rule be subject to an additional independent screen which required them "to adhere to an enforceable legal standard of commercially reasonable practices."[35] Furthermore, the NPRM sought comment on whether certain practices, such as "paid prioritization," should be barred altogether or permitted if they meet the "commercial reasonableness" legal standard.[36]

In addition, the NPRM proposed to enhance the transparency rule, which was upheld by the court; to ensure that consumers and edge providers have the needed information to understand the services received and monitor practices; and to establish a multifaceted dispute resolution process including the creation of an ombudsperson to represent the interests of consumers, start-ups, and small businesses.

President Obama, in a statement released on November 10, 2014, urged the FCC to establish rules that would reclassify consumer broadband service under Title II of the 1934 Communications Act with forbearance. More specifically, the statement called for regulations that prohibit blocking; prohibit throttling; ban paid prioritization; and increase transparency. It was also stated that these rules should also be fully applicable to mobile broadband and if necessary to interconnection points. Monitored exceptions for reasonable network management and specialized services and forbearance from Title II

---

[34] *In the Matter of Protecting and Promoting the Open Internet,* para. 97.
[35] *In the Matter of Protecting and Promoting the Open Internet,* para. 116.
[36] *In the Matter of Protecting and Promoting the Open Internet,* para. 138.

regulations "that are not needed to implement the principles above" were also included in the statement.[37]

While the FCC evaluates all comments, including those of sitting Presidents, as an independent regulatory agency it has the sole responsibility to adopt the final proposal. New rules addressing this issue were adopted by the FCC in February 2015. (See "The FCC 2015 Open Internet Order," below).

The FCC 2015 Open Internet Order

The FCC, in its February 26, 2015, open meeting, voted 3-2, along party lines, to adopt new open internet rules (Open Internet Order) and subsequently released these rules on March 12, 2015.[38] The order applies to mobile as well as fixed broadband internet access service and relies on Title II of the Communications Act and Section 706 of the Telecommunications Act of 1996 and, for mobile broadband, Title III for its legal authority. The order includes among its provisions the following:

- reclassifies "broadband Internet access service" (that is the retail broadband service Americans buy from cable, phone, and wireless providers) as a telecommunications service under Title II;
- bans blocking, throttling, and paid prioritization;
- creates a general conduct standard that internet service providers cannot harm consumers or edge providers (e.g., Google, Netflix) and gives the FCC the authority to address questionable practices on a case-by-case basis (reasonable network management will not be considered a violation of this rule);
- enhances existing transparency rules for both end users and edge providers (a temporary exemption from the transparency enhancements is given for small fixed and mobile providers)[39] and creates a "safe harbor" process for the format and nature of the required disclosure for consumers;[40]

---

[37] A full copy of the text of the President's statement and is available at https://obamawhitehouse.archives.gov/net-neutrality.

[38] Federal Communications Commission, "FCC Releases Open Internet Order," http://www.fcc.gov/document/fcc-releases-open-internet-order.

[39] The 2015 Open Internet Order granted a temporary exemption from the enhanced transparency requirements for fixed and mobile providers with 100,000 or fewer subscribers. Subsequently, on February 23, 2017, the FCC adopted (2-1) an Order raising the exemption level from 100,000 or fewer subscribers to 250,000 or fewer subscribers and granting the exemption from enhanced transparency requirements for five years. Available at https://apps.fcc.gov/ edocs_public/attachmatch/FCC-17-17A1.pdf.

[40] The FCC on April 4, 2016, issued a public notice containing proposed formats, i.e., "consumer broadband labels," to provide consumers with the information needed under the transparency rules. These formats while recommended, are not required by the FCC, but will serve as a "safe harbor" to meet broadband internet service provider transparency requirements under the 2015 Open Internet Order. Available at http:// transition.fcc.gov/Daily_Releases/ Daily_Business/2016/db0404/DA-16-357A1.pdf.

- permits an internet service provider to engage in "reasonable network management" (other than paid prioritization) and will take into account the specific network management needs of mobile networks and other technologies such as unlicensed Wi-Fi networks;
- does not apply the open internet rules to "non-BIAS data services," (aka, specialized services) a category of services defined by the FCC as those that "do not provide access to the Internet generally" (e.g., heart monitors or energy consumption sensors);
- does not apply the open internet rules to interconnection but does gives the FCC authority to hear complaints and take enforcement action, if necessary, on a case-by-case basis, under Sections 201 and 202, regarding interconnection activities of internet service providers if deemed unjust and unreasonable;
- applies major provisions of Title II such as no unjust and unreasonable practices or discrimination, consumer privacy,[41] disability access, consumer complaint and enforcement processes, and fair access to poles and conduits; and
- forbears, without any further proceedings, from various Title II provisions (e.g., cost accounting rules, tariffs, and last-mile unbundling) resulting in forbearance from 30 statutory provisions and over 700 codified rules.

With limited exceptions, the rules went into effect on June 12, 2015, which was 60 days after their publication in the *Federal Register*.[42] Various trade groups and selected individual providers filed appeals to the courts challenging the 2015 Open Internet Order's legality;[43] the appeals were consolidated in the U.S. Court of Appeals for the D.C. Circuit.[44] A motion to stay the effective date of the Order was denied, allowing the rules to go into effect as scheduled on June 12, 2015.[45]

Subsequently, the U.S. Court of Appeals for the D.C. Circuit, in a June 14, 2016, ruling, voted (2-1) to uphold the legality of all aspects the 2015 FCC Order.[46] A petition

---

[41] On October 27, 2016, the FCC adopted rules to implement the customer privacy requirements for broadband internet service providers under the 2015 Open Internet Order. *In the Matter of Protecting the Privacy of Customers of Broadband and other Telecommunications Services, WC Docket No. 16-106*. Available at https://apps.fcc.gov/edocs_public/
attachmatch/FCC-16-148A1.pdf. However, these rules were revoked, by congressional action, prior to their complete implementation. (See "115th Congress" below).

[42] Federal Communications Commission, "Protecting and Promoting the Open Internet; Final Rule," 80 *Federal Register*, 19738-19850, April 13, 2015.

[43] For a discussion of selected legal issues surrounding the 2015 Order see CRS Report R43971, *Net Neutrality: Selected Legal Issues Raised by the FCC's 2015 Open Internet Order*, by Kathleen Ann Ruane.

[44] Parties have 60 days after publication in the *Federal Register* to file suit to challenge the FCC order. (Consolidated under U.S. Telecom Association v. Federal Communications Commission, D.C. Cir., 15-1063, order 4/14/15).

[45] U.S. Telecom Association v. FCC, D.C. Cir. No. 15-1063, Order Denying Motion for Stay and Granting Motion for Expedited Review (June 11, 2015). Available at http://docs.techfreedom.org/oiostaydenial.pdf.

[46] U.S. Telecom Association v. FCC, D.C. Cir., No. 15-1063, decision rendered 6/14/16. Available at http://pdfserver.amlaw.com/nlj/6-14-16%20DC%20Circuit%20net%20
neutrality%20opinion.pdf.

requesting en banc review[47] of the decision was denied on May 1, 2017, by the majority of the judges.[48] Various parties filed on September 28, 2017, petitions asking the U.S. Supreme Court for review;[49] but the U.S. Supreme Court, on November 5, 2018, denied review.

The FCC 2017 Open Internet Notice of Proposed Rulemaking

On May 18, 2017, the FCC adopted (2-1) a Notice of Proposed Rulemaking (NPR) to reexamine the regulatory framework established in the 2015 Open Internet Order and embrace a "light-touch" regulatory approach.[50] The NPR returned internet broadband access service to a Title I classification and sought comment on the existing rules governing internet service providers. More specifically, the NPR proposed to

- reinstate the information service classification of broadband internet access service (both fixed and mobile), thereby removing the service from the Title II, common carrier classification imposed by the 2015 Open Internet Order and placing it under Title I;
- reinstate that mobile broadband internet access service is not a commercial mobile service;
- eliminate the general conduct standard;
- seek comment on the need to "keep, modify, or eliminate" the "bright line" (no blocking, no throttling, and no paid prioritization) and transparency rules;
- return authority to the Federal Trade Commission to oversee and enforce the privacy practices of internet service providers;
- reevaluate the FCC's enforcement regime with respect to the necessity for ex ante regulatory intervention; and
- conduct a cost-benefit analysis as part of the proceeding.

The NPR comment and reply comment periods closed, and a draft order, largely upholding the NPR and overturning the 2015 Order, was approved (3-2) by the FCC on December 14, 2017. [See "WC Docket No. 17-108 (The FCC 2017 Order)," below].

---

[47] An en banc review calls for a review by the full court of all the active appeals judges in the jurisdiction instead of the previous three-judge quorum. U.S. Telecom Association v. FCC, 825 F.3d 674 (D.C. Cir. 2016), *petition for reh'g en banc filed*.

[48] U.S. Telecom Assoc. v. FCC, 825 F.3d 674 (D.C. Cir. 2016), *reh'g en banc denied* by U.S. Telecom. Assoc. v. FCC, No. 15-1063, slip op. (D.C. Cir. May 1, 2017).

[49] U.S. Telecom. Ass'n v. FCC, 825 F.3d 674 (D.C. Cir. 2016), *petition for cert. filed*, No. 17-498, No. 17-499, No. 17 500, No. 17-501, No. 17-502, No. 17-503, No. 17-504 (October 3, 2017).

[50] *In the Matter of Restoring Internet Freedom*, WC Docket No. 17-108. Available at http://transition.fcc.gov/Daily_Releases/Daily_Business/2017/db0523/FCC-17-60A1.pdf.

WC Docket No. 17-108 (The FCC 2017 Order)

The FCC, in its December 14, 2017, open meeting, voted 3-2, along party lines, to adopt a new framework for the provision of broadband internet access services that largely reversed the 2015 regulatory framework and shifted much of the oversight from the FCC to the Federal Trade Commission (FTC) and the Department of Justice (DOJ).[51] WC Docket No. 17-108 (The 2017 Order), among other things, removed broadband internet access services (BIAS) from the 2015 regulatory classification as telecommunications services subject to common carrier Title II classification; removed the "bright line" no blocking, no throttling, and no paid prioritization rules; and eliminated the general conduct standard; but expanded the public transparency rules.

More specifically, WC Docket No. 17-108 is divided into three parts: a Declaratory Ruling, a Report and Order, and an Order.[52]

The Declaratory Ruling

The Declaratory Ruling includes provisions that restore the classification of BIAS as an information service; reinstate the private mobile service classification of mobile BIAS; and restore broadband consumer protection authority to the FTC.

The Declaratory Ruling also finds that "... Title II regulation reduced Internet service provider (ISP) investment in networks, as well as hampered innovation, particularly among small ISPs serving rural customers"; and "... public policy, in addition to legal analysis, supports the information service classification, because it is more likely to encourage broadband investment and innovation, thereby furthering the closing of the digital divide and benefitting the entire Internet ecosystem."

The Report and Order

The Report and Order includes provisions that enhance transparency requirements by requiring internet service providers to publicly disclose information about their practices including blocking, throttling, and paid prioritization; and eliminates the internet conduct standard.

The Report and Order also finds that "... transparency, combined with market forces as well as antitrust and consumer protection laws, achieve benefits comparable to those of the 2015 'bright line' rules at lower cost."

---

[51] *FCC Acts to Restore Internet Freedom*, WC Docket No. 17-108, adopted December 14, 2017. Available at https://www.fcc.gov/document/fcc-takes-action-restore-internet-freedom.

[52] *FCC Acts to Restore Internet Freedom*, press release. Available at http://transition.fcc.gov/Daily_Releases/Daily_Business/2017/db1214/DOC-348261A1.pdf.

The Order

The Order finds that adding to the record in this proceeding is not in the public interest.

Reaction to the 2017 Order has been mixed. Some see the 2015 FCC rules as regulatory overreach and welcome a less regulatory approach, which they feel will stimulate broadband investment, deployment, and innovation. Others support the 2015 regulations and feel that their reversal will result in a concentration of power to the detriment of content, services, and applications providers, as well as consumers, and refute the claim that these regulations have had a negative impact on broadband investment, expansion, or innovation.

The 2017 Order, Restoring Internet Freedom, was released by the FCC on January 4, 2018,[53] and published in the *Federal Register* on February 22, 2018.[54] Publication in the *Federal Register* triggered timelines for both court challenges and Congressional Review Act consideration. Petitions challenging the legality of the 2017 Order have been consolidated in the U.S. Court of Appeals, D.C. Circuit[55] with oral arguments held on February 1, 2019. CRA resolutions to overturn the 2017 Order were introduced in the 115th Congress. The Senate measure (S.J.Res. 52) passed (52-47) the Senate on May 16, 2018, but the House measure (H.J.Res. 129) was not considered. (See "Congressional Activity," below).

Implementation of the 2017 Order

The 2017 Order went into effect on June 11, 2018.[56] As a result the 2015 Order has been revoked and replaced by the provisions contained in the 2017 Order. According to the FCC[57] the 2017 Order framework has three parts

- The FTC will assume the major role and will take action against internet service providers that undertake anticompetitive acts or unfair and deceptive practices;
- Internet service providers will be subject to enhanced transparency requirements and must publically disclose, via a publically available, easily accessible company website or through the FCC's website, information regarding their network management practices, performance, and commercial terms of service; and

---

[53] Restoring Internet Freedom, Federal Communications Commission, released January 4, 2018. Available at https://apps.fcc.gov/edocs_public/Query.do?docket=17-108.

[54] Federal Communications Commission, "Restoring Internet Freedom," *Federal Register*, vol. 83, no. 36, Thursday, February 22, 2018, Rules and Regulations. Available at https://www.gpo.gov/fdsys/pkg/FR-2018-02-22/pdf/2018- 03464.pdf.

[55] *Mozilla v. FCC, et al.*, No. 18-1051 (D.C. Cir. filed February 22, 2018).

[56] Federal Communications Commission, "Restoring Internet Freedom," *Federal Register*, vol. 83, no. 92, Friday, May 11, 2018, Rules and Regulations. Available at https://www.gpo.gov/fdsys/pkg/FR-2018-05-11/pdf/2018-10063.pdf.

[57] Federal Communications Commission, "Restoring Internet Freedom Order." Available at https://www.fcc.gov/restoring-internet-freedom.

- Broadband internet access services are reclassified as information services, and regulations imposed by the 2015 Order are vacated, including the classification of broadband internet access services as telecommunications services subject to common carrier Title II classification; the "bright line" no blocking, no throttling, and no paid prioritization rules; and the general conduct standard.

NETWORK MANAGEMENT

As consumers expand their use of the internet and new multimedia and voice services become more commonplace, control over network quality and pricing is an issue. The ability of data bits to travel the network in a nondiscriminatory manner (subject to reasonable management practices), as well as the pricing structure established by broadband service providers for consumer access to that data, have become significant issues in the debate.

Prioritization

In the past, internet traffic has been delivered on a "best efforts" basis. The quality of service needed for the delivery of the most popular uses, such as email or surfing the web, is not as dependent on guaranteed quality. However, as internet use expands to include video, online gaming, and voice service, the need for uninterrupted streams of data becomes important. As the demand for such services continues to expand, a debate over the need to prioritize network traffic to ensure the quality of these services has formed.

The need to establish prioritized networks, although embraced by some, has led others to express policy concerns. Concern has been expressed that the ability of network providers to prioritize traffic may give them too much power over the operation of, and access to, the internet. If a multitiered internet develops where content providers pay for different service levels, some have expressed concern that the potential to limit competition exists if smaller, less financially secure content providers are unable to afford to pay for a higher level of access. Also, they state, if network providers have control over who is given priority access, the ability to discriminate among who gets such access is also present. If such a scenario were to develop, they claim, any potential benefits to consumers of a prioritized network would be lessened by a decrease in consumer choice and/or increased costs, if the fees charged for premium access are passed on to the consumer.

Others state that prioritization will benefit consumers by ensuring faster delivery and quality of service and may be necessary to ensure the proper functioning of expanded service options. They claim that the marketplace for the provision of such services is expanding and any potential abuse is significantly decreased in a marketplace where multiple, competing broadband providers exist. Under such conditions, they claim that if a network broadband provider blocks access to content or charges unreasonable fees content providers and consumers could obtain their access from other network providers. As consumers and content providers migrate to these competitors, market share and profits of the offending network provider will decrease, they state, leading to corrective action or failure. Furthermore, any abuses that may occur, they state, can be addressed by existing enforcement agencies at the federal and state level.

Deep Packet Inspection

The use of one management tool, deep packet inspection (DPI), illustrates the complexity of the net neutrality debate. DPI refers to a network management technique that enables network operators to inspect, in real time, both the header and the data field of the packets.[58] As a result, DPI not only can allow network operators to identify the origin and destination points of the data packet, but also enables the network operator to determine the application used and content of that packet. The information that DPI provides enables the network operator to differentiate, or discriminate, among the packets travelling over its network. The ability to discriminate among packets enables the network operator to treat packets differently. This ability itself is not necessarily viewed in a negative light. Network managers use DPI to assist them in performing various functions that are necessary for network management and that contribute to a positive user experience. For example, DPI technology is used in filters and firewalls to detect and prevent spam, viruses, worms, and malware. DPI is also used to gain information to help plan network capacity and diagnostics, as well as to respond to law enforcement requests.[59] However, some claim that the ability to discriminate based on the information gained via DPI also has the potential to be misused.[60] It is the potential negative impact that DPI use could have on consumers and suppliers that raises concern for some policymakers. For example, concern has been expressed that the information gained

[58] The header contains the processing information which includes the source and destination addresses, and the data field includes the message content and the identity of the source application.

[59] For a further discussion of the positive uses, by network operators, of DPI technologies see testimony of Kyle McSlarrow, President and CEO National Cable and Telecommunications Association, hearings on "Communications Networks and Consumer Privacy: Recent Developments," House Committee on Energy and Commerce, Subcommittee on Communications, Technology, and the Internet, April 23, 2009. Available at https://www.gpo.gov/fdsys/pkg/ CHRG-111hhrg72880/html/CHRG-111hhrg72880.htm.

[60] For a further discussion of the potential abuses associated with DPI technology see testimony of Ben Scott, Policy Director, Free Press, hearings on "Communications Networks and Consumer Privacy: Recent Developments," House Committee on Energy and Commerce, Subcommittee on Communications, Technology, and the Internet, April 23, 2009. Available at https://www.gpo.gov/fdsys/pkg/CHRG-111hhrg72880/html/CHRG-111hhrg72880.htm.

could be used to discriminate against a competing service, causing harm to both the competitor and consumer choice by, for example, routing a network operator's own, or other preferred content, along a faster priority path, or selectively slowing down competitors' traffic. DPI's potential to extract personal information about the data that it inspects has also generated concerns, by some, about consumer privacy.[61]

Therefore it is not the management tool itself that is under scrutiny, but the behavior that potentially may occur as a result of the information that DPI provides. How to develop a policy that permits some types of discrimination (i.e., "good" discrimination) that may be beneficial to network operation and improve the user experience while protecting against what would be considered "harmful" or anticompetitive discrimination becomes the crux of the policy debate.

Metered/Usage-Based Billing

The move by some network broadband operators toward the use of metered or usage-based billing has caused considerable controversy. Under such a plan, users subscribe to a set monthly bandwidth cap, for an established fee, and are charged additional fees or could be denied service if that usage level is exceeded. The use of such billing practices, on both a trial and permanent basis, is becoming more commonplace.

Reaction to the imposition of usage-based billing has been mixed. Supporters of such billing models state that a small percentage of users consume a disproportionately high percentage of bandwidth and that some form of usage-based pricing may benefit the majority of subscribers, particularly those who are light users.[62] Furthermore, they state that offering a range of service tiers at varying prices offers consumers more choice and control over their usage and subsequent costs. The major growth in bandwidth usage, they also claim, places financial pressure on existing networks for both maintenance and expansion, and establishing a pricing system which charges high bandwidth users is more equitable.

Opponents of such billing plans claim that such practices will stifle innovation in high bandwidth applications and are likely to discourage the experimentation with and adoption of new applications and services. Some concerns have also been expressed that a move to metered/usagebased pricing will help to protect the market share for video

[61] For example, concern that information can be gathered, without permission, based on consumer use of the internet to develop user profiles to provide targeted online advertising, also known as "behavioral advertising," has raised privacy issues. For an examination of this issue see testimony from hearings "Communications Networks and Consumer Privacy: Recent Developments," held April 23, 2009, by the House Energy and Commerce Subcommittee on Communications, Technology, and the Internet. Available at https://www.gpo.gov/fdsys/pkg/CHRG-111hhrg72880/ html/CHRG-111hhrg72880.htm.

[62] For example, Time Warner states that the top 25% of its users consume 100 times more bandwidth than the bottom 25% and 30% of its high speed internet service (i.e., Road Runner) customers use less than 1 GB (Gigabyte) per month. See *Consumption Based Billing FAQs*. Available at http://www.time warnercable.com/corporate/announcements/cbb_faq.html.

services offered in packaged bundles by network broadband service providers, who compete with new applications, and that if such caps must exist, they should be applied to all online video sources. The move to usage-based pricing, they state, will unfairly disadvantage competing online video services and stifle a nascent market, since video applications are more bandwidth-intensive. Opponents have also questioned the accuracy of meters, and specific usage limits and overage fees established in specific trials, stating that the former seem to be "arbitrarily low" and the latter "arbitrarily high."[63]

Furthermore, they state that since network congestion only occurs in specific locations and is temporary, monthly data caps are not a good measure of congestion causation. Citing the generally falling costs of network equipment and the stability of profit margins, they also question the claims of network broadband operators that increased revenue streams are needed to supply the necessary capital to invest in new infrastructure to meet the growing demand for high bandwidth applications.[64]

Zero Rating/Sponsored Data Plans

Zero rating plans refer to the practice by internet service providers of allowing their subscribers to consume specific content or services without incurring charges against the subscriber's usage limits. In the case of sponsored data a third party (e.g., an interested edge provider) pays the charges that the customer would otherwise incur. Many different variations of these services are being implemented in the marketplace by wireless carriers.

Supporters of such plans claim that they can improve consumer choice and encourage consumers to try new and perhaps usage-heavy services and can improve competition among edge providers. Those critical of such plans state that they can be used for anticompetitive purposes to favor one edge provider over another, particularly those edge providers that are affiliated with the internet service provider, or those that are entrenched and well financed.

Whether or not such activities should, or should not, be considered a violation of the terms of the FCC's 2015 Open Internet Order's general conduct rule remains controversial. The FCC, under former Chairman Wheeler, requested in a December 2015 letter that AT&T, T-Mobile, and Comcast[65] individually meet with FCC staff to discuss their various plans to enable the FCC to "have all the facts to understand how this service

[63] See Free Press letter to House Energy and Commerce Committee, April 22, 2009. Available at http://www. Freepress.net/files/FP_metering_letter.pdf.

[64] *As Costs Fall, Companies Push to Raise Internet Price*, New York Times, April 20, 2009. Available at http://www.nytimes.com/2009/04/20/business/20isp.html?_r=1.

[65] AT&T's Sponsored Data and Data Perks Programs, Comcast's Stream TV service, and T-Mobile's Binge On program were specifically referred to in the FCC letters. Is should be noted that Comcast states that its Stream TV service is not a zero-rated service since it does not travel across the public internet and is only available to its subscribers at their residence.

relates to the Commission's goal of maintaining a free and open Internet while incentivizing innovation and investment from all sources."[66] A similar letter was sent, at a later date (January 2016), to Verizon inquiring about its FreeBee Data 360 offering. The FCC concluded in a January 11, 2017, Policy Review Report that the specific approaches taken by AT&T and Verizon may harm competition and put forward a draft framework for evaluating such data offering plans.[67] Subsequently, however, the FCC, under current Chairman Pai, issued an Order on February 3, 2017, that retracted the report and closed the inquiries, further stating that the report "will have no legal or other effect or meaning going forward."[68]

CONGRESSIONAL ACTIVITY

116th Congress

Debate over what the appropriate regulatory framework should be for broadband access has continued in the 116th Congress. Six bills (H.R. 1006, H.R. 1096, H.R. 1101, H.R. 1644, H.R. 1860, S. 682) have been introduced to date. An amended version of H.R. 1644 passed (232-190) the House on April 10, 2019.

H.R. 1644 (and its companion measure, S. 682) add a new title to the Communications Act that negates the 2017 Order and restores the 2015 Order and its subsequent regulations. H.R. 1644, introduced on March 8, 2019, by Representative Doyle and S. 682, introduced on March 6, 2019, by Senator Markey, add a new title to the Communications Act that repeals the 2017 Order, stating that the rule "shall have no force or effect," and prohibits the FCC, in most circumstances, from reissuing the rule or enacting a new rule that is substantially the same. The bills also restore the 2015 Order and its subsequent regulations, thereby once again classifying both mobile and fixed broadband internet access service as a telecommunications service under Title II of the Communications Act and restoring regulations, including those that prohibit blocking, throttling, and paid prioritization and establish a general conduct standard.

An amended version of H.R. 1644 was passed (30-22) by the House Energy and Commerce Committee on April 3, 2019. A revised version of the subcommittee-passed H.R. 1644, a manager's amendment in the nature of a substitute (AINS) offered by Representative Doyle, was considered by the Committee. The AINS contained a

---

66 Copies of the three letters which were sent by the FCC on December 16, 2015, are available at http://src.bna.com/bCh; http://src.bna.com/bCj; http://src.bna.com/bCi.

67 *Policy Review of Mobile Broadband Operators' Sponsored Data Offerings for Zero Rated Content and Services*, released January 11, 2017, Wireless Telecommunications Bureau, available at https://apps.fcc.gov/edocs_public/attachmatch/DOC-342987A1.pdf.

68 In the Matter of Wireless Telecommunications Bureau Report: Policy Review of Mobile Broadband Operators' Sponsored Data Offerings for Zero Rated Content and Services. Order adopted February 3, 2017, available at https://apps.fcc.gov/edocs_public/attachmatch/DA-17-127A1.pdf.

provision, in the definition section, to clarify the forbearance provisions in the subcommittee passed bill. One amendment to the AINS, containing a one-year exemption from the 2015 Order enhanced transparency requirements for small internet service providers (that is those with 100,000 or fewer subscribers), was also approved prior to Committee passage. An amended version of H.R. 1644 (containing 12 additional amendments considered on the floor) passed (232-190) the House on April 10, 2019.

Three bills (H.R. 1006, H.R. 1096, H.R. 1101) establish a regulatory framework by amending Title I of the Communications Act and H.R. 1006, introduced on February 6, 2019, by Representative Latta, amends Title I of the Communications Act to address potential negative behaviors of BIAS providers. Provisions include those that prohibit blocking and unjust and discriminatory behavior, subject to reasonable network management; establish transparency requirements; establish FCC enforcement authority, including the authority to issue fines and forfeitures up to $2 million; in general, prohibit the FCC from imposing regulations on BIAS services under Title II; protect the needs of emergency communications, law enforcement, public safety or national security, copyright infringement or other unlawful activity; and preserve the authority of the Department of Justice and the Federal Trade Commission.

H.R. 1096, introduced on February 7, 2019, by Representative Rodgers, amends Title I of the Communications Act to require rules applicable to BIAS providers that establish transparency requirements; prohibit blocking and degrading (throttling) lawful traffic, subject to reasonable network management; prohibit paid prioritization; and contain a savings clause relating to emergency communications, law enforcement, public safety or national security, copyright infringement or other unlawful activity.

H.R. 1101, introduced on February 7, 2019, by Representative Walden, amends Title I of the Communications Act to establish obligations for BIAS providers. These include no blocking or throttling, subject to reasonable network management; no paid prioritization; and establishment of transparency rules. Additional provisions require the FCC to enforce these rules through adjudication of complaints and establish, no later than 60 days after enactment, formal complaint procedures to address alleged violations; protect the needs of emergency communications, law enforcement, public safety or national security, copyright infringement or other unlawful activity; protect the ability of BIAS providers to offer specialized services and the right of consumers to choice of service plans or control over their chosen BIAS; and establish that BIAS or any other mass-market retail service providing advanced telecommunications capability shall be considered an information service and that Section 706 may not be relied upon as a grant of authority.

A more narrowly focused measure, H.R. 1860, introduced on March 25, 2019, by Representative Kinzinger, prohibits the FCC from regulating the rates charged for BIAS.

115th Congress

Congressional activity in the 115th Congress sought to address a wide range of issues directly related to the debate over the appropriate framework for the provision of and access to broadband networks. Legislation included Congressional Review Act (CRA) resolutions to overturn the 2017 Order (H.J.Res. 129 and S.J.Res. 52); a measure (S. 993) to nullify the 2015 Order; comprehensive legislation (H.R. 4682, H.R. 6393, S. 2510, and S. 2853) to provide a regulatory framework to outline FCC authority over broadband internet access services; and measures to address the privacy and transparency regulations of the 2015 Order.

Publication of the 2017 Order in the *Federal Register* on February 22, 2018, triggered timelines for CRA consideration.[69] CRA resolutions to overturn the 2017 Order were introduced on February 27, 2018, in both the House (H.J.Res. 129) by Representative Doyle, and the Senate (S.J.Res. 52) by Senator Markey. Both measures stated that Congress disapproves "the rule" (in this case the FCC 2017 Order) and that the rule "shall have no force or effect." If the CRA joint resolution is enacted the rule "shall be treated as though such rule had never taken effect."

Additionally, the agency, in this case the FCC, may not, in most circumstances, promulgate the same rule again. S.J.Res. 52 passed (52-47) the Senate on May 16, 2018, and was sent to the House and held at the desk. On May 17, 2018, Representative Doyle filed a discharge petition to bring a House floor vote on H.J.Res. 129 but the House measure did not come up for consideration.

On the other hand, legislation (S. 993) to nullify the FCC's 2015 Open Internet Order was introduced on May 1, 2017, by Senator Lee. S. 993 nullified the FCC's 2015 Order, prohibited the FCC from reclassifying broadband internet access service as a telecommunications service, and prohibited the FCC from issuing a substantially similar rule absent congressional authorization. No further action was taken on the measure.

The FCC's adoption (3-2) of the 2017 Order that largely reversed the 2015 regulatory framework (see "WC Docket No. 17-108 (The FCC 2017 Order)" above) reopened debate over whether Congress should take broader action to amend existing law to provide guidance and more stability to FCC authority. Four measures (H.R. 4682, H.R. 6393, S. 2510, and S. 2853) to provide a regulatory framework to outline FCC authority over broadband internet access services were introduced.

---

69 Under the Congressional Review Act (CRA; 5 U.S.C. 801-808), when an agency final rule has been published in the *Federal Register* and submitted to the House and Senate, it begins a 60-day Senate session period during which Congress can consider a joint resolution disapproving that rule under special "fast track" parliamentary procedures which preclude a filibuster of the legislation. For a further discussion of the CRA, see CRS In Focus IF10023, *The Congressional Review Act (CRA)*, by Maeve P. Carey and Christopher M. Davis; and CRS Report R43992, *The Congressional Review Act (CRA): Frequently Asked Questions*, by Maeve P. Carey and Christopher M. Davis.

H.R. 4682, introduced on December 19, 2017, by Representative Blackburn, and S. 2510, introduced by Senator Kennedy on March 7, 2018, amended the Communications Act of 1934 to address a broad range of issues. More specifically, provisions contained in both measures included those which prohibited broadband internet access service (BIAS) providers from blocking lawful content or degrading (throttling) lawful internet traffic, subject to reasonable network management; granted FCC enforcement authority and required the FCC to establish formal complaint procedures to address alleged violations; preserved the ability of BIAS providers to offer, with a prohibition on certain practices, specialized services; established BIAS as an information service; and preempted state and local authority over "internet openness obligations" for the provision of BIAS with exceptions for emergency communications or law enforcement, public safety, or national security obligations. Provisions in H.R. 4682 also established eligibility of BIAS services for Federal Universal Service Fund program support. H.R. 4682 and S. 2510 were referred to the House Committee on Energy and Commerce and the Senate Committee on Commerce, Science, and Transportation, respectively, but no further action was taken. S. 2853, introduced on May 16, 2018, by Senator Thune, also amended the Communications Act of 1934 to establish a framework to address broadband internet access services. S. 2853 is similar to H.R. 4682 and S. 2510 in that it classified broadband as an information service, prohibited both blocking and throttling subject to reasonable network management, permitted with limitations the ability to provide specialized services, and provided for a similar role for the FCC. However, unlike the two previous measures, S. 2853 contained provisions that prohibit paid prioritization and contain transparency obligations; did not contain specific provisions to preempt state and local authority; and clarified that Section 706 of the Telecommunications Act of 1996 may not be used as a grant of regulatory authority. S. 2853 was referred to the Senate Committee on Commerce, Science, and Transportation but no further action was taken. H.R. 6393, introduced on July 17, 2018, by Representative Coffman, established a new title, Title VIII, in the 1934 Communications Act to provide a regulatory framework for broadband internet access providers. Provisions included those that banned blocking, throttling, and "paid preferential treatment" subject to reasonable network management, and established a general conduct standard. Included among the additional provisions were those that established transparency requirements, granted the FCC oversight over interconnection, permitted specialized services, stated that internet service providers are eligible to receive funds from, and may be required to contribute to, the Universal Service Fund, ensured disability access to broadband equipment and services, and exempted Title VIII provisions from FCC forebearance authority.

H.R. 6393 was referred to the House Committee on Energy and Commerce but no further action was taken.

Congressional action in the 115th Congress also focused on two specific aspects of the 2015 Open Internet Order rules: privacy (S.J.Res. 34, S. 878, S. 964, H.J.Res. 86,

H.Res. 230, H.R. 1754, H.R. 1868, H.R. 2520, H.R. 3175) and transparency (S. 228, H.R. 288).

Privacy

Congress successfully used the Congressional Review Act (CRA; 5 U.S.C. paras. 801-808) to revoke the customer privacy rules adopted by the FCC under the 2015 Open Internet Order.[70] Legislation (S.J.Res. 34, H.J.Res. 86), in the form of a joint resolution, was introduced by Senator Flake and Representative Blackburn, respectively, to overturn the FCC's customer privacy rules. The identical joint resolutions stated "that Congress disapproves the rule submitted by the Federal Communications Commission relating to 'Protecting the Privacy of Customers of Broadband and Other Telecommunications Services' (81 *Federal Register* 87274 (December 2, 2016) and such rule will have no force or effect." S.J.Res. 34 passed the Senate (50-48) on March 23, 2017, and the House (215-205) on March 28, 2017, and was signed by the President on April 3, 2017 (P.L. 115-22). This action prevents any new rule subject to the joint resolution from taking effect and invalidates any rules that have already been in effect. Additionally, it prevents the agency (in this case the FCC) from reissuing the rule in "substantially the same form" or issuing a "new rule that is substantially the same" as the disapproved rule unless specifically authorized by a law enacted after the approved resolution (CRA. 5 U.S.C. para 801(b)(2)).

Additional measures (H.R. 1754, H.R. 1868, H.R. 2520, H.R. 3175, S. 878, and S. 964) addressing other aspects of the privacy issue, as it relates to protection of broadband user data privacy, were introduced but received no further action.

Transparency

Legislation (H.R. 288, S. 228) addressing the transparency requirements contained in the 2015 Open Internet Order was under consideration. Transparency requirements refer to the disclosures that internet service providers are required to provide to their end users and edge providers and include, among other things, network management practices, performance, and commercial terms. These requirements were expanded upon or "enhanced" in the 2015 Open Internet Order, and small internet service providers (i.e., those with 100,000 or fewer subscribers) were given a temporary exemption from these enhanced requirements. H.R. 288, introduced on January 4, 2017, by Representative Walden, addresses the transparency requirements contained in the 2015 Order. H.R. 288 expanded the exemption to include internet service providers with 250,000 or fewer subscribers and sunset the transparency exemption five years from the bill's enactment.

---

[70] The FCC on October 27, 2016, adopted rules to implement the customer privacy requirements for broadband internet service providers under the 2015 Open Internet Order. *In the Matter of Protecting the Privacy of Customers of Broadband and Other Telecommunications Services,* WC Docket No. 16-106. Available at https://apps.fcc.gov/edocs_public/attach
match/FCC-16-148A1.pdf.

The bill also required the FCC to report to Congress 180 days after the bill's enactment on whether the exemption should be made permanent and whether the definition of "small" for exemption purposes should be modified. H.R. 288 passed the House, by voice vote, on January 10, 2017, but received no further action.

S. 228, introduced on January 24, 2017, by Senator Daines, also addressed the transparency exemption and was largely identical to H.R. 288. S. 228 provided for an exemption of the enhanced transparency rule contained in the 2015 Open Internet Order for small businesses. The term "small business" was defined for purposes of the exemption as any provider of broadband internet access service that has not more than 250,000 subscribers. The bill also required the FCC to report to Congress 180 days after the bill's enactment, on whether the exemption should be made permanent and whether the definition of "small business" for exemption purposes should be modified. The exemption was to last for five years after enactment or until the FCC completed the above report and a rulemaking to implement those recommendations. S. 228 was referred to the Senate Commerce, Science, and Transportation Committee, but no further action was taken.

114th Congress

Ten measures (S. 40, S. 2283, S. 2602, S.J.Res. 14, H.R. 196, H.R. 279, H.R. 1212, H.R. 2666, H.R. 4596, and H.J.Res. 42) addressing broadband regulation were introduced in the 114th Congress. Amended versions of H.R. 4596, dealing with transparency, and H.R. 2666, dealing with rate regulation, passed the House on March 16, 2016, and April 15, 2016, respectively. Draft legislation, offered by the House Energy and Commerce Committee and the Senate Committee on Commerce, Science, and Transportation, had also been a focal point of hearings. However, no final action was taken on any of these measures.

S. 40, the Online Competition and Consumer Choice Act of 2015, and its companion measure H.R. 196, introduced on January 7, 2015, by Senator Leahy and Representative Matsui, respectively, addressed the relationship between a broadband internet access provider and a content provider. Both bills directed the FCC to establish/adopt regulations, within 90 days of enactment, to prohibit broadband internet access providers from entering into agreements with content providers, for pay, to give preferential treatment or priority to their content (often termed "paid prioritization"), and prohibited broadband providers from giving preferential treatment to their own or affiliated content. These rules applied to the traffic/content that travels between the access provider and the end user, often termed "the last mile." Exceptions were given to address the needs of emergency communications or law enforcement, public safety, or national security authorities.

H.R. 279, introduced on January 12, 2015, by Representative Latta, prohibited the FCC from regulating the provision of broadband internet access as a telecommunications service. More specifically, the bill included provisions that classified broadband internet access service as an "information service," not a telecommunications service, and clarified that a provider of broadband internet access service may not be treated as a telecommunications carrier when engaged in the provision of an information service. This measure prevented the FCC from regulating providers of broadband internet access services under Title II of the Communications Act. Representative Blackburn, in direct response to the FCC's February 26, 2015, adoption of the Open Internet Order, introduced H.R. 1212, the "Internet Freedom Act," on March 3, 2015. H.R. 1212 blocked the implementation of the FCC's adopted Open Internet Order (GN Docket No. 14-28) by stating that it "shall have no force or effect." Furthermore, it prohibited the FCC from reissuing a rule in substantially the same form or issuing a new rule that is substantially the same, unless the reissued or new rule is specifically authorized by a law enacted after the date of the enactment of this act. Exceptions were granted to protect national security or public safety, or to assist or facilitate actions taken by federal or state law enforcement agencies. Similarly S. 2602, the "Restoring Internet Freedom Act," introduced by Senator Lee, on February 25, 2016, also negated the FCC's 2015 Open Internet Order. S. 2602 was identical to H.R. 1212 but did not contain the provisions relating to exceptions.

A more targeted measure, H.R. 2666, the No Rate Regulation of Broadband Internet Access Act, introduced by Representative Kinzinger, on June 4, 2015, prohibited the FCC from regulating the rates charged for broadband internet access as defined by the Open Internet Order. H.R. 2666 was approved (15-11) by the House Communications Subcommittee by a party line vote on February 11, 2016. An amended version of H.R. 2666 was approved (29-19) by the House Energy and Commerce Committee on March 15, 2016. Prior to full committee passage of H.R. 2666, an amendment stating that the bill would not affect the FCC's authority over data roaming, interconnection, truth-in-billing, paid prioritization, and rates charged for services that receive universal service support was approved. H.R. 2666 passed the House (241-173) without further amendment, on April 15, 2016.

Another approach, using the Congressional Review Act (CRA; 5 U.S.C. paras. 801-808) to overturn the 2015 Order, was also under consideration. H.J.Res. 42, introduced on April 13, 2015, by Representative Doug Collins, contained a resolution stating that "Congress disapproves the rule submitted by the Federal Communications Commission relating to the matter of protecting and promoting the open Internet ... adopted by the Commission on February 26, 2015 ... and such rule shall have no force or effect." A similar measure, S.J.Res. 14, introduced on April 28, 2015, by Senator Rand Paul stated that "Congress disapproves the rule submitted by the Federal Communications Commission relating to regulating broadband Internet access ..., and such rule will have no force and effect." The CRA empowers Congress to review, under an expedited legislative process, new federal regulations and, if a joint resolution is passed and signed into law, to invalidate such regulations.[71]

Attempts were also made, through the appropriations process, to add language that would have delayed the FCC from using its funds to implement the Open Internet Order until the courts address its legality and/or regulate rates. Language attached to the House Financial Services FY2017 appropriation measure (H.R. 5485) contained among its provisions those that prevented the FCC from using any funds "... to implement, administer or enforce" the Open Internet Order until the legal challenges to the Order have been resolved (Title VI §632). The bill also contained a provision that prohibited the FCC from using FY2017 funds to directly or indirectly regulate the prices, fees, or data caps and allowances charged or imposed by providers of broadband internet access services (Title VI §631). The funding bill was passed (30-17) by the House Appropriations Committee on June 9, 2016, and passed the House (239-185) on July 7, 2016.

Separately, legislation (H.R. 4596, S. 2283) addressing the transparency requirements contained in the 2015 Open Internet Order was also under consideration. Transparency requirements refer to the disclosures that internet service providers are required to provide to their end users and edge providers and includes, among other things, network management practices, performance, and commercial terms. These requirements were expanded upon or "enhanced" in the 2015 Open Internet Order, and small internet service providers were given a temporary exemption from these enhanced requirements. H.R. 4596, introduced on February 24, 2016, by Representative Walden, addressed the transparency requirements contained in the 2015 Order. This measure was amended and passed, by voice vote, by the House Energy and Commerce Committee on February 25, 2016, and subsequently passed (411-0) the House on March 16, 2016. H.R. 4596, as passed by the House, sunset the transparency exemption five years from the bill's enactment and defined small internet service providers as those who have 250,000 subscribers or less. The bill also required the FCC to report to Congress 180 days after the bill's enactment on whether the exemption should be made permanent and whether the definition of "small" for exemption purposes should be modified. S. 2283, introduced on November 16, 2015, by Senator Daines, also addressed the transparency exemption. An amended version of S. 2283 passed the Senate Commerce Committee, by voice vote, on June 15, 2016. The bill, as amended, provided for an exemption of the enhanced transparency rule contained in the 2015 Open Internet Order for small businesses. The term "small business" was defined for purposes of the exemption as any provider of

---

[71] Under the Congressional Review Act (CRA; 5 U.S.C. paras. 801-808) Congress is given 60 in-session days from publication in the *Federal Register* or submission to Congress, whichever is later, to review and potentially overturn major federal agency rulemakings. For a further discussion of the CRA, see CRS In Focus IF10023, *The Congressional Review Act (CRA)*, by Maeve P. Carey and Christopher M. Davis; and CRS Report R43992, *The Congressional Review Act (CRA): Frequently Asked Questions*, by Maeve P. Carey and Christopher M. Davis.

broadband internet access service that has not more than 250,000 subscribers. The bill also required the FCC to report to Congress 180 days after the bill's enactment on whether the exemption should be made permanent and whether the definition of "small" for exemption purposes should be modified. The exemption would have lasted for three years after enactment or until the FCC completed the above report and a rulemaking to implement those recommendations.

To some degree the debate in the 114th Congress over broadband regulation became more nuanced. Some looked to the FCC to address this issue using current provisions in the 1934 Communications Act to protect the marketplace from potential abuses that could threaten the net neutrality concept. Others felt that existing laws are outdated and limited, cannot be used to establish regulations to address current issues, and would not stand up to court review. They advocated that the FCC should look to Congress for guidance to amend current law to update FCC authority before action is taken. Senator Thune released a list of 11 principles that he felt should be used as a guide to develop legislation. These principles are as follows: prohibit blocking; prohibit throttling; prohibit paid prioritization; require transparency; apply rules to both wireline and wireless; allow for reasonable network management; allow for specialized services; protect consumer choice; classify broadband internet access as an information service under the Communications Act; clarify that Section 706 of the Telecommunications Act may not be used as a grant of regulatory authority; and direct the FCC to enforce and abide by these principles.[72]

Draft legislation, guided by these principles, was offered by the House Energy and Commerce Committee and the Senate Committee on Commerce, Science, and Transportation.[73] The draft amended the Communications Act of 1934 to prohibit blocking lawful content and nonharmful devices (subject to reasonable network management), throttling data (subject to reasonable network management), and paid prioritization; and required transparency of network management practices. The FCC was directed to enforce these provisions through the establishment of a formal complaint procedure. The draft permitted, within certain guidelines, the offering of specialized services. The provision of broadband internet access service (as well as other mass market retail services providing advanced telecommunications capability) was classified as an information service. The draft also prohibited the FCC, or any state commission, from using Section 706 of the Telecommunications Act of 1996 as a grant of authority. This draft legislation was the focus of hearings, held on January 21, 2015, in the Senate Commerce Committee and the House Subcommittee on Communications and Technology.

---

[72] *Thune Announces Hearing on FCC Internet Authority*, U.S. Senate Committee on Commerce, Science, and Transportation, press release issued January 14, 2015.

[73] *Congressional Leaders Unveil Draft Legislation Ensuring Consumer Protections and Innovative Internet*, January 16, 2015, available at http://energycommerce.house.gov.

Additional hearings focusing on a wide range of issues related to the net neutrality/broadband regulation debate were held by the Senate Commerce Committee, the House Judiciary Committee, the House Subcommittee on Communications and Technology, the House Committee on Oversight and Government Reform, and the House Financial Services Subcommittee.

113th Congress

Seven measures (H.R. 3982, H.R. 4070, H.R. 4752, H.R. 4880, H.R. 5429, S. 1981, and S. 2476) were introduced in direct response to the January 2014 decision issued by the U.S. Court of Appeals, D.C. Circuit, which struck down the antiblocking and nondiscrimination rules adopted by the FCC in its Open Internet Order (*Verizon Communications Inc. v. Federal Communications Commission*, D.C. Cir., No.11-1355). H.R. 3982, the Open Internet Preservation Act of 2014, and its companion measure S. 1981, introduced on February 3, 2014, restored the antiblocking and nondiscrimination rules struck down by the court until the FCC takes final action, based on Section 706 authority, upheld by the court, to establish new rules in its current Open Internet proceeding. The FCC was also given the authority to adjudicate cases under those rules that occurred during that period. H.R. 4880, the "Online Competition and Consumer Choice Act of 2014," and its companion measure S. 2476, introduced on June 17, 2014, directed the FCC to establish regulations that prohibit paid prioritization agreements between internet service providers and content providers on the internet connection between the internet service provider and the consumer and prohibit broadband providers from prioritizing or giving preferential treatment to their own traffic, or the traffic of their affiliates, over the traffic of others. H.R. 5429, the "Open Internet Act of 2014," introduced on September 9, 2014, restored the authority of the FCC to adopt the rules vacated by the U.S. D.C. Court of Appeals in *Verizon v. Federal Communications Commission* (the FCC's 2010 Open Internet Order).

On the other hand, H.R. 4070, the Internet Freedom Act, introduced on February 21, 2014, stated that the FCC's 2010 Open Internet rules shall have "no force or effect" and prohibited the FCC from reissuing regulations in the same or substantially the same form unless they were specifically authorized by a law enacted after the date of the enactment of the act. Exceptions were made for regulations determined by the FCC to be necessary to prevent damage to U.S. national security; ensure the public safety; or assist or facilitate actions taken by a federal or state law enforcement agency. H.R. 4752, introduced on May 28, 2014, amended the Communications Act of 1934 to prohibit the FCC from reclassifying broadband networks under Title II of the Communications Act. The bill included, among other provisions, that the term "information service" is not a telecommunications service but includes broadband internet access service and that a provider of an information service may not be treated as a telecommunications carrier when engaged in the provision of an information service.

The House Judiciary Committee, Subcommittee on Regulatory Reform, held a hearing on June 20, 2014, examining the role of antitrust law and regulation as it related to the broadband access debate. The Senate Judiciary Committee held a field hearing in Vermont on July 1, 2014, and a hearing on September 17, 2014, to address issues related to an open internet.

112th Congress

A consensus on the net neutrality issue remained elusive and support for the FCC's Open Internet Order was mixed. (See "The FCC 2010 Open Internet Order," above.) While some Members of Congress supported the action and in some cases would have supported an even stronger approach, others felt that the FCC had overstepped its authority and that the regulation of the internet is not only unnecessary, but harmful. Internet regulation and the FCC's authority to implement such regulations was a topic of legislation (H.R. 96, H.R. 166, S. 74, H.R. 2434, H.R. 1, H.R. 3630, H.J.Res. 37, S.J.Res. 6) and hearings (Senate Commerce Committee, House Communications Subcommittee, and House Intellectual Property, Competition, and the Internet Subcommittee) in the 112th Congress.

Legislation to limit FCC regulation was introduced. H.R. 96, the Internet Freedom Act, introduced, on January 5, 2011, by Representative Blackburn and 59 additional original cosponsors, prohibited, with exceptions, the FCC from proposing, promulgating, or issuing any regulations regarding the internet or IP-enabled services, effective the date of the bill's enactment. Exceptions were made for regulations that the FCC determined were necessary to prevent damage to national security, to ensure the public safety, or to assist or facilitate actions taken by a federal or state law enforcement agency. The bill also contained a finding that the internet and IP-enabled services are services affecting interstate commerce and are not subject to state or municipality jurisdiction. Another measure, H.R. 166, the "Internet Investment, Innovation, and Competition Preservation Act," introduced on January 5, 2011, by Representative Stearns, required the FCC to prove the existence of a "market failure" before regulating information services or internet access services. The FCC must also conclude that the "market failure" is causing "specific, identified harm to consumers" and that regulations are necessary to ameliorate that harm. The bill also contained provisions that required any FCC regulation to be the "least restrictive," determine that the benefits exceed the cost, permit network management, not prohibit managed services, be reviewed every two years, and be subject to sunset. Any such regulation was required to be enforced on a nondiscriminatory basis between and among broadband network, service, application, and content providers. A more narrowly focused limitation was contained within H.R. 3630, the "Middle Class Tax Relief and Job Creation Act of 2011," as passed (234-193) by the House on December 13, 2011. Section 4105 of Title IV (spectrum provisions) of the bill prohibited the FCC from imposing network access/management requirements on licensees. More specifically, the provision prohibited the promulgation of auction service rules that restrict a licensee's ability to manage network traffic or prioritize the traffic on its network, or that would require providing network access on a wholesale basis. However, the provision was removed from the bill prior to final passage (P.L. 112-96).

Legislation to strengthen the FCC's ability to regulate open access by amending Title II of the 1934 Communications Act was also introduced. S. 74, the Internet Freedom, Broadband Promotion, and Consumer Protection Act of 2011, introduced January 25, 2011, by Senator Cantwell, provided for strengthened open access protections. More specifically, the bill contained among its provisions those that codify the four FCC principles issued in 2005 as well as those to require internet service providers to be nondiscriminatory regarding access and transparent in their network management practices. The bill also required internet service providers to provide service to end users upon "reasonable request" and offer stand-alone broadband access at "reasonable rates, terms, and conditions" and prohibited internet service providers from requiring paid prioritization. The bill's requirements applied to both wireline and wireless platforms; however, the FCC was allowed to take into consideration differences in network technologies when applying requirements. The FCC was tasked with establishing the necessary rules, and injured parties could be awarded damages by the FCC or a federal district court.

Other measures, which proved unsuccessful, were considered to prevent, or at least delay, implementation of the FCC's Open Internet Order. Attempts were made, through the appropriations process, to add language that would prevent the FCC from using its funds to implement the Open Internet Order. Language attached to the FY2011 appropriation measure, H.R. 1, to prevent the use of FCC FY2011 funds for implementation of the order was passed by the House. The Continuing Appropriations Act, 2011 (H.R. 1), passed (235-189) by the House on February 19, 2011, contained an amendment, introduced by Representative Walden and passed by the House (244-181), to prohibit the FCC from using any funds made available by the act to implement the FCC's Open Internet Order adopted on December 21, 2010. No such provision, however, was included in the final FY2011 appropriations bill, H.R. 1473, passed by Congress and signed by the President (P.L. 112-10). Similarly, language included in the FY2012 Financial Services and General Government Appropriations bill (H.R. 2434), which includes funding for the FCC, contained a provision that barred the FCC from using any funds to implement its Open Internet Order adopted December 21, 2010. This measure passed the House Appropriations Committee on June 23, 2011 (H.Rept. 112-136),[74] but no such provision was included in the final FY2012 consolidated appropriations bill, H.R. 2055, which was signed by President Obama (P.L. 112-74) on December 23, 2011.

Another approach, using the Congressional Review Act to overturn the order,[75] was also considered. Identical resolutions of disapproval were introduced, on February 16, 2011, in both the House (H.J.Res. 37) and Senate (S.J.Res. 6). These measures stated that Congress disapproves of the rule submitted by the FCC's report and order relating to the matter of preserving the open internet and broadband industry practices adopted by the FCC on December 21, 2010, and further stated that "such rule would have no force or effect." A hearing on H.J.Res. 37 was held by the House Energy and Commerce Communications and Technology Subcommittee on March 9, 2011, and the subcommittee passed the measure (15-8) on a party-line vote immediately following the hearing. On March 25, 2011, the House Energy and Commerce Committee passed (30-23) H.J.Res. 37. On April 8, 2011, the full House considered and passed (240-179) H.J.Res. 37. However, an identical resolution of disapproval (S.J.Res. 6) failed to pass the Senate on November 10, 2011, by a 52-46 vote.

Legislation addressing the issue of data usage caps was also introduced. The Data Cap Integrity Act of 2012 (S. 3703), introduced on December 20, 2012, by Senator Wyden, addressed the usage of data caps by internet service providers (ISPs) and their implementation. Included among the bill's provisions were those that required that an ISP that imposes data caps must be certified by the FCC as to accuracy of data cap measurement; that the cap "functions to reasonably limit network congestion without unnecessarily restricting Internet use"; and that the cap does not discriminate (that is, for purposes of measuring does not provide "preferential treatment of data that is based on the source or content of the data"). The bill also required ISPs that apply data caps to provide data tools, or identify commercially available data measurement tools, to consumers for monitoring and management. Civil penalties for violations were to be used to reimburse those violated, and unobligated funds in excess of $5 million (annually) were to be transferred from the newly created "Data Cap Integrity Fund" to the U.S. Department of the Treasury for deficit reduction.

111th Congress

Although the 111th Congress saw considerable activity addressing the net neutrality debate, no final action was taken. One stand-alone measure (H.R. 3458) that

---

[74] The Senate Appropriations subcommittee-passed (September 14, 2011) appropriations measure, S. 1573, did contain a provision to prohibit the FCC from using funds to implement the Open Internet Order, but it did not remain in the full committee passed (September 15, 2011) version (S.Rept. 112-79).

[75] Under the Congressional Review Act (CRA; 5 U.S.C. paras. 801-808) Congress is given 60 in-session-days, from publication in the *Federal Register* or submission to Congress, whichever is later, to review and potentially overturn federal agency major rulemakings.

comprehensively addressed the net neutrality debate was introduced in the 111th Congress. H.R. 3458, the Internet Freedom Preservation Act of 2009, introduced by Representative Edward Markey, and also supported by then-House Energy and Commerce Committee Chairman Waxman, sought to establish a national policy of nondiscrimination and openness with respect to internet access offered to the public. The bill also required the offering of unbundled, or stand-alone, internet access service as well as transparency for the consuming public with respect to speed, nature, and limitations on service offerings and the public disclosure of network management practices. The FCC was tasked with promulgating the rules relating to the enforcement and implementation of the legislation. Then-House Communications, Technology, and the Internet Subcommittee Chairman Boucher stated that he continued to work with broadband providers and content providers to seek common ground on network management practices, and chose to pursue that approach.[76] Furthermore, the Senate Commerce and House Energy and Commerce Committees and Communications Subcommittees held a series of staff-led sessions with industry stakeholders to discuss a range of communications policies including broadband regulation and FCC authority.[77]

Two bills (S. 1836, H.R. 3924) were introduced in response to the adoption by the FCC of a NPR on preserving the open internet. S. 1836, introduced on October 22, 2009, by Senator McCain, prohibited, with some exceptions, the FCC from proposing, promulgating, or issuing any further regulations regarding the internet or IP-enabled services. Exceptions included those relating to national security, public safety, federal or state law enforcement, and Universal Service Fund solvency.[78] Additional provisions reaffirmed that existing regulations, including those relating to CALEA, remain in force and stated as a general principle that the internet and all IP-enabled services are services affecting interstate commerce and are not subject to state or municipal locality jurisdiction. H.R. 3924, introduced by Representative Blackburn on October 26, 2009, was identical to S. 1836, except for title and the omission of the reference to the Universal Service Fund. H.Con.Res. 311, introduced by Representative Gene Green and 49 other House Members on July 30, 2010, affirmed that it is the responsibility of Congress to determine the regulatory authority of the FCC with respect to broadband internet services and called upon the FCC to suspend any further action on its proceedings until such time as Congress delegates such authority to the FCC.

---

[76] "Boucher Opts for Talks, Not Legislation, on Net Neutrality," *National Journal*, Congress Daily, February 26, 2009. "Boucher, Stakeholders Working on Network Management Issues," *Telecommunications Reports*, March 15, 2009, p. 19.

[77] *Bicameral Bipartisan Telecommunications Update Statement*, U.S. Senate Committee on Commerce, Science and Transportation, press release, June 18, 2010, available at http://democrats.energycommerce.house.gov/index.php?q=news/bicameral-bipartisan-telecommunications-update-statement.

[78] For a discussion and analysis of issues regarding the Universal Service Fund see CRS Report RL33979, *Universal Service Fund: Background and Options for Reform*, by Angele A. Gilroy.

Another measure (H.R. 5257), introduced by Representative Stearns, addressed the possible reclassification of broadband service and would have required, among other provisions, that the FCC prove the existence of a "market failure" before regulating information services or internet access services. Furthermore, the bill required, among other provisions, that the FCC conclude that the market failure is causing "specific, identified harm to consumers" and if devising regulations must adopt those that are the "least restrictive," permit network management, and are subject to sunset. Still another measure (S. 3624), introduced by Senator DeMint, contained provisions that required the FCC to prove consumers are being substantially harmed by a lack of marketplace choice before imposing new regulations and to weigh the potential cost of action against any benefits to consumers or competition. The FCC was given the authority to hear complaints for violations and award damages to injured parties. The bill also required that any rules the FCC adopted would sunset in five years unless it could make the same finding again.

The net neutrality issue was also narrowly addressed within the context of the American Recovery and Reinvestment Act of 2009 (ARRA, P.L. 111-5). The ARRA contains provisions that require the National Telecommunications and Information Administration (NTIA), in consultation with the FCC, to establish "nondiscrimination and network interconnection obligations" as a requirement for grant participants in the Broadband Technology Opportunities Program (BTOP). The law further directs that the FCC's four broadband policy principles, issued in August 2005, are the minimum obligations to be imposed.[79] These obligations were issued July 1, 2009, in conjunction with the release of the notice of funds availability (NOFA) soliciting applications for the program. The FCC's *National Broadband Plan* (NBP), which was required to be written in compliance with provisions contained in the ARRA, while making no recommendations, did contain discussions regarding the open internet and the classification of information services.

Concern over the move by some broadband network providers to expand their implementation of metered or consumption-based billing prompted the introduction of legislation (H.R. 2902) to provide for oversight of volume usage service plans. H.R. 2902, the Broadband Internet Fairness Act, introduced by former Representative Massa, required, among its provisions, that any broadband internet service provider serving 2 million or more subscribers submit any volume usage based service plan that the provider is proposing or offering to the Federal Trade Commission (FTC) for approval. The FTC, in consultation with the FCC, was required to review such plans "to ensure that such plans are fairly based on cost." Such plans were subject to agency review and public hearings. Plans determined by the FTC to impose "rates, terms, and conditions that are

[79] For a further more detailed discussion of the broadband infrastructure programs contained in P.L. 111-5 see CRS Report R40436, *Broadband Infrastructure Programs in the American Recovery and Reinvestment Act*, by Lennard G. Kruger.

unjust, unreasonable, or unreasonably discriminatory" were to be declared unlawful. Violators were subject to injunctive relief requiring the suspension, termination, or revision of such plans and were subject to a fine of not more than $1 million.

In: Broadband
Editor: Orlando Harrison
ISBN: 978-1-53616-409-1

Chapter 5

BROADBAND INTERNET ACCESS AND THE DIGITAL DIVIDE: FEDERAL ASSISTANCE PROGRAMS (UPDATED)*

Lennard G. Kruger and Angele A. Gilroy

ABSTRACT

The "digital divide" is a term that has been used to characterize a gap between "information haves and have-nots," or in other words, between those Americans who use or have access to telecommunications and information technologies and those who do not. One important subset of the digital divide debate concerns high-speed internet access and advanced telecommunications services, also known as broadband. Broadband is provided by a series of technologies (e.g., cable, telephone wire, fiber, satellite, mobile and fixed wireless) that give users the ability to send and receive data at volumes and speeds necessary to support a number of applications including voice communications, entertainment, telemedicine, distance education, telework, ecommerce, civic engagement, public safety, and energy conservation.

Broadband technologies are currently being deployed primarily by the private sector throughout the United States. While the numbers of new broadband subscribers continue to grow, studies and data suggest that the rate of broadband deployment in urban/suburban and high-income areas is outpacing deployment in rural and low-income areas. Some policymakers, believing that disparities in broadband access across American society could have adverse economic and social consequences on those left behind, assert that the federal government should play a more active role to address the "digital divide" in broadband access.

With the conclusion of the grant and loan awards established by the American Recovery and Reinvestment Act of 2009 (P.L. 111-5), there remain two primary ongoing

---

* This is an edited, reformatted and augmented version of Congressional Research Service, Publication No. RL30719, dated January 9, 2019.

federal vehicles which direct federal money to fund broadband infrastructure: the broadband and telecommunications programs at the Rural Utilities Service (RUS) of the U.S. Department of Agriculture and the Universal Service Fund (USF) programs under the Federal Communications Commission (FCC). RUS broadband programs were reauthorized and modified by the 2018 farm bill. The USF High Cost Fund is undergoing a major transition to the Connect America Fund, which is targeted to the deployment, adoption, and utilization of both fixed and mobile broadband.

Meanwhile, the Consolidated Appropriations Act, 2018 (P.L. 115-141) appropriated $600 million to RUS to conduct a new broadband loan and grant pilot program (called the ReConnect Program), and appropriated $7.5 million to the National Telecommunications and Information Administration to update the national broadband availability map in coordination with the FCC. Additionally, P.L. 115-141 contained provisions seeking to facilitate deployment of broadband infrastructure on federal property, as well as making more spectrum available for wireless broadband.

To the extent that the 116th Congress may consider various options for further encouraging broadband deployment and adoption, a key issue is how to strike a balance between providing federal assistance for unserved and underserved areas where the private sector may not be providing acceptable levels of broadband service, while at the same time minimizing any deleterious effects that government intervention in the marketplace may have on competition and private sector investment.

INTRODUCTION

The "digital divide" is a term used to describe a gap between "information haves and have-nots," or in other words, between those Americans who use or have adequate access to telecommunications and information technologies and those who do not.[1] Whether or not individuals or communities fall into the "information haves" category depends on a number of factors, ranging from the presence of computers in the home, to training and education, to the availability of affordable internet access.

Broadband technologies are currently being deployed primarily by the private sector throughout the United States. While the numbers of new broadband subscribers continue to grow, studies and data suggest that the rate of broadband deployment in urban/suburban and high-income areas is outpacing deployment in rural and low-income areas.

STATUS OF BROADBAND IN THE UNITED STATES

Prior to the late 1990s, American homes accessed the internet at maximum speeds of 56 kilobits per second by dialing up an Internet Service Provider over the same copper telephone line used for traditional voice service. A relatively small number of businesses

[1] The term "digital divide" can also refer to international disparities in access to communications and information technology. This chapter focuses on domestic issues only.

and institutions used broadband or high-speed connections through the installation of special "dedicated lines" typically provided by their local telephone company. Starting in the late 1990s, cable television companies began offering cable modem broadband service to homes and businesses. This was accompanied by telephone companies beginning to offer DSL service (broadband over existing copper telephone wireline). Growth in broadband service has been steep, rising from 2.8 million high-speed lines reported as of December 1999, to 409 million connections as of June 30, 2017.[2] Of the 409 million high-speed connections reported by the FCC, 353 million serve residential users.[3]

Table 1. Percentage of Broadband Technologies by Types of Connection

	Connections over 200 kbps in at least one direction	Residential connections over 200 kbps in at least one direction	Fixed Connections at least 10 Mbps downstream and 1 Mbps upstream	Residential fixed connections at least 10 Mbps downstream and 1 Mbps upstream	Fixed Connections at least 25 Mbps downstream and 3 Mbps upstream	Residential fixed connections at least 25 Mbps downstream and 3 Mbps upstream
Cable modem	15.7%	16.9%	71.0%	71.4%	80.3%	80.8%
DSL	6.2%	6.5%	13.0%	12.9%	3.7%	3.7%
Mobile wireless	74.0%	72.5%	—	—	—	—
Fiber	3.2%	3.3%	13.7%	13.4%	15.2%	14.8%
All other	0.9%	0.8%	2.4%	2.2%	0.8%	0.6%

Table 1 depicts the relative deployment of different types of broadband technologies. A distinction is often made between "current generation" and "next generation" broadband (commonly referred to as next generation networks or NGN). "Current generation" typically refers to initially deployed cable, DSL, and many wireless systems, while "next generation" refers to dramatically faster download and upload speeds offered by fiber technologies and also by successive generations of cable, DSL, and wireless technologies. In general, the greater the download and upload speeds offered by a broadband connection, the more sophisticated (and potentially valuable) the application that is enabled.

[2] Federal Communications Commission, *Internet Access Services: Status as of June 30, 2017*, released November 2018, p. 12, available at https://docs.fcc.gov/public/attachments/DOC-355166A1.pdf.

[3] Ibid.

Broadband Availability

FCC data indicate where fixed broadband service is and is not being deployed.[4] The FCC has set a speed benchmark of 25 Mbps (download speed)/3 Mbps (upload speed) as the measure by which it determines whether a fixed service provides advanced telecommunications capability. Table 2 shows recent percentages of Americans in urban, rural, and tribal areas with access to terrestrial fixed broadband at speeds of 25 Mbps/3Mbps, as presented in the FCC's *Communications Marketplace Report*.[5] According to the most recent FCC deployment data, as of December 2017, 94% of all Americans had access to fixed terrestrial broadband at speeds of at least 25 Mbps/3 Mbps, with over 19.4 million Americans still lacking fixed terrestrial broadband at those speeds.[6]

Table 2. Percentage of Americans with Access to Fixed Terrestrial Broadband at Minimum Speed of 25 Mbps/3 Mbps

All U.S.	**83.6%**	**89.4%**	**89.9%**	**91.9%**	**94.0%**
Urban	92.3%	96.4%	96.7%	97.7%	98.5%
Rural	47.6%	60.4%	61.5%	67.8%	75.7%
Tribal	37.1%	57.2%	57.8%	63.1%	68.0%

Source: FCC, Communications Marketplace Report, p. 132.

Table 3 shows the percentage of Americans as of December 2017 with access to fixed 25 Mbps/3Mbps terrestrial broadband by state.

Another important broadband availability metric is the extent to which there are multiple broadband providers offering competition and consumer choice. Typically, multiple providers are more prevalent in urban than in rural areas or tribal areas (see Table 4).

---

[4] See FCC, *Fixed Broadband Deployment*, available at https://broadbandmap.fcc.gov/#/.

[5] FCC, *Communications Marketplace Report*, GN Docket No. 18-231, FCC 18-181, p. 132, available at https://docs.fcc.gov/public/attachments/FCC-18-181A1.pdf.

[6] Fixed terrestrial broadband at speeds of 25 Mbps/3Mbps includes cable modem, fiber, and a limited number of DSL connections. It does not include satellite broadband, which is also a fixed (nonmobile) broadband service. According to the *Communications Marketplace Report*, "Form 477 deployment data for satellite broadband indicate that satellite service offering 25 Mbps/3 Mbps speeds is available to all but 0.03% of the population." However, the FCC acknowledges that "these data could overstate the deployment of these services" and that "while satellite operators may be able to offer service to wide swaths of the country, overall satellite capacity may limit the number of consumers that can actually subscribe to satellite service at any one time." Ibid, p. 129, footnotes no. 778 and 779.

Table 3. Percentage of Americans with access to fixed terrestrial broadband by state (December 2017 data, minimum speed of 25 Mbps/3 Mbps)

	% of population with access, all areas	% of population with access, rural areas	% of population with access, urban areas
United States	94.0%	75.7%	98.5%
Alabama	86.1	69.7	97.7
Alaska	80.5	51.6	96.4
Arizona	86.9	39.8	93.3
Arkansas	77.4	55.9	94.4
California	97.0	67.3	98.9
Colorado	92.9	63.2	98.4
Connecticut	100	100	100
Delaware	97.6	93.8	98.5
Dist. of Columbia	100	N/A	100
Florida	96.2	77.9	98.2
Georgia	92.5	77.6	97.3
Hawaii	96.1	71.5	98.6
Idaho	85.3	58.6	97.3
Illinois	94.7	61.1	99.0
Indiana	89.9	67.4	98.5
Iowa	90.7	77.5	98.1
Kansas	91.2	71.9	97.9
Kentucky	90.9	79.7	98.7
Louisiana	87.6	63.3	96.5
Maine	93.3	89.6	99.4
Maryland	100	100	100
Massachusetts	97.9	92.3	98.4
Michigan	92.0	73.1	98.6
Minnesota	94.8	83.7	98.9
Mississippi	79.6	62.6	97.0
Missouri	88.7	65.1	98.8
Montana	86.1	72.6	97.5
Nebraska	87.3	58.0	97.9
Nevada	92.7	46.5	96.1
New Hampshire	94.7	89.1	98.3
New Jersey	100	100	100
New Mexico	83.4	47.4	94.8
New York	100	100	100
North Carolina	94.8	84.8	99.8
North Dakota	93.1	87.3	97.6
Ohio	94.7	78.4	99.3
Oklahoma	79.0	48.3	95.0
Oregon	92.4	68.9	98.3
Pennsylvania	100	100	100
Rhode Island	100	100	100
South Carolina	89.9	73.7	98.2
South Dakota	88.9	76.1	99.2

Table 3. (Continued)

	% of population with access, all areas	% of population with access, rural areas	% of population with access, urban areas
Tennessee	91.3	77.0	98.5
Texas	92.7	68.9	97.4
Utah	94.2	64.0	98.5
Vermont	89.3	83.4	98.7
Virginia	100	100	100
Washington	97.3	88.9	99.0
West Virginia	84.6	72.5	97.2
Wisconsin	91.3	72.1	99.5
Wyoming	81.3	52.8	98.7

Source: FCC, *Communications Marketplace Report,* Appendix D-1, available at: https://docs.fcc.gov/public/attachments/FCC-18-181A9.pdf.

Table 4. Percentage of Americans with multiple options for fixed terrestrial broadband (25/3 Mbps)

	No provider	1 provider	2 providers	3 or more
Nationwide	6.0%	23.7%	42.4%	28.0%
Urban	1.5%	20.1%	45.8%	32.5%
Rural	24.3%	38.4%	28.0%	9.3%
Tribal	32.0%	36.1%	21.1%	10.8%

Source: FCC, *Communications Marketplace Report,* pp. 97-98, data as of December 31, 2017.

Broadband Adoption

In contrast to broadband *availability*, which refers to whether or not broadband service is offered, broadband *adoption* refers to the extent to which American households actually subscribe to and use broadband. According to Census data from the 2016 American Community Survey, 81.4% of American households have a broadband internet subscription.[7] Other Census data from July 2015 show that 68% of Americans use the internet at home.[8] The Census data also show that Americans increasingly are connecting to the internet through other devices in addition to desktop computers: 52% of Americans used two or more devices to connect, including tablets, laptops, mobile phones, and TV connected boxes (gaming consoles and streaming video players).[9]

---

[7] U.S. Census Bureau, 2016 American Community Survey, accessed at NTIA BroadbandUSA, "Data Section," at https://broadbandusa.ntia.doc.gov/ntia-map.

[8] For other metrics, see NTIA's Digital Nation Data Explorer tool at https://www.ntia.doc.gov/data/digital-nation-dataexplorer#sel=internetUser&disp=map.

[9] Guila McHenry, NTIA, "Majority of Americans Use Multiple Internet-Connected Devices, Data Shows," December 7, 2015, available at https://www.ntia.doc.gov/blog/2015/majority-americans-use-multiple-internet-connected-devicesdata-shows.

The most recent survey data from the Pew Research Center show that populations continuing to lag behind in internet adoption include people with low incomes, seniors, the less-educated, and households in rural areas (see Table 5). Pew has reported that the cost of monthly subscriptions is the main reason some people do not have broadband connections.[10]

In June 2015, GAO released a report (*Intended Outcomes and Effectiveness of Efforts to Address Adoption Barriers Are Unclear*) which found that affordability, lack of perceived relevance, and lack of computer skills are the principal barriers to broadband adoption.[11] GAO examined adoption efforts by NTIA and the FCC, and identified three key approaches used to address broadband adoption barriers: discounts on computer equipment and broadband subscriptions; outreach efforts to promote broadband availability and benefits; and training to help people develop skills in using computers and broadband.[12]

Table 5. Percentage of U.S. adults who do not use the internet in 2018

U.S. Adults	11%
Men	11%
Women	12%
White	11%
Black	13%
Hispanic	12%
18-29 age	2%
30-49	3%
50-64	13%
65+	34%
Less than $30K income	19%
$30K-$50K	7%
$50K-$75K	3%
$75K+	2%
Less than high school	35%
High school	16%
Some college	7%
College +	3%
Urban	8%
Suburban	10%
Rural	22%

Source: Pew Research Center, *11% of Americans don't use the internet. Who are they?* March 5, 2018, survey conducted January 3-10, 2018, available at http://www.pewresearch.org/fact-tank/2018/03/05/some-americansdont-use-the-internet-who-are-they/.

[10] John B. Horrigan and Maeve Duggan, Pew Research Center, *Home Broadband 2015*, December 21, 2015, available at http://www.pewinternet.org/2015/12/21/home-broadband-2015/.

[11] Government Accountability Office, *Intended Outcomes and Effectiveness of Efforts to Address Adoption Barriers Are Unclear*, GAO-15-473, June 2, 2015, p. 11, available at http://www.gao.gov/assets/680/670588.pdf.

[12] Ibid, p. 17.

Broadband in Rural Areas[13]

While the number of new broadband subscribers continues to grow, the rate of broadband deployment in urban areas has outpaced deployment in rural and tribal areas. While there are many examples of rural communities with state of the art telecommunications facilities,[14] recent surveys and studies have indicated that, in general, rural areas (and particularly tribal areas)[15] tend to lag behind urban and suburban areas in broadband deployment. For example

- According to the FCC's *Communications Marketplace Report*, "As of year-end 2017, 94% of the overall population had coverage [of fixed terrestrial broadband at speeds of 25 Mbps/3 Mbps], up from 91.9% in 2016. Nonetheless, the gap in rural and Tribal America remains notable: over 24% of Americans in rural areas and 32% of Americans in Tribal lands lack coverage from fixed terrestrial 25 Mbps/3 Mbps broadband, as compared to only 1.5% of Americans in urban areas."[16]
- Also according to the FCC's *Communications Marketplace Report*, rural areas continue to lag behind urban areas in mobile broadband deployment. Although evaluated urban areas saw an increase of 10 Mbps/3 Mbps mobile LTE (median speed) from 81.9% in 2014 to 92.6% in 2017, such deployment in evaluated rural areas remained relatively flat at about 70%.[17]
- According to January 2018 survey data from the Pew Research Center, 58% of adults in rural areas said they have a high-speed broadband connection at home, as opposed to 67% of adults in urban areas and 70% of adults in suburban areas.[18]
- A November 2017 Census Bureau survey reported by the National Telecommunications and Information Administration (NTIA) Digital Nation Data Explorer showed 72.9% of rural residents reporting using the internet, versus 78.5% of urban residents.[19] According to NTIA, the data "indicates a

---

[13] For more information on rural broadband and broadband programs at the Rural Utilities Service, see CRS Report RL33816, *Broadband Loan and Grant Programs in the USDA's Rural Utilities Service*, by Lennard G. Kruger.

[14] See for example: National Exchange Carrier Association (NECA), *Trends: A Report on Rural Telecom Technology,* December 2015, available at https://www.neca.org/WorkArea/linkit.aspx?LinkIdentifier=id&ItemID=12331&libID= 12351.

[15] For more information on tribal broadband, see CRS Report R44416, *Tribal Broadband: Status of Deployment and Federal Funding Programs*, by Lennard G. Kruger.

[16] FCC, *Communications Marketplace Report*, p. 131.

[17] Ibid., p. 133.

[18] Pew Research Center, *Internet/Broadband Fact Sheet*, February 5, 2018, available at http://www.pewinternet.org/fact-sheet/internet-broadband/.

[19] U.S. Department of Commerce, National Telecommunications and Information Administration, *Digital Nation Data Explorer*, June 6, 2018, available at https://www.ntia.doc.gov/data/digital-nation-data-explorer#sel=internetUser&demo=metro&pc=prop&disp=chart.

fairly constant 6-9 percentage point gap between rural and urban communities' internet use over time."[20]

The comparatively lower population density of rural areas is likely the major reason why broadband is less deployed than in more highly populated suburban and urban areas. Particularly for wireline broadband technologies—such as cable modem and fiber—the greater the geographical distances among customers, the larger the cost to serve those customers. Thus, there is often less incentive for companies to invest in broadband in rural areas than, for example, in an urban area where there is more demand (more customers with perhaps higher incomes) and less cost to wire the market area.

The terrain of rural areas can also be a hindrance, in that it is more expensive, for example, to deploy broadband technologies in a mountainous or heavily forested area. An additional added cost factor for remote areas can be the expense of "backhaul" (e.g., the "middle mile"), which refers to the installation of a dedicated line which transmits a signal to and from an internet backbone, which is typically located in or near an urban area. Some policymakers believe that disparities in broadband access across American society could have adverse consequences on those left behind, and that advanced telecommunications applications critical for businesses and consumers to engage in ecommerce are increasingly dependent on high-speed broadband connections to the internet. Thus, some say, communities and individuals without access to broadband could be at risk to the extent that connectivity becomes a critical factor in determining future economic development and prosperity. A February 2006 study done by the Massachusetts Institute of Technology for the Economic Development Administration of the Department of Commerce marked the first attempt to quantitatively measure the impact of broadband on economic growth. The study found that "between 1998 and 2002, communities in which mass-market broadband was available by December 1999 experienced more rapid growth in employment, the number of businesses overall, and businesses in IT-intensive sectors, relative to comparable communities without broadband at that time."[21]

Subsequently, other studies have attempted to assess the economic impact of broadband deployment. For example

- A June 2007 report from the Brookings Institution found that for every one percentage point increase in broadband penetration in a state, employment is

[20] U.S. Department of Commerce, National Telecommunications and Information Administration, "The State of the Urban/Rural Digital Divide," August 10, 2016, available at https://www.ntia.doc.gov/blog/2016/state-urbanruraldigital-divide.

[21] Gillett, Sharon E., Massachusetts Institute of Technology, *Measuring Broadband's Economic Impact*, report prepared for the Economic Development Administration, U.S. Department of Commerce, February 28, 2006, p. 4, available at http://cfp.mit.edu/publications/CFP_Papers/Measuring_bb_econ_impact-final.pdf.

projected to increase by 0.2% to 0.3% per year. For the entire U.S. private nonfarm economy, the study projected an increase of about 300,000 jobs.[22]

- A July 2009 study commissioned by the Internet Innovation Alliance found net consumer benefits of home broadband on the order of $32 billion per year, up from an estimated $20 billion in consumer benefits from home broadband in 2005.[23]
- A January 2009 study conducted by the Information Technology and Innovation Foundation (ITIF) found that investing an additional $10 billion in one year on broadband networks will create or retain 498,000 U.S. jobs for that year.[24]
- A study (first published in 2013)[25] funded by the National Agricultural and Rural Development Policy Center found that nonmetropolitan counties that had high levels of broadband adoption (greater than 60%) in 2010 had significantly higher growth in median household income—23.4% versus just over 22%—between 2001 and 2010 when compared to counties that had similar characteristics in the 1990s but were not as successful at adopting broadband.
- A 2016 study from the Hudson Institute found that rural broadband providers directly and indirectly added $24.1 billion to the U.S. economy in 2015. The rural broadband industry supported 69,595 jobs in 2015, both through its own employment and the employment that its purchases of goods and services generated.[26]

Broadband and the Federal Role

Section 706 of the Telecommunications Act of 1996

Section 706 of the Telecommunications Act of 1996 (P.L. 104-104) requires the FCC to regularly initiate an inquiry assessing the availability of broadband to all Americans

---

[22] Crandall, Robert, William Lehr, and Robert Litan, *The Effects of Broadband Deployment on Output and Employment: A Cross-sectional Analysis of U.S. Data*, June 2007, 20 pp., available at https://www.brookings.edu/ research/the-effects-of-broadband-deployment-on-output-and-employment-a-cross-sectional-analysis-of-u-s-data/.

[23] Mark Dutz, Jonathan Orszag, and Robert Willig, *The Substantial Consumer Benefits of Broadband Connectivity for U.S. Households*, Internet Innovation Alliance, July 2009, p. 4, available at http://internet innovation.org/files/specialreports/CONSUMER_BENEFITS_OF_BROADBAND.pdf.

[24] Robert D. Atkinson, Daniel Castro and Stephen Ezell, Information Technology and Innovation Foundation, *The Digital Road to Recovery: A Stimulus Plan to Create Jobs, Boost Productivity and Revitalize America*, January 2009, 22 p, available at https://www.itif.org/files/roadtorecovery.pdf.

[25] Brian Whitacre, Roberto Gallardo, and Sharon Strover, "Broadband's Contribution to Economic Health in Rural Areas," *Research & Policy Brief Series*, Community and Regional Development Institute, Cornell University, February 2015, available at https://cardi.cals.cornell.edu/sites/cardi.cals.cornell.edu/files/ shared/documents/ResearchPolicyBriefs/Policy-Brief-Feb15-draft03.pdf.

[26] Hanns Kuttner, Hudson Institute, *The Economic Impact of Rural Broadband*, April 2016, available at https://s3.amazonaws.com/media.hudson.org/files/publications/201604 19KuttnerTheEconomicImpactofRuralBroadban d.pdf.

and to determine whether broadband is "being deployed to all Americans in a reasonable and timely fashion." If the determination is negative, the act directs the FCC to "take immediate action to accelerate deployment of such capability by removing barriers to infrastructure investment and by promoting competition in the telecommunications market."

Starting in 1999, there have been eleven Section 706 reports, each providing a snapshot and assessment of broadband deployment.[27] As part of this assessment, and to help determine whether broadband is being deployed in "a reasonable and timely fashion," the FCC has set a minimum broadband speed that essentially serves as the benchmark the FCC uses to determine what it considers broadband service for the purposes of its Section 706 determination. In 2015 the FCC, citing changing broadband usage patterns and multiple devices using broadband within single households, raised its minimum fixed broadband benchmark speed from 4 Mbps (download)/1 Mbps (upload) to 25 Mbps/3 Mbps. The designation of minimum benchmark speeds for fixed broadband, and how mobile broadband speeds should be benchmarked and factored into an overall determination of broadband deployment, has proven controversial.[28]

On February 2, 2018, the FCC adopted and released its latest 706 report, the *2018 Broadband Deployment Report*. The FCC concluded that advanced telecommunications capability is being deployed to all Americans in a reasonable and timely fashion. This determination was based on evaluating progress—comparing deployment in the present year to deployment in previous years. According to the *Report*:

> We find that analyzing progress to determine whether deployment is occurring in a reasonable and timely fashion is the approach that is most consistent with the language of section 706, as the analysis of such progress enables the Commission to determine whether advanced telecommunications capability "is being deployed" in the manner that section 706 requires. The use of the present progressive tense—"is being deployed"—as well as the language requiring an evaluation of whether that deployment is "reasonable and timely" indicates that Congress intended that the Commission evaluate the current state of deployment to all Americans, not a rigid requirement that each and every American be served *at this moment*.[29]

The FCC's latest 706 determination that broadband is being deployed in a reasonable and timely fashion is a departure from the FCC's previous five determinations that broadband is not being deployed in a reasonable and timely fashion. The latest 706

---

[27] An archive of notices of inquiry and released broadband progress reports is available at https://www.fcc.gov/general/ archive-released-broadband-progress-notices-inquiry.

[28] See CRS Report R45039, *Defining Broadband: Minimum Threshold Speeds and Broadband Policy*, by Lennard G. Kruger.

[29] Federal Communications Commission, *2018 Broadband Deployment Report*, In the Matter of Inquiry Concerning Deployment of Advanced Telecommunications Capability to All Americans in a Reasonable and Timely Fashion, GN Docket no. 17-199, FCC 18-10, adopted and released February 2, 2018, p. 4, available at https://www.fcc.gov/document/fcc-releases-2018-broadband-deployment-report.

determination was approved by the three Republican FCC commissioners, with the remaining two Democratic commissioners dissenting. According to FCC Commissioner Rosenworcel's dissent:

> This chapter concludes that in the United States the deployment of broadband to all Americans is reasonable and timely. This is ridiculous—and irresponsible. Today there are 24 million Americans without access to broadband. There are 19 million Americans in rural areas who lack the ability to access high-speed services at home. There are 12 million school-aged children who are falling into the Homework Gap because they do not have the broadband at home they need for nightly schoolwork. Ask any one of them if they think the deployment of the most essential digital age infrastructure is reasonable and timely and you will get a resounding "No." To call these numbers a testament to our national success is insulting and not credible.[30]

In gathering data, information, and viewpoints for the *Report*, the August 8, 2017, *Notice of Inquiry*[31] proposed to maintain the 25 Mbps/3 Mbps benchmark for fixed broadband, while at the same time soliciting comments on whether to establish a lower benchmark speed specifically for mobile broadband. One proposal under consideration was whether the presence of fixed *or* mobile broadband should indicate that an area has adequate broadband service. Ultimately, the *Report* concluded that adoption of a single mobile benchmark is currently unworkable, given available data and the inherent variability of actual mobile speeds, and that mobile broadband service is not a full substitute for fixed service at this time.

On August 8, 2018, the FCC adopted the *Fourteenth Broadband Deployment Report Notice of Inquiry*, which is collecting comments in preparation for the next Section 706 report in 2019. The FCC proposes to maintain the 25 Mbps/3 Mbps benchmark and is seeking comments on "whether since the *2018 Report* there have been developments that would support a different conclusion about substitutability" between mobile and fixed broadband service, and "to the extent that mobile services are able to offer equivalent functionality as fixed services either now or in the future ... whether or not and in what circumstances, if any, mobile and fixed services should be considered substitutes."[32]

Broadband Data and Mapping

Obtaining an accurate snapshot of broadband deployment is problematic. Rapidly evolving technologies, the constant flux of the telecommunications industry, the

---

[30] Ibid, p. 89.

[31] FCC, *Thirteenth Section 706 Report Notice of Inquiry*, GN Docket No. 17-199, FCC 17-109, August 8, 2017, available at https://transition.fcc.gov/Daily_Releases/Daily_Business/2017/db0808/FCC-17-109A1.pdf.

[32] FCC, *Fourteenth Broadband Deployment Report Notice of Inquiry*, GN Docket No. 18-238, FCC 18-119, adopted August 8, 2018, para. 11, available at https://docs.fcc.gov/public/attachments/FCC-18-119A1.pdf.

uncertainty of consumer wants and needs, and the sheer diversity and size of the nation's economy and geography make the status of broadband deployment difficult to characterize. Improving the quality of broadband deployment data has become an issue of congressional interest, as policymakers recognize that more accurate broadband availability maps could help ensure that federal broadband programs target unserved areas of the country that are most in need of assistance.

Since the initial deployment of broadband in the late 1990s, two federal agencies have implemented broadband availability data collection and mapping initiatives: the NTIA's State Broadband Initiative,[33] which was funded by the American Recovery and Reinvestment Act of 2009 (ARRA) and used to develop the previous National Broadband Map; and the FCC's Form 477 Data Program, which is used to populate and update the current National Broadband Map.[34]

One of the major criticisms of the FCC's Form 477 National Broadband Map is that broadband availability can be overstated because fixed broadband deployment data are collected at the census block level. A census block is considered served if there is broadband service (or the strong potential of broadband service) to one or more locations. This is especially problematic in rural areas, which have large census blocks and may be considered served if, for example, a single neighborhood in that large census block has broadband service.

On August 3, 2017, the FCC adopted a *Further Notice of Proposed Rulemaking* to explore ways "to improve the quality, accuracy, and usefulness of the data it collects on fixed and mobile voice and broadband service," while at the same time examining how it can "reduce burdens on industry by eliminating unnecessary or onerous data filing requirements."[35]

The Administration requested $50 million for broadband mapping in FY2018. The Consolidated Appropriations Act, 2018 (P.L. 115-141) appropriated $7.5 million to NTIA to update the national broadband availability map in coordination with the FCC and using partnerships previously developed with the states. For FY2019, the House and Senate appropriations committees would provide an additional $7.5 million to NTIA for broadband mapping.

On May 30, 2018, NTIA issued a *Request for Comments* on actions it should take to improve the quality and accuracy of broadband availability data.[36] NTIA received

---

[33] Previously, in 2008, the Broadband Data Improvement Act (P.L. 110-385) had directed the Department of Commerce to establish a state broadband data and development grant program, and to use the data gathered by the states to create a broadband inventory map.

[34] For more information on NTIA and FCC broadband data and mapping activities, see CRS Insight IN10925, *Broadband Data and Mapping*, by Lennard G. Kruger.

[35] FCC, *Further Notice of Proposed Rulemaking*, "In the Matter of Modernizing the FCC Form 477 Data Program," WC Docket No. 11-19, FCC 17-103, adopted August 3, 2017, 44 pp., available at https://docs.fcc.gov/public/attachments/FCC-17-103A1.pdf.

[36] NTIA, "Improving the Quality and Accuracy of Broadband Availability Data," *Federal Register*, vol. 83, no. 104, May 30, 2018, pp. 24747-24749, available at https://www.ntia.doc.gov/files/ntia/publications/fr-05302018-rfc-improving-broadband-data.pdf.

comments[37] on the following issues: identifying additional broadband availability data; technology type, service areas, and bandwidth associated with such data; new approaches, tools, technologies, or methodologies to capture broadband availability data; validating broadband availability data; and identifying gaps in broadband availability. On October 25, 2018, NTIA issued a notice for public comment regarding its intention to collect broadband availability data at a more granular level than what the FCC currently collects.[38]

The National Broadband Plan

As mandated by the American Recovery and Reinvestment Act of 2009 (ARRA), on March 16, 2010, the FCC released its report, *Connecting America: The National Broadband Plan.*[39] The National Broadband Plan (NBP) sought to "create a high-performance America," which the FCC defined as "a more productive, creative, efficient America in which affordable broadband is available everywhere and everyone has the means and skills to use valuable broadband applications."[40] In order to achieve this mission, the NBP recommended that the country set six goals for 2020

- Goal No. 1: At least 100 million U.S. homes should have affordable access to actual download speeds of at least 100 megabits per second and actual upload speeds of at least 50 megabits per second.
- Goal No. 2: The United States should lead the world in mobile innovation, with the fastest and most extensive wireless networks of any nation.
- Goal No. 3: Every American should have affordable access to robust broadband service, and the means and skills to subscribe if they so choose.
- Goal No. 4: Every American community should have affordable access to at least 1 gigabit per second broadband service to anchor institutions such as schools, hospitals, and government buildings.
- Goal No. 5: To ensure the safety of the American people, every first responder should have access to a nationwide, wireless, interoperable broadband public safety network.

---

[37] Available at https://www.ntia.doc.gov/federal-register-notice/2018/comments-improving-quality-and-accuracy-broadband-availability-data.

[38] NTIA, "Agency Information Collection Activities; Proposed Information Collection; Comment Request; Broadband Availability Data," *Federal Register*, vol. 83, no. 207, October 25, 2018, pp. 53852-53853, available at https://www.govinfo.gov/content/pkg/FR-2018-10-25/pdf/2018-23296.pdf.

[39] Available at http://www.broadband.gov/plan/.

[40] Federal Communications Commission, Connecting America: *The National Broadband Plan*, March 17, 2010, p. 9.

- Goal No. 6: To ensure that America leads in the clean energy economy, every American should be able to use broadband to track and manage their real-time energy consumption.

The National Broadband Plan was categorized into three parts

- Part I (Innovation and Investment), which "discusses recommendations to maximize innovation, investment and consumer welfare, primarily through competition. It then recommends more efficient allocation and management of assets government controls or influences."[41] The recommendations address a number of issues, including spectrum policy, improved broadband data collection, broadband performance standards and disclosure, special access rates, interconnection, privacy and cybersecurity, child online safety, poles and rights-of-way, research and experimentation (R&E) tax credits, and R&D funding.
- Part II (Inclusion), which "makes recommendations to promote inclusion—to ensure that all Americans have access to the opportunities broadband can provide."[42] Issues include reforming the Universal Service Fund, intercarrier compensation, federal assistance for broadband in tribal lands, expanding existing broadband grant and loan programs at the Rural Utilities Service, enabling greater broadband connectivity in anchor institutions, and improved broadband adoption and utilization especially among disadvantaged and vulnerable populations.
- Part III (National Purposes), which "makes recommendations to maximize the use of broadband to address national priorities. This includes reforming laws, policies and incentives to maximize the benefits of broadband in areas where government plays a significant role."[43] National purposes include health care, education, energy and the environment, government performance, civic engagement, and public safety. Issues include telehealth and health IT, online learning and modernizing educational broadband infrastructure, digital literacy and job training, smart grid and smart buildings, federal support for broadband in small businesses, telework within the federal government, cybersecurity and protection of critical broadband infrastructure, copyright of public digital media, interoperable public safety communications, next generation 911 networks, and emergency alert systems.

---

[41] Ibid, p. 11.
[42] Ibid.
[43] Ibid.

FEDERAL BROADBAND PROGRAMS

With the conclusion of grant and loan awards established by the American Recovery and Reinvestment Act of 2009 (P.L. 111-5),[44] there remain two ongoing major federal vehicles which direct federal money to fund broadband: the Universal Service Fund (USF) programs under the Federal Communications Commission (FCC), and the broadband and telecommunications programs at the Rural Utilities Service (RUS) of the U.S. Department of Agriculture.

In June 2017, the National Telecommunications and Information Administration (NTIA) released an updated comprehensive *Guide to Federal Funding of Broadband Projects*.[45] The guide provides summary and contact information for a variety of federal programs that may fund projects involving broadband infrastructure, adoption, access, planning, or research.[46]

The Universal Service Concept and the FCC[47]

Since its creation in 1934 the Federal Communications Commission (FCC) has been tasked with "mak[ing] available, so far as possible, to all the people of the United States ... a rapid, efficient, Nation-wide, and world-wide wire and radio communications service with adequate facilities at reasonable charges."[48] This mandate led to the development of what has come to be known as the universal service concept.

The universal service concept, as originally designed, called for the establishment of policies to ensure that telecommunications services are available to all Americans, including those in rural, insular, and high cost areas, by ensuring that rates remain affordable. Over the years this concept has evolved and expanded, fostering the development of various FCC policies and programs that target both providers of and subscribers to telecommunications and, more recently, broadband services. Passage of the Telecommunications Act of 1996 (P.L. 104-104) codified the longstanding commitment by U.S. policymakers to ensure universal service in the provision of telecommunications services, and the FCC established, in 1997, a federal Universal Service Fund (USF) to meet the expanded objectives and principles contained in the act. The USF is

---

[44] See CRS Report R40436, *Broadband Infrastructure Programs in the American Recovery and Reinvestment Act*, by Lennard G. Kruger.

[45] U.S. Department of Commerce, National Telecommunications and Information Administration, *BroadbandUSA: Guide to Federal Funding of Broadband Projects*, June 2017, 44 pp., available at https://broadbandusa.ntia.doc.gov/sites/default/files/resource-files/ntia_guidetofedfunding_062317.pdf.

[46] NTIA also provides an online broadband federal funding search tool, available at https://broadbandusa.ntia.doc.gov/ funding-list.

[47] The section on universal service was prepared by Angele Gilroy, Specialist in Telecommunications, Resources, Science and Industry Division.

[48] Communications Act of 1934, As Amended, Title I §1 (47 U.S.C. 151).

administered by the Universal Service Administrative Company (USAC), an independent not-for-profit organization, under the direction of the FCC. The USF is being transformed in stages, over a multiyear period, from a mechanism to support voice telecommunications services to one that supports the deployment, adoption, and utilization of both fixed and mobile broadband. The USF currently administers four programs: the High Cost/Connect America Fund Program, the Schools and Libraries Program, the Rural Health Care Program, and the Low Income Program. The USF disbursed $8.8 billion in 2017 with all 50 states, the District of Columbia, and all territories receiving some benefit.[49]

Universal Service and Broadband

One of the major policy debates surrounding universal service in the last decade was whether access to advanced telecommunications services (i.e., broadband) should be incorporated into universal service objectives. The 1996 Telecommunications Act tasked the federal-state Joint Board with defining the services which should be included in the definition of services to be eligible for universal service support. The Joint Board's recommendation, which was adopted by the FCC in May 1997, largely limited the definition to voice telecommunications services. Some policymakers expressed concern that the FCC-adopted definition was too limited and did not take into account the importance and growing acceptance of advanced services such as broadband and internet access. They pointed to a number of provisions contained in the universal service principles of the 1996 act to support their claim. Universal service principles contained in Section 254(b)(2) state that "Access to advanced telecommunications services should be provided to all regions of the Nation." The subsequent principle (b)(3) calls for consumers in all regions of the nation, including "low-income" and those in "rural, insular, and high cost areas," to have access to telecommunications and information services including "advanced services" at a comparable level and a comparable rate charged for similar services in urban areas. Such provisions, they state, dictate that the FCC expand its universal service definition.

The 1996 act does take into consideration the changing nature of the telecommunications sector and allows, if future conditions warrant, for the modification of the universal service definition. Section 254(c) of the act states that "universal service is an evolving level of telecommunications services" and that the FCC is tasked with "periodically" reevaluating this definition "taking into account advances in telecommunications and information technologies and services." Furthermore, the Joint

[49] *2017 Annual Report, Universal Service Administrative Company*, p. 2. Total funding approved for disbursement for the months of January-December 2017, available at https://www.usac.org/about/tools/publications/annual-reports/ default.aspx.

Board is given specific authority to recommend "from time to time" to the FCC modification in the definition of the services to be included for federal universal service support. The Joint Board, on November 19, 2007, concluded such an inquiry and recommended that the FCC change the mix of services eligible for universal support. The Joint Board recommended, among other things, that "the universal availability of broadband Internet services" be included in the nation's communications goals and hence be supported by federal universal service funds.[50]

This debate was put to rest when provisions contained in the American Recovery and Reinvestment Act of 2009 (ARRA) called for the FCC to develop, and submit to Congress, a national broadband plan to ensure that every American has "access to broadband capability."[51]

The FCC in its national broadband plan, *Connecting America: the National Broadband Plan*, recommended that access to and adoption of broadband be a national goal. Furthermore the national broadband plan proposed that the Universal Service Fund be restructured to become a vehicle to help reach this goal. The FCC, in an October 2011 decision, adopted an Order that calls for the USF to be transformed, in stages, over a multiyear period, from a mechanism to support voice telephone service to one that supports the deployment, adoption, and utilization of both fixed and mobile broadband. This transformation includes the phaseout of the USF's legacy High Cost Program and the creation of a new fund, the Connect America Fund, to replace it as well as an expansion and modification of the Schools and Libraries, Rural Health Care, and Low Income programs.[52]

The High Cost/Connect America Fund Program

Historically the High Cost Program provided support for eligible telecommunications carriers to help offset the higher-than-average costs of providing voice telephone service in rural, insular, or other high cost areas. This mechanism has been the largest USF program based on disbursements and has been particularly important to rural areas due to the lack of subscriber density often combined with higher costs. The High Cost Program is undergoing a transition from one that primarily supports voice communications to one that supports a broadband platform that enables multiple applications, including voice. The High Cost program is being phased out in stages and is being replaced by the Connect America Fund (CAF), which will support the provision of affordable voice and

[50] The Joint Board recommended that the definition of those services that qualify for universal service support be expanded and that the nation's communications goals include the universal availability of: mobility services (i.e., wireless); broadband internet services; and voice services at affordable and comparable rates for all rural and nonrural areas. For a copy of this recommendation see http://hraunfoss.fcc.gov/edocs_public/attachmatch/FCC-07j-4A1.pdf.

[51] American Recovery and Reinvestment Act of 2009, P.L. 111-5, Section 6001 (k)(2)(D).

[52] For a detailed discussion of this Order and USF transition see *In the Matter of the Connect America Fund, et. al.*, WC Docket No. 10-90 et. al., Report and Order and Further Notice of Proposed Rulemaking, FCC 11-161, adopted October 27, 2011, and released November 18, 2011, available at http://hraunfoss.fcc.gov/edocs_public/attachmatch/FCC-11- 161A1.pdf.

broadband services, both fixed and mobile, in high cost areas. The CAF will eventually replace all of the existing support mechanisms in the High Cost Program and contains a Mobility Fund and a Remote Areas Fund to meet these needs. According to data released by program administrators, approximately $4.67 billion in funding was disbursed in 2017.[53]

The Schools and Libraries, and Rural Health Care Programs

Congress, through the 1996 act, not only codified, but also expanded the concept of universal service to include, among other principles, that elementary and secondary schools and classrooms, libraries, and rural health care providers have access to telecommunications services for specific purposes at discounted rates. (See §§254(b)(6) and 254(h) of the 1996 Telecommunications Act, 47 U.S.C. 254.)

The Schools and Libraries (E-Rate) Program

Under universal service provisions contained in the 1996 act, elementary and secondary schools and classrooms and libraries are designated as beneficiaries of universal service discounts. Universal service principles detailed in Section 254(b)(6) state that "Elementary and secondary schools and classrooms ... and libraries should have access to advanced telecommunications services." The act further requires in Section 254(h)(1)(B) that services within the definition of universal service be provided to elementary and secondary schools and libraries for education purposes at discounts, that is at "rates less than the amounts charged for similar services to other parties."

The FCC established the Schools and Libraries Division within USAC to administer the schools and libraries or "E (education)-rate" program to comply with these provisions. Under this program, eligible schools and libraries receive discounts ranging from 20% to 90% for telecommunications services depending on the poverty level of the school's (or school district's) population and its location in a high cost (i.e., rural) telecommunications area. Two categories of services are eligible for discounts: category one services (telecommunications, telecommunications services, and internet access), and category two services that deliver internet access within schools and libraries (internal connections, basic maintenance of internal connections, and managed internal broadband services). The funding cap for funding year 2018 (July 1, 2018, to June 30, 2019) is $4.1 billion. According to data released by program administrators, approximately $2.62 billion in funding was disbursed in 2017.[54]

---

[53] For additional information and data on this program see *2017 Annual Report, Universal Service Administrative Company*, pp. 2 and 10, available at https://www.usac.org/about/tools/publications/annual-reports/default.aspx.

[54] For additional information and data on this program see *2017 Annual Report, Universal Service Administrative Company*, pp. 2 and 7, available at https://www.usac.org/about/tools/publications/annual-reports/default.aspx.

The Rural Health Care Program

Section 254(h) of the 1996 act requires that public and nonprofit rural health care providers have access to telecommunications services necessary for the provision of health care services at rates comparable to those paid for similar services in urban areas. Subsection 254(h)(1) further specifies that "to the extent technically feasible and economically reasonable" health care providers should have access to advanced telecommunications and information services. The FCC established the Rural Health Care Division (RHCD) within USAC to administer the universal support program to comply with these provisions. The Rural Health Care Program provides funding through three programs: the Telecommunications Program, the Healthcare Connect Fund, and the rural Health Care Pilot Program. The goal of these programs is to improve the quality of health care for those living in rural areas by ensuring access to broadband and telecommunications services. Under FCC established rules only public or nonprofit health care providers are eligible to receive funding.

The Telecommunications Program, established in 1997, provides discounts for telecommunications services to ensure that eligible rural health care providers pay no more than urban providers for telecommunications services. The primary use of the funding is to provide reduced rates for telecommunications and information services necessary for the provision of health care.[55]

The Rural Health Care Pilot Program was established in 2006 to help public and nonprofit health care providers build state- and region-wide broadband networks dedicated to the provision of health care services. The program provides funding up to 85% of eligible costs. No new funding is available under this program and current participants that need additional support will transfer to the most recently created program, the Healthcare Connect Fund.

The FCC in December 2012 created the Healthcare Connect Fund,[56] a program to expand health care provider access to broadband, particularly in rural areas, and replace the Rural Health Care Pilot Program with a permanent program. The Healthcare Connect Fund program supports high-capacity broadband connectivity and encourages the development of state and regional networks. This program provides a 65% discount on eligible expenses related to broadband connectivity and is available to individual rural health care providers and consortia. Consortia can include nonrural providers but at least 50% of providers must be located in a rural area. The total annual funding cap for all of the above mentioned USF rural health care programs is $400 million.[57] According to data

---

[55] For additional information on this program see the RHCD website: http://www.universalservice.org/rhc/.

[56] Title II (the Rural Healthcare Connectivity Act of 2016) of P.L. 114-182 includes skilled nursing facilities to the list of health care providers eligible to receive RHC program support. This change became effective June 21, 2017.

[57] For more details on the USF rural health care support mechanism and the Healthcare Connect Fund see *In the Matter of Rural Health Care Support Mechanism*, WC Docket No. 02-60, Federal Communications Commission, adopted December 12, 2012, available at http://hraunfoss.fcc.gov/edocs_public/attachmatch/FCC-12-150A1.pdf.

released by program administrators, approximately $261.5 million was disbursed in 2017.[58]

The Low Income Program

As initially designed the Low Income Program provided a discount for voice telephony service for eligible low-income consumers. The major program has two subprograms, Lifeline and Link Up,[59] with the Lifeline Program providing the vast majority of support. In March 2016 the FCC adopted an Order to expand the Lifeline Program to support mobile and fixed broadband internet access services on a stand-alone basis, or with a bundled voice service.[60] Households must meet a needs-based criteria for eligibility. The Lifeline Program provides assistance to only one line per eligible household either wired or wireless, in the form of a monthly subsidy of, in most cases, $9.25.[61] Support is not given directly to the subscriber but to the designated service provider. According to data released by program administrators, approximately $1.27 billion was disbursed in 2017.[62]

Rural Utilities Service Programs

RUS implements four programs specifically targeted at providing assistance for broadband connectivity infrastructure deployment in rural areas: the Rural Broadband Access Loan and Loan Guarantee Program (also referred to as the Farm Bill Broadband Loans), the Telecommunications Infrastructure Loans and Loan Guarantees (previously the rural telephone loan program dating back to 1949), the ReConnect Program (pilot broadband loan and grant program), and the Community Connect Grant Program.[63] Additionally, RUS houses the Distance Learning and Telemedicine Grant Program, which supports broadband-based distance learning and telemedicine applications. The 115th Congress reauthorized and modified RUS broadband programs as part of the 2018 farm bill (P.L. 115-334, Agriculture Improvement Act of 2018).

---

[58] For additional information and data on this program see *2017 Annual Report, Universal Service Administrative Company*, pp. 2 and 8, available at https://www.usac.org/about/tools/publications/annual-reports/default.aspx.

[59] The Link Up program assists eligible low-income subscribers to pay the costs associated with the initiation of service and is no longer available except for on Tribal Lands.

[60] For additional information on the Lifeline Program see CRS Report R44487, *Federal Lifeline Program: Frequently Asked Questions*, by Angele A. Gilroy.

[61] Tribal Lands Lifeline provides an additional discount of up to $25 for eligible low-income consumers living on Tribal Lands for a total discount of up to $34.25.

[62] For additional information and data on this program see *2017 Annual Report, Universal Service Administrative Company*, pp. 2 and 9, available at https://www.usac.org/about/tools/publications/annual-reports/default.aspx. http://www.usac.org/_res/documents/about/pdf/annual-reports/usac-annual-report-interactive-2016.pdf.

[63] For more information on these programs, see CRS Report RL33816, *Broadband Loan and Grant Programs in the USDA's Rural Utilities Service*, by Lennard G. Kruger.

P.L. 111-5: The American Recovery and Reinvestment Act of 2009

On February 17, 2009, President Obama signed P.L. 111-5, the American Recovery and Reinvestment Act (ARRA). Broadband provisions of the ARRA provided a total of $7.2 billion, for broadband grants, loans, and loan/grant combinations. The total consisted of $4.7 billion to NTIA/DOC for a newly established Broadband Technology Opportunities Program (grants) and $2.5 billion to the RUS/USDA Broadband Initiatives Program (grants, loans, and grant/loan combinations).

Regarding the $2.5 billion to RUS/USDA broadband programs, the ARRA specified that at least 75% of the area to be served by a project receiving funds shall be in a rural area without sufficient access to high-speed broadband service to facilitate economic development, as determined by the Secretary of Agriculture. Priority was given to projects that provide service to the most rural residents that do not have access to broadband services. Priority was also given to borrowers and former borrowers of rural telephone loans.

Of the $4.7 billion appropriated to NTIA

- $4.35 billion was directed to a competitive broadband grant program, of which not less than $200 million shall be available for competitive grants for expanding public computer center capacity (including at community colleges and public libraries); not less than $250 million to encourage sustainable adoption of broadband service; and $10 million transferred to the Department of Commerce Office of Inspector General for audits and oversight; and
- $350 million was directed for funding the Broadband Data Improvement Act (P.L. 110-385) and for the purpose of developing and maintaining a broadband inventory map, which shall be made accessible to the public no later than two years after enactment. Funds deemed necessary and appropriate by the Secretary of Commerce may be transferred to the FCC for the purposes of developing a national broadband plan, which shall be completed one year after enactment.

Final BTOP and BIP program awards were announced by September 30, 2010. With a few exceptions, all ARRA broadband projects were concluded as of September 30, 2015.[64]

---

[64] For more information on implementation of the broadband provisions of the ARRA, see CRS Report R40436, *Broadband Infrastructure Programs in the American Recovery and Reinvestment Act*, by Lennard G. Kruger. For information on the distribution and oversight of ARRA broadband grants and loans, see CRS Report R41775, *Background and Issues for Congressional Oversight of ARRA Broadband Awards*, by Lennard G. Kruger.

Infrastructure Initiative and Broadband

On February 12, 2018, the Trump Administration released its *Legislative Outline for Rebuilding Infrastructure in America*.[65] The plan does not dedicate any funding exclusively for broadband, but does include rural broadband among the types of infrastructure projects that would be eligible for funding. Proposed funding streams include the following:

- *$50 billion* for a Rural Infrastructure Program. Funding would be block-granted to the states under a formula distribution for infrastructure projects including transportation, water and waste, power and electric, water resources, and broadband (including other high-speed data and communication conduits). Governors "would have the discretion to choose investments to respond to the unique rural needs of their states."[66] Eligible infrastructure projects would serve rural areas with populations of less than 50,000. An unspecified portion of the Rural Infrastructure Program funds would be set aside for tribal infrastructure and territorial infrastructure.
- *$20 billion* for a Transformative Projects Program, which would "provide Federal funding and technical assistance for bold, innovative, and transformative infrastructure projects that could dramatically improve infrastructure."[67] Funding would be awarded on a competitive basis to projects "that are likely to be commercially viable, but that possess unique technical and risk characteristics that otherwise deter private sector investment."[68] The program would be led by the Department of Commerce, which would chair an interagency project selection and evaluation committee. Federal funding would be available for up to 30% of eligible costs for project demonstration, up to 50% of eligible costs for project planning, and up to 80% of eligible costs for capital construction.
- *$14 billion* for expanding existing federal credit programs that address infrastructure. This would include additional budget authority to the USDA's Rural Utilities Service for RUS loan programs (which include telecommunications and broadband loans and loan guarantees).
- *$6 billion* for expanding the scope of Public Activity Bonds (PABs). The proposal would expand and modify eligible exempt facilities for PABs to include a number of new categories, including rural broadband service facilities.

---

[65] The White House, *Legislative Outline for Rebuilding Infrastructure in America*, released February 12, 2018, 53 pages, available at https://www.whitehouse.gov/wp-content/uploads/2018/02/INFRASTRUCTURE-211.pdf.

[66] Ibid. p. 6.

[67] Ibid, p. 8.

[68] Ibid.

It will be up to Congress to determine the extent to which the Administration infrastructure proposal will be implemented, and how an infrastructure initiative will be legislated. Whether dedicated rural broadband funding should or should not be part of the legislative response will likely be debated.[69] Meanwhile, rural broadband was included in the $20 billion carved out for infrastructure in the two-year budget agreement ($10 billion per year for two years) reached between the House and Senate in February 2018 (P.L. 115-123). However, the amount of targeted funding specifically for broadband was not specified.

The Trump Administration's *Legislative Outline* also contains many recommendations for reducing the costs and improving the time-effectiveness of infrastructure deployment by streamlining permitting regulations and procedures. In the 115th Congress, the House and Senate have considered legislation that would streamline permitting for broadband deployment.[70] The FCC has also begun a process to develop recommendations for lowering or removing regulatory barriers to broadband deployment (see section below, "FCC's Broadband Deployment Advisory Committee").

Other Federal Programs and Initiatives

FCC's Broadband Deployment Advisory Committee

Aside from funding, another way the federal government can facilitate broadband deployment is by taking steps to lower or remove regulatory barriers to broadband deployment facing private sector providers. On January 31, 2017, FCC Chairman Ajit Pai announced the formation of a new federal advisory committee, the Broadband Deployment Advisory Committee (BDAC),[71] which will provide advice and recommendations for the FCC on how to accelerate the deployment of broadband by reducing and/or removing regulatory barriers to infrastructure investment.[72] The BDAC is composed of stakeholders, appointed by the FCC chairman, representing industry, states, localities, tribes, academia, and others. Five working groups have been formed; these are Model Code for Municipalities, Model Code for States, Competitive Access to

---

[69] Democrats in Congress have released infrastructure proposals that include dedicated funding for broadband. See for example: the Senate Democratic Caucus proposal, *A Better Deal*, which would invest at least $40 billion in direct federal funding to provide universal high-speed internet (https://www.democrats. senate.gov/imo/media/doc/A%20Better%20Deal%20Universal%20High%20Speed%20Internet%20FINAL.pdf); the Lift America Act, which would provide $40 billion over five years to deploy broadband (https://democrats-energycommerce.house.gov/ newsroom/press-releases/ec-democrats-unveil-comprehensive-infrastructure-package); and the 21st Century New Deal for Jobs Act, which would provide $100 billion for broadband over ten years (https://cpc-grijalva.house.gov/21stcentury-new-deal-for-jobs/).

[70] See for example Consolidated Appropriations Act, 2018 (P.L. 115-141), Division P, Title V (Ray Baum's Act of 2018) and Title VI (MOBILE NOW).

[71] https://www.fcc.gov/broadband-deployment-advisory-committee.

[72] See https://www.fcc.gov/broadband-deployment-advisory-committee.

Broadband Infrastructure, Removing State and Local Regulatory Barriers, and Streamlining Federal Siting.[73]

The FCC has also initiated proceedings and adopted orders addressing the issue of reducing regulatory barriers for the deployment of wireless and wireline broadband.[74]

BroadbandUSA

BroadbandUSA is housed at the Department of Commerce's National Telecommunications and Information Administration (NTIA). Using the expertise gained during administration of the ARRA Broadband Technology Opportunities Program (BTOP), the BroadbandUSA program offers one-to-one technical assistance to communities seeking to plan and implement broadband initiatives. BroadbandUSA will leverage knowledge of federal funding and its network of contacts to help communities identify and leverage funding opportunities; provide support to communities seeking public-private partnerships; review, analyze, and provide recommendations and guidance associated with community-level reports, studies, and procurements; and provide background information and training to organizations that need assistance navigating the broadband landscape.[75] BroadbandUSA also organizes regional events and workshops bringing together broadband stakeholders and publishes guides and tools[76] that can serve as resources for communities seeking to launch broadband initiatives.

Additionally, NTIA serves as cochair of the Broadband Interagency Working Group (BIWG)[77] alongside the Department of Agriculture's Rural Utilities Service (RUS). Through the BIWG,

NTIA works with other federal agencies to improve coordination across programs, reduce regulatory barriers to broadband deployment, promote awareness of the importance of federal support for broadband investment and digital inclusion programs, and collect and share information with communities about available federal resources for broadband deployment and digital inclusion efforts. The BIWG was formed in January 2017, in response to the *Broadband Opportunity Council Agency's Progress Report*.[78]

---

[73] FCC, *Public Notice*, "FCC Announced Membership and First Meeting of the Broadband Deployment Advisory Committee," GN Docket No. 17-83, April 6, 2017, available at https://apps.fcc.gov/edocs_public/attachmatch/DA-17- 328A1.pdf.

[74] See FCC, *Fourteenth Broadband Deployment Report Notice of Inquiry*, GN Docket No. 18-238, FCC 18-119, adopted August 8, 2018, para. 24, available at https://docs.fcc.gov/public/attachments/FCC-18-119A1.pdf.

[75] For more information on the types of technical assistance BroadbandUSA offers, see http://www2.ntia.doc.gov/technical_assistance.

[76] See http://www2.ntia.doc.gov/publications.

[77] NTIA, Broadband Interagency Working Group (BIWG) Members, March 2018, available at NTIA, Broadband Interagency Working Group (BIWG) Members, March 29, 2018, available at https://www.ntia.doc.gov/other-publication/2018/broadband-interagency-working-group-biwg-members.

[78] Department of Commerce and Department of Agriculture, *Broadband Opportunity Council Agency's Progress Report*, January 2017, available at https://www.ntia.doc.gov/files/ntia/publications/broadband_opportunity_council_agencies_progress_report_jan2017.pdf. The report was issued pursuant to the March 23, 2015, Presidential Memorandum, "Expanding Broadband Deployment and Adoption by Addressing Regulatory Barriers and Encouraging Investment and Training," and the subsequent August 20, 2015,

BroadbandUSA also coordinates the State Broadband Leaders Network (SBLN), which includes state level office representatives involved in broadband efforts. The SBLN shares priorities and best practices; discusses emerging telecommunications policy issues; links states and local jurisdictions to federal agencies and funding sources; and addresses barriers to collaboration across states and agencies.[79]

Appalachian Regional Commission

Section 1436 of the Fixing America's Surface Transportation Act (FAST Act, P.L. 114-94) authorized a high-speed broadband deployment initiative for the 13-state Appalachian region consisting of $10 million in available broadband grants annually through FY2020. In August 2016, ARC published a *Broadband Planning Primer and Toolkit*.[80]

Economic Development Administration

Broadband projects are eligible for funding under the Economic Development Assistance programs of the Economic Development Administration (EDA) in the Department of Commerce. The Explanatory Statement that accompanied the FY2018 Consolidated Appropriations Act (P.L. 115-141) stated that funding provided under EDA's Public Works, Economic Adjustment Assistance, and other programs may be used to support broadband infrastructure projects, and that EDA is encouraged to prioritize unserved areas. The Explanatory Statement directed that "EDA shall submit a report to the Committees within 30 days of the end of fiscal year 2018 describing the number and value of broadband projects supported with fiscal year 2018 funds."[81]

CONSOLIDATED APPROPRIATIONS ACT, 2018

The Consolidated Appropriations Act, 2018 (P.L. 115-141) contained many broadband-related provisions

---

Broadband Opportunity Council Report and Recommendations, available at https://www.ntia.doc.gov/files/ntia/publications/ broadband_opportunity_council_report_final.pdf.

[79] For more information on the SBLN, see https://broadbandusa.ntia.doc.gov/ntia-resources/state-broadband-leadersnetwork-sbln. A fact sheet is available at https://broadbandusa.ntia.doc.gov/sites/default/files/resource-files/ sbln_fact_sheet_007262017_0.pdf.

[80] Available at https://www.arc.gov/images/programs/telecom/ARCBroadbandPlanningPrimerToolkit.pdf.

[81] Explanatory Statement accompanying FY2018 Consolidated Appropriations Act, Division B, p. 3, available at http://docs.house.gov/billsthisweek/20180319/DIV%20B%20CJS%20SOM-%20FY18-OMNI.OCR.pdf.

- Title VII, Section 779 of Division A (Agriculture, Rural Development, Food and Drug Administration, and Related Agencies Appropriations Act, 2018) appropriates $600 million to RUS to "conduct a new broadband loan and grant pilot program." The law states that the funding is to "remain available until expended," and that at least 90% of the households to be served by a project receiving a loan or grant under the pilot program shall be in a rural area without sufficient access to broadband, defined for this pilot program as 10 Mbps downstream, and 1 Mbps upstream, which shall be reevaluated and redetermined, as necessary, on an annual basis by the Secretary of Agriculture. The pilot broadband loan and grant program is being implemented as the ReConnect Program, which is offering loans, grants, and loan/grant combinations. More information on the ReConnect Program is available at https://reconnect.usda.gov.
- Title I of Division B (Commerce, Justice, Science, and Related Agencies Appropriations Act, 2018) appropriates $7.5 million to update the national broadband availability map in coordination with the FCC and using partnerships previously developed with the states.
- Title V of Division P (Ray Baum's Act of 2018). Section 101 amends the Communications Act of 1934 to provide for the deposits of bidders in auctions of spectrum frequencies to be deposited in the Treasury. Section 504 directs the FCC to submit a report to Congress on promoting broadband for veterans, in particular low-income veterans and veterans residing in rural areas. Section 505 directs the FCC to promulgate regulations to establish a methodology that shall apply to the collection of mobile service coverage data for the purposes of the Universal Service program. Section 508 requires the FCC to submit a report to Congress evaluating broadband coverage in Indian country and on land held by a Native Corporation pursuant to the Alaska Native Claims Settlement Act, with the FCC required to complete a proceeding to address the unserved areas identified in the report.
- Title VI (MOBILE NOW) of Division P seeks to make more spectrum available for wireless broadband, facilitate broadband infrastructure deployment on federal lands, include communications facility installation data in the federal real property database, and require consultation between telecommunications providers and state highway authorities receiving federal highway money. Additionally, Section 615 directs GAO to conduct a study to evaluate the availability of broadband access using unlicensed spectrum and wireless networks in low-income neighborhoods.

CONCLUDING OBSERVATIONS

To the extent that Congress may consider various options for encouraging broadband deployment and adoption, a key issue is how to strike a balance between providing federal assistance for unserved and underserved areas where the private sector may not be providing acceptable levels of broadband service, while at the same time minimizing any deleterious effects that government intervention in the marketplace may have on competition and private sector investment.

The 116th Congress may address the digital divide issue by considering various approaches to providing support for infrastructure deployment, including support for rural broadband. In addition to loans, loan guarantees, and grants for broadband infrastructure deployment, a wide array of policy instruments are available to policymakers, including universal service reform, tax incentives to encourage private sector deployment, broadband bonds, demand-side incentives (such as assistance to low-income families for purchasing computers), reducing regulatory barriers to broadband deployment, and spectrum policy to spur rollout of wireless broadband services. In assessing federal incentives for broadband deployment, Congress may consider the appropriate mix of broadband deployment incentives to create jobs in the short and long term, the extent to which incentives should target next-generation broadband technologies, and the extent to which unserved and underserved areas with existing broadband providers should receive federal assistance.

APPENDIX. BROADBAND LEGISLATION IN THE 115TH CONGRESS

Aside from the 2018 farm bill (P.L. 115-334) and annual appropriations legislation, the following broadband-related bills were introduced into the 115th Congress:

Funding and Federal Assistance

H.R. 547 (DeLauro), introduced on January 13, 2017, would have facilitated efficient investments and financing of infrastructure projects (including broadband projects) through the establishment of a National Infrastructure Development Bank. Referred to multiple committees.

H.R. 800 (Huffman), introduced on February 1, 2017, as the New Deal Rural Broadband Act of 2017, would have established an Office of Rural Broadband within USDA; authorize a “Breaking Ground on Rural Broadband Program” to make grants, loans, or loan guarantees to eligible entities for serving rural and underserved areas ($20 billion to remain available until September 30, 2022); establish a Tribal Broadband Assistance Program ($25 million for each of fiscal years 2017 through 2022); establish a broadband grant program to accompany the Rural Broadband Loan program; and modify the Telecommunications Infrastructure Loan program by raising the threshold for an eligible rural area from 5,000 to 20,000 population and by permitting RUS to give preference to loan applications that support regional telecommunications development. Referred to the Committee on Agriculture, and in addition to the Committees on Natural Resources and Energy and Commerce.

H.R. 1139 (Cramer), introduced February 16, 2017, as the Preserving State Commission Oversight Act of 2017, would have amended the Communications Act of 1934 to protect low-income Lifeline subscribers by mandating a continued role for states in designating eligible telecommunications carriers for participation in the Universal Service Program, and for other purposes. Referred to the House Committee on Energy and Commerce.

H.R. 1581 (Ruiz), introduced on March 16, 2017, as the Tribal Digital Access Act of 2017, would have amended the Communications Act of 1934 to add access to telecommunications and information services in Indian country and areas with high populations of Indian people to the universal service principle relating to access to such services in rural, insular, and high cost areas. Referred to the Committee on Energy and Commerce.

H.R. 1591 (Welch), introduced on March 16, 2017, would have directed the FCC to adopt rules and conduct outreach to offer recipients of assistance under the Lifeline Assistance Program mobile devices that are capable of receiving a WiFi signal and are capable of tethering with other WiFi compatible hardware or devices. Referred to the Committee on Energy and Commerce.

H.R. 2479 (Pallone), introduced on May 17, 2017, as the LIFT America Act, would have provided $40 billion over five years to deploy secure and resilient broadband to expand access for communities nationwide while promoting security by design. Three-quarters of this funding will be used to deploy broadband in unserved areas of the country through a national reverse auction. The remaining funds will be given to states to distribute through separate statewide reverse auctions. If there are no unserved areas in a state, the state may use the funding to deploy broadband in underserved areas, to deploy broadband or connective technology to schools and libraries, or to fund the deployment of Next Generation 9-1-1. Requires that grant recipients offer a service tier of 25 Mbps (download)/3Mbps (upload) at $60 per month. Referred to multiple committees.

H.R. 3314 (Polis), introduced on July 19, 2017, as the 100 by '50 Act, would have included broadband grants and loans under a community need-based economic transition assistance program. Referred to multiple committees.

H.R. 3546 (Austin Scott of Georgia), introduced on July 28, 2017, as the End Taxpayer Funded Cell Phones Act of 2017, would have prohibited universal service support of commercial mobile service and commercial mobile data service through the Lifeline program. Referred to the Committee on Energy and Commerce.

H.R. 3621 (Russell), introduced on July 28, 2017, as the Rechecking Eligibility of Applicants to the Lifeline Program to Prevent Losses Yearly Act of 2017 (REAPPLY Act), would have required Lifeline subscribers to reapply for such services on an annual basis. Referred to Committee on Energy and Commerce.

H.R. 3912 (Walorski), introduced on October 2, 2017, as the Move America Act of 2017, would have included rural broadband service infrastructure as eligible for funding under Move America bonds. Referred to the Committee on Ways and Means.

H.R. 3994 (Tonko), introduced on October 6, 2017, as the ACCESS Broadband Act, would have established the Office of Internet Connectivity and Growth within NTIA at the Department of Commerce. Reported by the Committee on Energy and Commerce on July 18, 2018 (H.Rept. 115-841). Passed by House on July 23, 2018.

H.R. 4209 (Larson), introduced on November 1, 2017, as the American Wins Act, would have established a Build America Trust Fund in the Department of the Treasury, which would provide $3 billion to the Department of Commerce to carry out a program to expand access to broadband to communities throughout the United States, with an emphasis on communities unserved by broadband. Referred to multiple committees.

H.R. 4232 (Pocan), introduced on November 2, 2017, as the Broadband Connections for Rural Opportunities Program (BCROP) Act, would have amended Section 601 of the Rural Electrification Act of 1936 (7 U.S.C. 950bb) to establish a broadband grant program to accompany the Rural Broadband Loan program. Also would have raised the broadband loan program authorization from $25 million to $50 million. Referred to the Committees on Energy and Commerce and on Agriculture.

H.R. 4287 (Ben Ray Lujan), introduced on November 7, 2017, as the Broadband Infrastructure Finance and Innovation Act of 2017, would have established in the Department of Commerce a broadband infrastructure finance and innovation program to make available loans, loan guarantees, and lines of credit for the construction and deployment of broadband infrastructure. Referred to the Committee on Energy and Commerce.

H.R. 4291 (Stefanik), introduced on November 7, 2017, as the Precision Farming Act, would have utilized Rural Utilities Service loans and loan guarantees under the rural broadband access program to provide broadband service for agricultural producers, and would have provided universal service support for installation charges for broadband service for agricultural producers in order to improve precision farming and ranching. Referred to the Committees on Energy and Commerce and on Agriculture.

H.R. 4308 (Lujan Grisham), introduced on November 8, 2017, as the Rural Broadband Expansion Act, would have authorized the Rural Utility Service's Community Connect broadband grant program at $100 million for each of fiscal years 2019 through 2023. Referred to the Committees on Agriculture and on Energy and Commerce.

H.R. 4677 (Moulton), introduced on December 18, 2017, as the Small Business Broadband and Emerging Information Technology Enhancement Act of 2017, would have improved certain programs of the Small Business Administration to better assist small business customers in accessing broadband technology. Referred to the Committee on Small Business.

H.R. 4817 (Long), introduced on January 17, 2018, as the Promoting Exchanges for Enhanced Routing of Information so Networks are Great Act of 2018 (PEERING Act of 2018), would have directed NTIA to make grants for the establishment or expansion of internet exchange facilities. Referred to the Committee on Energy and Commerce.

H.R. 4832 (Cramer), introduced on January 18, 2018, as the Restoring Economic Strength and Telecommunications Operations by Releasing Expected Dollars Act of 2018 (RESTORED Act of 2018), would have amended the Communications Act of 1934 to clarify that an eligible telecommunications carrier may use high cost universal service support to aid in the restoration of telecommunications capabilities in an area in which the President has declared a major disaster or emergency and may elect to receive an advance payment of such support. Referred to the Committee on Energy and Commerce.

H.R. 4986 (Blackburn), introduced on February 8, 2018, as the Repack Airwaves Yielding Better Access for Users of Modern Services Act of 2018 (RAY BAUM's Act). Reauthorizes FCC. Section 505 would have directed the FCC to promulgate regulations to establish a methodology that shall apply to the collection of mobile service coverage data for the purposes of the Universal Service program. Referred to Committee on Energy and Commerce, and in addition to the Committees on Transportation and Infrastructure, and Oversight and Government Reform. Reported (amended) by Committee on Energy and Commerce on March 6, 2018 (H.Rept. 115- 587). Passed House on March 6, 2018.

H.R. 5016 (Abraham), introduced on February 14, 2018, as the Revitalize Rural America Act, would have directed the Secretary of Transportation to establish a $2.1 billion Revitalize Rural America Grant Program that would fund infrastructure projects, including rural broadband. Referred to Committee on Transportation and Infrastructure, and in addition to the Committee on Energy and Commerce.

H.R. 5172 (O'Halleran), introduced on March 6, 2018, would have assisted Indian tribes in maintaining, expanding, and deploying broadband systems. Referred to the Committee on Agriculture, and in addition to the Committee on Energy and Commerce.

H.R. 5213 (Hartzler), introduced on March 8, 2018, as the Expanding Rural Access to Broadband Act, would have prohibited the Rural Utilities Service from providing assistance for the provision of broadband service with a download speed of less than 25 megabits per second or an upload speed of less than 3 megabits per second, and clarify the broadband loan and loan guarantee authority provided in Section 601 of the Rural Electrification Act of 1936. Referred to the Committee on Agriculture, and in addition to the Committee on Energy and Commerce.

H.R. 5294 (Barletta), introduced on March 15, 2018, as the Treating Barriers to Prosperity Act of 2018, would have authorized the Appalachian Regional Commission to make grants and provide technical assistance to develop relevant infrastructure, including broadband infrastructure that supports the use of telemedicine. Reported by the Committee on Transportation and Infrastructure on June 12, 2018 (H.Rept. 115-749); passed House on June 13, 2018.

H.R. 5318 (Huffman), introduced on March 15, 2018, as the Investing for Tomorrow's Schools Act, would have authorized the Secretary of the Treasury, in consultation with the Secretary of Education, to enter into cooperative agreements with states for the establishment of state infrastructure banks and multistate infrastructure banks for making loans to community learning centers to connect and improve broadband services. Referred to Committee on Education and the Workforce.

H.R. 5497 (Peterson), introduced on April 12, 2018, as the Office of Rural Telecommunications Act, which would have directed the FCC to establish the Office of Rural Telecommunications which would coordinate with RUS, NTIA, and other federal broadband programs. Referred to the Committee on Energy and Commerce.

H.R. 6073(Cramer), introduced on June 12, 2018, as the RURAL Broadband Act of 2018, would have prohibited USDA from providing broadband loans or grants for projects that overbuild or otherwise duplicate broadband networks operated by another provider that have received universal service support from the FCC or previous broadband assistance from RUS. Referred to the Committee on Agriculture, and in addition to the Committee on Energy and Commerce.

H.R. 6442 (Kilmer), introduced on July 19, 2018, as the Broadband for All Act of 2018, would have amended the Internal Revenue Code of 1986 to provide a tax credit to consumers to reimburse a portion of the cost of broadband infrastructure serving limited broadband districts. Referred to Committee on Ways and Means.

H.R. 6781 (Marino), introduced September 25, 2018, as the Rural Broadband Connectivity Act of 2018, would have provided tax credits for broadband providers. Referred to the Committee on Ways and Means.

S. 277 (Manchin), introduced on February 2, 2017, as the Rural Telecommunications and Broadband Service Act of 2017, would have established a Rural Telecommunications and Broadband Advisory Committee within the Federal Communications Commission. Referred to the Committee on Commerce, Science, and Transportation.

S. 421 (Fischer), introduced February 16, 2017, as the Preserving State Commission Oversight Act of 2017, would have amended the Communications Act of 1934 to protect low-income Lifeline subscribers by mandating a continued role for states in designating eligible telecommunications carriers for participation in the Universal service program, and for other purposes. Referred to the Committee on Commerce, Science, and Transportation.

S. 987 (Merkley), introduced on April 27, 2017, as the 100 by '50 Act, would have included broadband grants and loans under a community need-based economic transition assistance program. Referred to the Committee on Finance.

S. 1229 (Hoeven), introduced on May 25, 2017, as the Move America Act of 2017, would have included rural broadband service infrastructure as eligible for funding under Move America bonds. Referred to the Committee on Finance.

S. 1377 (Wicker), introduced on June 19, 2017, as the Reaching Underserved Rural Areas to Lead Telehealth Act, would have removed the limitation on certain amounts for which large nonrural hospitals may be reimbursed under the Healthcare Connect Fund of the Federal Communications Commission, and for other purposes. Referred to the Committee on Commerce, Science, and Transportation.

S. 1676 (Gillibrand), introduced on July 31, 2017, as the Broadband Connections for Rural Opportunities Program (BCROP) Act, would have amended Section 601 of the Rural Electrification Act of 1936 (7 U.S.C. 950bb) to establish a broadband grant program to accompany the Rural Broadband Loan program. Also would have raised the broadband loan program authorization from $25 million to $50 million. Referred to the Committee on Agriculture, Nutrition, and Forestry.

S. 2165 (Sanders), introduced on November 28, 2017, as the Puerto Rico and Virgin Islands Equitable Rebuild Act of 2017, would have provided $300 million in FY2018 to Department of Agriculture and Department of Commerce broadband programs to expand access to, and the quality of, broadband service across Puerto Rico and the Virgin Islands. Referred to the Committee on Finance.

S. 2205 (Heinrich), introduced on December 7, 2017, as the Tribal Connect Act of 2017, would have improved access by Indian tribes to support from the Schools and Libraries Universal Service Support program (E-rate) of the Federal Communications Commission. Referred to the Committee on Indian Affairs.

S. 2654 (Smith), introduced on April 12, 2018, as the Community Connect Grant Program Act of 2018, would have amended the Rural Electrification Act of 1936 to authorize the Community Connect Grant Program at an annual level of $50 million per year. Defines "eligible broadband service" as operating at or above the applicable minimum download and upload speeds established by the FCC in defining the term "advanced telecommunications capability." Referred to Committee on Agriculture, Nutrition, and Forestry.

S. 2955 (Wicker), introduced on May 24, 2018, as the Mobile Accuracy and Precision (MAP) Broadband Act of 2018, would have reformed the Mobility Fund Phase II challenge process conducted by the FCC. Referred to the Committee on Commerce, Science, and Transportation.

S. 2958 (Udall), introduced on May 24, 2018, would have required the FCC to make the provision of Wi-Fi access on school buses eligible for E-rate support. Referred to the Committee on Commerce, Science, and Transportation.

S. 2959 (Hoeven), introduced on May 24, 2018, as the Office of Rural Broadband Act, would have directed the FCC to establish the Office of Rural Broadband which would coordinate broadband efforts between the FCC, RUS, NTIA, and other agencies. Referred to the Committee on Commerce, Science, and Transportation.

S. 2970 (Daines), introduced on May 24, 2018, as the RURAL Broadband Act of 2018, would have prohibited USDA from providing broadband loans or grants for projects that overbuild or otherwise duplicate broadband networks operated by another provider that have received universal service support from the FCC or previous broadband assistance from RUS. Referred to the Committee on Agriculture, Nutrition, and Forestry.

S. 3080 (Murkowski), introduced on June 18, 2018, as the Food Security, Housing, and Sanitation Improvements in Rural, Remote, and Frontier Areas Act of 2018, would have amended the Rural Electrification Act of 1936 to include a satellite project or technology within the definition of broadband service. Referred to the Committee on Agriculture, Nutrition, and Forestry.

S. 3255 (Cruz), introduced on July 23, 2018, as the E-FRONTIER Act, would have prohibited the President or a federal agency from constructing, operating, or offering wholesale or retail services on broadband networks without authorization from Congress. Referred to the Committee on Commerce, Science, and Transportation.

S. 3346 (Cortez Masto), introduced on August 1, 2018, as the ACCESS Broadband Act, would have established the Office of Internet Connectivity and Growth within NTIA at the Department of Commerce. Referred to the Committee on Commerce, Science, and Transportation.

S. 3360 (Wyden), introduced August 21, 2018, as the Broadband Internet for Small Ports Act, would have established priority for small harbors to receive RUS broadband funding. Referred to the Committee on Agriculture, Nutrition, and Forestry.

Broadband Data, Studies, Reports

H.R. 1084 (Kelly of Illinois), introduced on February 15, 2017, as the Today's American Dream Act, would have directed GAO to submit to Congress a report on the efficiency and effectiveness of efforts by federal agencies to expand access to broadband service. Referred to multiple committees.

H.R. 1546 (Loebsack), introduced on March 15, 2017, as the Rural Wireless Access Act of 2017, would have directed the FCC to establish a methodology for the collection by the commission of mobile service coverage data. Referred to the Committee on Energy and Commerce.

H.R. 2903 (McKinley), introduced on June 15, 2017, as the Rural Reasonable and Comparable Wireless Access Act of 2017, would have directed the FCC to promulgate regulations that establish a national standard for determining whether mobile and broadband services available in rural areas are reasonably comparable to those services provided in urban areas. Referred to the Committee on Energy and Commerce.

H.R. 3523 (Young of Alaska), introduced on July 27, 2017, would have required the Comptroller General of the United States to conduct a study and submit a report on filing requirements under the Universal Service Fund programs. Referred to the Committee on Energy and Commerce.

H.R. 3839 (Kelly of Illinois), introduced on September 26, 2017, as the Today's American Dream Act, would have directed GAO to submit to Congress a report on the efficiency and effectiveness of efforts by federal agencies to expand access to broadband service. Referred to multiple committees.

H.R. 3995 (McNerney), introduced on October 10, 2017, as the Improving Broadband Access for Veterans Act of 2017, would have required the FCC to submit to Congress a report on promoting broadband internet access service for veterans. Referred to the Committee on Energy and Commerce.

H.R. 4506 (Torres), introduced on November 30, 2017, as the Jobs for Tribes Act, would have directed GAO to conduct a study assessing a range of federal programs (including broadband and telecommunications programs) available to assist Indian communities with business and economic development. Referred to the Committees on Natural Resources; Foreign Affairs; and Education and the Workforce.

H.R. 4810 (Johnson of Ohio), introduced on January 17, 2018, as the Making Available Plans to Promote Investment in Next Generation Networks without Overbuilding and Waste Act of 2018 (MAPPING NOW Act of 2018), would have directed the Department of Commerce to carry out activities relating to the development and maintenance of a broadband inventory map through NTIA and not through an agreement with any other agency. Referred to the Committee on Energy and Commerce.

H.R. 4876 (Rush), introduced on January 22, 2018, as the Connecting Broadband Deserts Act of 2018, would have amended the Communications Act of 1934 to direct the FCC to conduct an annual inquiry on the availability of advanced telecommunications capability in broadband deserts. Referred to the Committee on Energy and Commerce.

H.R. 4881 (Latta), introduced on January 25, 2018, as the Precision Agriculture Connectivity Act of 2018, would have required the FCC to establish a task force for meeting the connectivity and technology needs of precision agriculture in the United States. Reported by the Committee on Energy and Commerce on July 18, 2018 (H.Rept. 115-837). Passed by House on June 23, 2018.

H.R. 4986 (Blackburn), introduced on February 8, 2018, as the Repack Airwaves Yielding Better Access for Users of Modern Services Act of 2018 (RAY BAUM's Act). Reauthorizes FCC. Section 504 would have directed the FCC to submit a report to Congress on promoting broadband for veterans, in particular low-income veterans and veterans residing in rural areas. Section 508 would have required the FCC to submit a report to Congress evaluating broadband coverage in Indian country and on land held by a Native Corporation pursuant to the Alaska Native Claims Settlement Act; the FCC shall complete a proceeding to address the unserved areas identified in the report. Section 711 would have directed GAO to conduct a study to evaluate the availability of broadband access using unlicensed spectrum and wireless networks in low-income neighborhoods. Referred to Committee on Energy and Commerce, and in addition to the Committees on Transportation and Infrastructure, and Oversight and Government Reform. Reported (amended) by Committee on Energy and Commerce on March 6, 2018 (H.Rept. 115-587). Passed House on March 6, 2018.

H.R. 5007 (Ruiz), introduced on February 13, 2018, would have directed the FCC to submit to Congress a report evaluating broadband coverage in Indian country and on land held by a Native Corporation and to complete a proceeding to address the unserved areas identified in the report. Referred to the Committee on Energy and Commerce.

H.R. 5213 (Hartzler), introduced on March 8, 2018, as the Expanding Rural Access to Broadband Act, would have required RUS to submit a report to Congress identifying administrative and legislative options for incentivizing private investment by utilizing RUS loan guarantee programs for the purpose of expanding broadband to rural areas; referred to the Committee on Agriculture, and in addition to the Committee on Energy and Commerce.

S. 645 (Klobuchar), introduced on March 15, 2017, as the Measuring the Economic Impact of Broadband Act of 2018, would have required the Secretary of Commerce to conduct an assessment and analysis of the effects of the digital economy on the economy of the United States. Referred to the Committee on Commerce, Science, and Transportation; reported by the Committee on October 2, 2018 (S.Rept. 115-341). Passed Senate with an amendment by Unanimous Consent on December 13, 2018.

S. 875 (Sullivan), introduced on April 6, 2017, would have required the Comptroller General of the United States to conduct a study and submit a report on filing requirements under the Universal Service Fund programs. Referred to the Committee on Commerce, Science, and Transportation and ordered to be reported with an amendment in the nature of a substitute favorably (S.Rept. 115-192).

S. 1104 (Manchin), introduced on May 11, 2017, as the Rural Wireless Access Act of 2017, would have required the FCC to establish a methodology for the collection by the commission of information about commercial mobile service and commercial mobile data service. Referred to the Committee on Commerce, Science, and Transportation.

S. 1116 (Hoeven), introduced on May 11, 2017, as the Indian Community Economic Enhancement Act of 2017, would have directed GAO to conduct a study assessing a range of federal programs (including broadband and telecommunications programs) available to assist Indian communities with business and economic development. Referred to the Committee on Senate Indian Affairs; reported by Committee on October 17, 2017 (S.Rept. 115-174).

S. 1621 (Wicker), introduced on July 24, 2017, as the Rural Wireless Access Act of 2017, would have required the FCC to establish a methodology for the collection by the commission of information about commercial mobile service and commercial mobile data service. Referred to the Committee on Commerce, Science, and Transportation. Reported by the Committee on February 7, 2018 (S.Rept. 115-206).

S. 1950 (Blumenthal), introduced on October 5, 2017, as the Improving Broadband Access for Veterans Act of 2017, would have required the FCC to submit to Congress a report on promoting broadband internet access service for veterans. Referred to the Committee on Commerce, Science, and Transportation.

S. 2343 (Wicker), introduced on January 25, 2018, as the Precision Agriculture Connectivity Act of 2018, would have required the FCC to establish a task force for meeting the connectivity and technology needs of precision agriculture in the United States. Referred to the committee on Commerce, Science, and Transportation; reported by Committee on October 2, 2018 (S.Rept. 115-342). Passed in Senate with an amendment by Unanimous Consent on December 6, 2018.

S. 2418 (Hassan), introduced on February 13, 2018, as the Rural Reasonable and Comparable Wireless Access Act of 2017, would have directed the FCC to promulgate regulations that establish a national standard for determining whether mobile and broadband services available in rural areas are reasonably comparable to those services provided in urban areas. Referred to the Committee on Commerce, Science, and Transportation; ordered to be reported with an amendment on May 22, 2018.

Spectrum for Wireless Broadband

H.R. 686 (Paulsen), introduced on January 24, 2017, as the DIGIT Act, would have ensured appropriate spectrum planning and interagency coordination to support the Internet of Things. Referred to the Committee on Energy and Commerce.

H.R. 1814 (Kinzinger), introduced on March 30, 2017, would have encouraged spectrum licensees to make unused spectrum available for use by rural and smaller carriers in order to expand wireless coverage. Referred to the Committee on Energy and Commerce.

H.R. 1888 (Guthrie), introduced on April 4, 2017, as the Federal Spectrum Incentive Act of 2017, would have amended the National Telecommunications and Information Administration Organization Act to provide incentives for the reallocation of federal government spectrum for commercial use. Referred to the Committee on Energy and Commerce, and in addition to the Committee on Armed Services.

H.R. 4109 (Guthrie), introduced on October 24, 2017, as the Spectrum Auction Deposits Act of 2017, would have amended the Communications Act of 1934 to provide for the deposits of bidders in auctions of spectrum frequencies to be deposited in the Treasury. Referred to the Committee on Energy and Commerce.

H.R. 4813 (Costello of Pennsylvania), introduced on January 17, 2018, as the Wireless Internet Focus on Innovation in Spectrum Technology for Unlicensed Deployment Act (WIFI STUDy Act), would have directed GAO to conduct a study to evaluate the role of unlicensed spectrum in offloading broadband traffic. Referred to the Committee on Energy and Commerce.

H.R. 4986 (Blackburn), introduced on February 8, 2018, as the Repack Airwaves Yielding Better Access for Users of Modern Services Act of 2018 (RAY BAUM's Act). Reauthorizes FCC. Section 101 would have amended the Communications Act of 1934 to provide for the deposits of bidders in auctions of spectrum frequencies to be deposited in the Treasury. Title VII (MOBILE Now) would make more spectrum available for wireless broadband. Referred to Committee on Energy and Commerce, and in addition to the Committees on Transportation and Infrastructure, and Oversight and Government Reform. Reported (amended) by Committee on Energy and Commerce on March 6, 2018 (H.Rept. 115-587). Passed House on March 6, 2018.

H.R. 4953 (Lance), introduced on February 6, 2018, as the AIRWAVES Act, would have facilitated a national pipeline of spectrum for commercial use. Referred to the Committee on Energy and Commerce.

H.R. 6017 (Guthrie), introduced on June 6, 2018, as the SPECTRUM NOW Act, would have amended the National Telecommunications and Information Administration Organization Act to provide for necessary payments from the Spectrum Relocation Fund for costs of spectrum research and development and planning activities. Referred to the Committee on Energy and Commerce.

S. 19 (Thune), introduced on June 3, 2017, as the MOBILE Now Act, would have made more spectrum available for wireless broadband, facilitate broadband infrastructure deployment on federal lands, establish a national broadband facilities asset database, and encourage consultation between telecommunications providers and state highway authorities receiving federal highway money. Reported (S.Rept. 115-4) by the Committee on March 21, 2017. Passed Senate on August 3, 2017.

S. 88 (Fischer), introduced on January 10, 2017, as the DIGIT Act, would have ensured appropriate spectrum planning and interagency coordination to support the Internet of Things. Referred to the Committee on Commerce, Science, and Transportation. Reported (S.Rept. 115- 90) by the Committee on June 5, 2017. Passed Senate on August 3, 2017.

S. 1682 (Gardner), introduced on August 1, 2017, as the AIRWAVES Act, would have facilitated a national pipeline of spectrum for commercial use, including wireless broadband internet access. Referred to the Committee on Commerce, Science, and Transportation.

S. 3010 (Wicker), introduced on June 6, 2018, as the SPECTRUM NOW Act, would have amended the National Telecommunications and Information Administration Organization Act to provide for necessary payments from the Spectrum Relocation Fund for costs of spectrum research and development and planning activities. Referred to the Committee on Commerce, Science, and Transportation.

S. 3347 (Markey), introduced on August 1, 2018, a bill to repeal the section of the Middle Class Tax Relief and Job Creation Act of 2012 that requires the Federal Communications Commission to reallocate and auction the T-Band spectrum. Referred to the Committee on Commerce, Science, and Transportation.

Addressing Barriers to Broadband Deployment

H.R. 800 (Huffman), introduced on February 1, 2017, as the New Deal Rural Broadband Act of 2017, includes language that would have directed USDA to establish and maintain an inventory of any real property that is owned, leased, or otherwise managed by the federal government on which a broadband facility could be constructed, as determined by the Under Secretary for Rural Broadband Initiatives. Referred to the Committee on Agriculture, and in addition to the Committees on Natural Resources and Energy and Commerce.

H.R. 2425 (Huffman), introduced on May 17, 2017, as the Public Lands Telecommunications Act, would have supported the establishment and improvement of communications sites on or adjacent to federal lands under the jurisdiction of the Secretary of the Interior or the Secretary of Agriculture through the retention and use of rental fees associated with such sites. Referred to the Committee on Natural Resources and in addition to the Committee on Agriculture. Ordered to be reported (amended) by the Committee on Natural Resources on June 27, 2017.

H.R. 2870 (Collins), introduced on June 12, 2017, as the Gigabit Opportunity Act, would have provided tax incentives for low-income communities in states that adopt Uniform Model Broadband Deployment laws developed by FCC and that have been designated by state as gigabit opportunity zones. Referred to the Committees on Energy and Commerce and Ways and Means.

H.R. 4682 (Blackburn), introduced on December 19, 2017, as the Open Internet Preservation Act, would have amended the Communications Act of 1934 to ensure internet openness, to prohibit blocking of lawful content, applications, services, and nonharmful devices, to prohibit impairment or degradation of lawful internet traffic, to limit the authority of the FCC and to preempt state law with respect to internet openness obligations, to provide that broadband internet access service shall be considered to be an information service. Referred to the Committee on Energy and Commerce. Reported (amended) by Committee on Energy and Commerce on March 6, 2018 (H.Rept. 115-587). Passed House on March 6, 2018.

H.R. 4795 (Walters of California), introduced on January 16, 2018, as the Communications Facilities Deployment on Federal Property Act of 2018, would have streamlined communications facilities deployment on federal property. Referred to the Committee on Transportation and Infrastructure, and in addition to the Committee on Energy and Commerce.

H.R. 4800 (Eshoo), introduced on January 17, 2018, as the Broadband Conduit Deployment Act of 2018, would have provided for the inclusion of broadband conduit installation in certain highway construction projects. Referred to the Committee on Transportation and Infrastructure.

H.R. 4802 (Kinzinger), introduced on January 16, 2018, as the Streamlining and Expediting Approval for Communications Technologies Act, would have tracked applications to locate or modify communications facilities on federal real property. Referred to the Committee on Transportation and Infrastructure, and in addition to the Committees on Oversight and Government Reform, and Energy and Commerce.

H.R. 4814 (Eshoo), introduced on January 17, 2018, as the Community Broadband Act of 2018, would have amended the Telecommunications Act of 1996 to preserve and protect the ability of local governments to provide broadband capability and services. Referred to the Committee on Energy and Commerce.

H.R. 4824 (Curtis), introduced on January 18, 2018, as the Rural Broadband Permitting Efficiency Act of 2018, would have allowed certain state permitting authority to encourage expansion of broadband service to rural communities. Referred to the Committee on Natural Resources, and in addition to the Committee on Agriculture; reported by Committee on Natural Resources on August 3, 2018 (H.Rept. 115-881). Passed by House September 12, 2018.

H.R. 4839 (Ben Ray Lujan of New Mexico), introduced on January 18, 2018, as the Broadband Inventory Infrastructure Act of 2018, would have provided for the establishment of an inventory of federal assets to provide information to entities that construct or operate communications facilities or provide communications service. Referred to the Committee on Transportation and Infrastructure, and in addition to the Committee on Energy and Commerce.

H.R. 4842 (Shimkus), introduced on January 18, 2018, as the Streamlining Permitting to Enable Efficient Deployment of Broadband Infrastructure Act of 2018, would have amended the Communications Act of 1934 to provide that the FCC is not required to perform any review under the National Environmental Policy Act of 1969 or division A of subtitle III of Title 54, United States Code, as a condition of permitting the placement and installation of a communications facility. Referred to the Committee on Energy and Commerce, and in addition to the Committee on Natural Resources.

H.R. 4845 (Olson), introduced on January 19, 2018, as the Connecting Communities Post Disasters Act of 2018, would have provided that the FCC and communications service providers regulated by the FCC shall not be subject to certain provisions of the National Environmental Policy Act of 1969 and the National Historic Preservation Act with respect to the construction, rebuilding, or hardening of communications facilities following a major disaster or an emergency declared by the President. Referred to the Committee on Energy and Commerce, and in addition to the Committee on Natural Resources.

H.R. 4847 (Brooks of Indiana), introduced on January 19, 2018, as the Broadband Deployment Streamlining Act, would have streamlined the process for consideration of applications for the placement of communications facilities on certain federal lands. Referred to the Committee on Transportation and Infrastructure, and in addition to the Committees on Agriculture, Natural Resources, and Energy and Commerce.

H.R. 4858 (Eshoo), introduced on January 19, 2018, as the Clearing Local Impediments Makes Broadband Open to New Competition and Enhancements (CLIMB ONCE Act), would have clarified Section 224 of the Communications Act of 1934 as not limiting the ability of a state to adopt a one touch make ready policy for pole attachments. Referred to the Committee on Energy and Commerce.

H.J.Res. 131 (Doyle), introduced on March 28, 2018, would have provided for congressional disapproval under Chapter 8 of Title 5, United States Code, of the rule submitted by the FCC relating to "Restoring Internet Freedom." Referred to the Committee on Energy and Commerce.

H.R. 5969 (Pocan), introduced on May 24, 2018, as the Speed Up Broadband Access Act of 2018, would have prohibited the use of federal funds for the provision of broadband service in any State that has in effect a law, regulation, or other requirement that prohibits, limits, places conditions on, or regulates the provision of broadband service by public, cooperative, or nonprofit broadband providers. Referred to the Committee on Energy and Commerce.

H.R. 6393 (Coffman), introduced on July 17, 2018, as the 21st Century Internet Act, would have amended the Communications Act of 1934 to provide for internet openness requirements for broadband internet access service providers. Referred to the Committee on Energy and Commerce.

S. 19 (Thune), introduced on June 3, 2017, as the MOBILE Now Act, would have made more spectrum available for wireless broadband, facilitate broadband infrastructure deployment on federal lands, establish a national broadband facilities asset database, and encourage consultation between telecommunications providers and state highway authorities receiving federal highway money. Reported (S.Rept. 115-4) by the Committee on Commerce, Science, and Transportation on March 21, 2017.

S. 604 (Hatch), introduced on March 9, 2017, as the Highway Rights-of-Way Permitting Efficiency Act of 2017, would have allowed certain state permitting authority to encourage expansion of broadband service to rural communities. Referred to the Committee on Environment and Public Works.

S. 742 (Booker), introduced on March 28, 2017, as the Community Broadband Act of 2017, would have removed state barriers for constructing municipal broadband networks and encourage public-private partnerships. Referred to the Committee on Commerce, Science, and Transportation.

S. 1013 (Moore), introduced on May 3, 2017, as the Gigabit Opportunity Act, would have provided tax incentives for low income communities in states that adopt Uniform Model

Broadband Deployment laws developed by FCC and that have been designated by state as gigabit opportunity zones. Referred to the Committee on Finance.

S. 1363 (Heller), introduced on June 15, 2017, as the Rural Broadband Deployment Streamlining Act, would have streamlined the process for broadband facility location applications on Department of Interior and Forest Service land. Referred to the Committee on Energy and Natural Resources.

S. 1988 (Wicker), introduced on October 19, 2017, as the Streamlining Permitting to Enable Efficient Deployment of Broadband Infrastructure Act of 2017 (the SPEED Act), would have streamlined broadband infrastructure permitting on established public rights-of-way. Referred to the Committee on Environment and Public Works.

S. 2381 (Klobuchar), introduced on February 6, 2018, as the Streamlining and Investing in Broadband Infrastructure Act, would have directed the Secretary of Transportation to require that broadband conduits be installed as a part of certain highway construction projects. Referred to the Committee on Environment and Public Works.

S.J.Res. 52 (Markey), introduced on February 27, 2018, would have provided for congressional disapproval under Chapter 8 of Title 5, United States Code, of the rule submitted by the FCC relating to “Restoring Internet Freedom.” Passed by Senate on May 16, 2018.

S. 2853 (Thune), introduced on May 16, 2018, to amend the Communications Act of 1934 to ensure internet openness, to prohibit blocking lawful content and nonharmful devices, to prohibit throttling data, to prohibit paid prioritization, to require transparency of network management practices, to provide that broadband shall be considered to be an information service, and to prohibit the commission or a State commission from relying on Section 706 of the Telecommunications Act of 1996 as a grant of authority. Referred to the Committee on Commerce, Science, and Transportation.

S. 3157 (Thune), introduced on June 28, 2018, as the STREAMLINE Small Cell Deployment Act, would have streamlined siting processes for small cell deployment. Referred to the Committee on Commerce, Science, and Transportation.

In: Broadband
Editor: Orlando Harrison

ISBN: 978-1-53616-409-1

Chapter 6

BROADBAND LOAN AND GRANT PROGRAMS IN THE USDA'S RURAL UTILITIES SERVICE (UPDATED)*

Lennard G. Kruger and Alyssa R. Casey

ABSTRACT

Given the large potential impact broadband access may have on the economic development of rural America, concern has been raised over a "digital divide" between rural and urban or suburban areas with respect to broadband deployment. While there are many examples of rural communities with state-of-the-art telecommunications facilities, recent surveys and studies have indicated that, in general, rural areas tend to lag behind urban and suburban areas in broadband deployment.

According to the Federal Communications Commission's *Communications Marketplace Report*, as of 2017, 24% of Americans in rural areas lacked coverage from fixed terrestrial 25 Mbps/3 Mbps broadband, as compared to only 1.5% of Americans in urban areas. The comparatively lower population density of rural areas is likely a major reason why broadband is less deployed than in more highly populated suburban and urban areas. Particularly for wireline broadband technologies—such as cable modem and fiber—the greater the geographical distances among customers, the larger the cost to serve those customers.

The Rural Utilities Service (RUS) at the U.S. Department of Agriculture (USDA) houses three ongoing assistance programs exclusively created and dedicated to financing broadband deployment: the Rural Broadband Access Loan and Loan Guarantee Program, the Community Connect Grant Program, and the ReConnect Program. Additionally, the Telecommunications Infrastructure Loan and Loan Guarantee Program (previously the Telephone Loan Program) funds broadband deployment in rural areas. Distance Learning

---

* This is an edited, reformatted and augmented version of Congressional Research Service, Publication No. RL33816, dated March 22, 2019.

and Telemedicine (DLT) grants—while not principally supporting connectivity—fund equipment and software that operate via telecommunications to rural end-users of telemedicine and distance learning applications.

The Consolidated Appropriations Act, 2019 (P.L. 116-6) provided $5.83 million to subsidize a rural broadband loan level of $29.851 million, $30 million to Community Connect broadband grants, $47 million for DLT grants, and $1.725 million in loan subsidies for a total loan level of $690 million for the Telecommunications Infrastructure Loan and Loan Guarantee Program. P.L. 116-6 also provided $550 million for the ReConnect Program, which is in addition to the $600 million provided in the 2018 Consolidated Appropriations Act.

The Administration's FY2020 budget proposal requested zero funding for Rural Broadband Access Loans, $200 million for the ReConnect Program, $1.933 million in budget authority to subsidize a loan level of $690 million for Telecommunications Infrastructure Loans and Loan Guarantees, $30 million for Community Connect Grants, and $43.6 million for Distance Learning and Telemedicine Grants.

On December 20, 2018, the President signed the 2018 farm bill (P.L. 115-334, Agriculture Improvement Act of 2018). Regarding the RUS broadband programs, the act includes provisions authorizing a grant component in combination with the broadband loan program; increasing the annual authorization level from $25 million to $350 million; raising the proposed service area eligibility threshold of unserved households from 15% to 50% for broadband loans; authorizing grants, loans, and loan guarantees for middle mile infrastructure; directing improved federal agency broadband program coordination; and providing eligible applicants with technical assistance and training to prepare applications. In the 116th Congress, appropriations will determine the extent to which these programs will be funded.

BACKGROUND: BROADBAND AND RURAL AMERICA

The broadband loan and grant programs at RUS are intended to accelerate the deployment of broadband services in rural America. "Broadband" refers to high-speed internet access and advanced telecommunications services for private homes, commercial establishments, schools, and public institutions. Currently in the United States, residential broadband is primarily provided via cable modem (from the local provider of cable television service), fiber-optic cable, mobile wireless (e.g., smartphones), or over the copper telephone line (digital subscriber line or "DSL"). Other broadband technologies include fixed wireless and satellite.

Broadband access enables a number of beneficial applications to individual users and to communities. These include ecommerce, telecommuting, voice service (voice over the internet protocol or "VOIP"), distance learning, telemedicine, public safety, and others. It is becoming generally accepted that broadband access in a community can play an important role in economic development.

Access to affordable broadband is viewed as particularly important for the economic development of rural areas because it enables individuals and businesses to participate fully in the online economy regardless of geographical location. For example, aside from enabling existing businesses to remain in their rural locations, broadband access could attract new business enterprises drawn by lower costs and a more desirable lifestyle. Essentially, broadband potentially allows businesses and individuals in rural America to live locally while competing globally in an online environment. A 2016 study from the Hudson Institute found that rural broadband providers directly and indirectly added $24.1 billion to the U.S. economy in 2015. The rural broadband industry supported 69,595 jobs in 2015, both through its own employment and the employment that its purchases of goods and services generated.[1]

Given the large potential impact broadband may have on the economic development of rural America, concern has been raised over a "digital divide" between rural and urban or suburban areas with respect to broadband deployment. While there are many examples of rural communities with state-of-the-art telecommunications facilities,[2] recent surveys and studies have indicated that, in general, rural areas tend to lag behind urban and suburban areas in broadband deployment. For example

- According to the Federal Communications Commission's (FCC's) *Communications Marketplace Report*, "As of year-end 2017, 94% of the overall population had coverage [of fixed terrestrial broadband at speeds of 25 Mbps/3 Mbps], up from 91.9% in 2016. Nonetheless, the gap in rural and Tribal America remains notable: 24% of Americans in rural areas and 32% of Americans in Tribal lands lack coverage from fixed terrestrial 25 Mbps/3 Mbps broadband, as compared to only 1.5% of Americans in urban areas. The data demonstrate, however, that the gap between urban and rural or Tribal areas has narrowed each year over the last five years."[3]
- Also according to the FCC's *Communications Market Report*, rural areas continue to lag behind urban areas in mobile broadband deployment. Although evaluated urban areas saw an increase of 10 Mbps/3 Mbps mobile LTE from 81.9% in 2014 to 92.6% in 2017, such deployment in evaluated rural areas remained relatively flat at about 70%.[4]
- According to January 2018 survey data from the Pew Research Center, 58% of adults in rural areas said they have a high-speed broadband connection at home, as opposed to 67% of adults in urban areas and 70% of adults in suburban areas.[5]

---

[1] Hanns Kuttner, Hudson Institute, *The Economic Impact of Rural Broadband*, April 2016, available at https://s3.amazonaws.com/media.hudson.org/files/publications/20160419 KuttnerTheEconomicImpactofRuralBroadband.pdf.

[2] See for example, National Exchange Carrier Association (NECA), *Trends: A Report on Rural Telecom Technology*, 18 pages, December 2015, available at https://www.neca.org/WorkArea/linkit.aspx?LinkIdentifier=id&ItemID=12331&libID=12351.

[3] FCC, *Communications Marketplace Report*, GN Docket No. 18-231, FCC 18-181, p. 131, available at https://docs.fcc.gov/public/attachments/FCC-18-181A1.pdf.

[4] Ibid., Figure G-2b, p. 133.

[5] Pew Research Center, *Internet/Broadband Fact Sheet*, February 5, 2018, available at http://www.pewinternet.org/fact- sheet/internet-broadband/.

- A November 2017 Census Bureau survey reported by the National Telecommunications and Information Administration (NTIA) Digital Nation Data Explorer showed 72.9% of rural residents reporting using the internet, versus 78.5% of urban residents.[6] According to NTIA, the data "indicates a fairly constant 6-9 percentage point gap between rural and urban communities' internet use over time."[7]

The comparatively lower population density of rural areas is likely the major reason why broadband is less deployed than in more highly populated suburban and urban areas. Particularly for wireline broadband technologies—such as cable modem, fiber, and DSL—the greater the geographical distances among customers, the larger the cost to serve those customers. Thus, there is often less incentive for companies to invest in broadband in rural areas than, for example, in an urban area where there is more demand (more customers with perhaps higher incomes) and less cost to wire the market area.

The terrain of rural areas can also be a hindrance, in that it is more expensive to deploy broadband technologies in a mountainous or heavily forested area. An additional added cost factor for remote areas can be the expense of "backhaul" (e.g., the "middle mile"), which refers to the installation of a dedicated line that transmits a signal to and from an internet backbone, which is typically located in or near an urban area.

Another important broadband availability issue is the extent to which there are multiple broadband providers offering competition and consumer choice. Typically, multiple providers are more prevalent in urban than in rural areas.[8]

RURAL BROADBAND PROGRAMS AT THE RURAL UTILITIES SERVICE

Because private providers are unlikely to earn enough revenue to cover the costs of deploying and operating broadband networks in many unserved rural areas, it is unlikely that private investment alone will bring service to these areas.[9] In 2000, given the lagging deployment of broadband in rural areas, Congress and the Administration acted to initiate pilot broadband loan and grant programs within the Rural Utilities Service of the U.S. Department of Agriculture. While RUS had long maintained telecommunications loan

[6] U.S. Department of Commerce, National Telecommunications and Information Administration, *Digital Nation Data Explorer*, June 6, 2018, available at https://www.ntia.doc.gov/data/digital-nation-data-explorer#sel=internetUser&demo=metro&pc=prop&disp=chart.

[7] U.S. Department of Commerce, National Telecommunications and Information Administration, "The State of the Urban/Rural Digital Divide," August 10, 2016, available at https://www.ntia.doc.gov/blog/2016/state-urbanrural- digital-divide.

[8] See Table 2 in CRS In Focus IF10441, *Broadband Deployment: Status and Federal Programs*, by Lennard G. Kruger.

[9] Government Accountability Office, *Projects and Policies Related to Deploying Broadband in Unserved and Underserved Areas*, GAO-14-409, April 2014, p. 9, available at http://www.gao.gov/assets/670/662711.pdf.

and grant programs (Rural Telephone Loans and Loan Guarantees, Rural Telephone Bank, and more recently, the Distance Learning and Telemedicine Loans and Grants), none were exclusively dedicated to financing rural broadband deployment. Title III of the FY2001 agriculture appropriations bill (P.L. 106-387) directed USDA/RUS to conduct a "pilot program to finance broadband transmission and local dial-up Internet service in areas that meet the definition of 'rural area' used for the Distance Learning and Telemedicine Program."

Subsequently, on December 5, 2000, RUS announced the availability of $100 million in loan funding through a one-year pilot program "to finance the construction and installation of broadband telecommunications services in rural America."[10] The broadband pilot loan program was authorized under the authority of the Distance Learning and Telemedicine Program (7 U.S.C. 950aaa), and was available to "legally organized entities" not located within the boundaries of a city or town having a population in excess of 20,000.

The FY2002 agriculture appropriations bill (P.L. 107-76) designated a loan level of $80 million for broadband loans, and on January 23, 2002, RUS announced that the pilot program would be extended into FY2002, with $80 million in loans made available to fund many of the applications that did not receive funding during the previous year.[11]

Meanwhile, the FY2002 agriculture appropriations bill (P.L. 107-76) allocated $20 million for a pilot broadband grant program, also authorized under the Distance Learning and Telemedicine Program. On July 8, 2002, RUS announced the availability of $20 million for a pilot grant program for the provision of broadband service in rural America. The program was specifically targeted to economically challenged rural communities with no existing broadband service.

Grants were made available to entities providing "community-oriented connectivity," which the RUS defined as those entities "who will connect the critical community facilities including the local schools, libraries, hospitals, police, fire and rescue services and who will operate a community center that provides free and open access to residents."[12]

The pilot program was extended into FY2003, as the Consolidated Appropriations Resolution of 2003 (P.L. 108-7) allocated $10 million for broadband grants.

Currently, RUS has four ongoing programs that have been established to incentivize and subsidize broadband infrastructure investment in unserved and underserved rural areas. These include the following:

---

[10] Rural Utilities Service, USDA, "Construction and Installation of Broadband Telecommunications Services in Rural America; Availability of Loan Funds," *Federal Register*, Vol. 65, No. 234, December 5, 2000, p. 75920.

[11] Rural Utilities Service, USDA, "Broadband Pilot Loan Program," *Federal Register*, Vol. 67, No. 15, January 23, 2002, p. 3140.

[12] Rural Utilities Service, USDA, "Broadband Pilot Grant Program," *Federal Register*, Vol. 67, No. 130, July 8, 2002, p. 45080.

- *Rural Broadband Access Loans*—funds the costs of construction, improvement, or acquisition of facilities and equipment needed to provide service in eligible rural areas.
- *Community Connect Grants*—funds broadband deployment into rural communities where it is not yet economically viable for private sector providers to deliver service.
- *Telecommunications Infrastructure Loans and Loan Guarantees*—funds the construction, maintenance, improvement, and expansion of telephone service and broadband in extremely rural areas with a population of 5,000 or less.
- *Distance Learning and Telemedicine Grants*—principally funds end-user equipment to help rural communities use telecommunications to link teachers and medical service providers in one area to students and patients in another.

In addition, a new broadband loan and grant pilot program—the ReConnect Program—has been established and funded at $600 million by the Consolidated Appropriations Act, 2018 (P.L. 115- 141).

Table A-1 in the Appendix shows the total amount and number of awards provided by the RUS broadband programs for each state between FY2009 and FY2016.

In its April 2017 report, *Rural Broadband Deployment: Improved Consistency with Leading Practices Could Enhance Management of Loan and Grant Programs*, GAO reported that (according to RUS data) since FY2004, RUS has approved 704 broadband projects totaling almost $8.6 billion in loans and $144.8 million in grants to deploy telecommunications or broadband infrastructure networks in rural areas.[13]

Rural Broadband Access Loan and Loan Guarantee Program

Building on the pilot broadband loan program at RUS, Section 6103 of the Farm Security and Rural Investment Act of 2002 (P.L. 107-171) amended the Rural Electrification Act of 1936 to authorize a loan and loan guarantee program to provide funds for the costs of the construction, improvement, and acquisition of facilities and equipment for broadband service in eligible rural communities.[14] Section 6103 made available, from the funds of the Commodity Credit Corporation (CCC), a total of $100 million through FY2007. P.L. 107-171 also authorized any other funds appropriated for the broadband loan program. The program was subsequently reauthorized by Section

---

[13] Government Accountability Office, *Rural Broadband Deployment: Improved Consistency with Leading Practices Could Enhance Management of Loan and Grant Program*, GAO-17-301, April 2017, p. 2, available at https://www.gao.gov/assets/690/684093.pdf.

[14] Title VI of the Rural Electrification Act of 1936 (7 U.S.C. 950bb).

6110 of the Food, Conservation, and Energy Act of 2008 (P.L. 110-246), and by Section 6104 of the Agricultural Act of 2014 (P.L. 113-79).

Beginning in FY2004, Congress annually blocked mandatory funding from the CCC. Thus— starting in FY2004—the program was funded as part of annual appropriations in the Distance Learning and Telemedicine account within the Department of Agriculture appropriations bill. Every fiscal year, Congress approves an appropriation (loan subsidy) and a specific loan level (lending authority) for the Rural Broadband Access Loan and Loan Guarantee Program. Table 1 shows—for the life of the program to date—loan subsidies and loan levels (lending authority) set by Congress in annual appropriations bills.

The Rural Broadband Access Loan and Loan Guarantee Program is codified as 7 U.S.C. 950bb. On July 30, 2015, the RUS published in the *Federal Register* the interim rule (7 C.F.R. part 1738) implementing the Rural Broadband Access Loan and Loan Guarantee Program as reauthorized by the enactment of the Agricultural Act of 2014 (P.L. 113-79),[15] and the interim rule was made final on June 9, 2016. Entities eligible to receive loans include corporations, limited liability companies, cooperative or mutual organizations, Indian tribes or tribal organizations, and state or local governments. Eligible areas for funding must be completely contained within a rural area (or composed of multiple rural areas). Additionally, at least 15% of the households in the proposed funded service areas must be unserved, no part of the proposed service area can have three or more incumbent service providers, and no part of the proposed service area can overlap with the service area of current RUS borrowers or of grantees that were funded by RUS.

The latest Notice of Solicitation of Applications (NOSA) announced that RUS is now accepting applications on a rolling basis through September 30, 2019, which will give RUS the ability to request additional information and modifications to submitted applications if necessary. RUS will evaluate the submitted applications every 90 days, and anticipates at least two evaluation periods for FY2019. The minimum loan amount is $100,000, while the maximum loan amount is $25 million. The NOSA has maintained its definition of broadband service and broadband lending speed at no less than 25 Mbps download and 3 Mbps upload for both mobile and fixed services.[16]

The 2018 farm bill, which was signed by the President on December 20, 2018 (P.L. 115-334, Agriculture Improvement Act of 2018), adds a grant component to the broadband loan program, increases the annual authorization level from $25 million to $350 million, and changes the proposed service area threshold from 15% to 50%. RUS

[15] Department of Agriculture, Rural Utilities Service, "Rural Broadband Access Loans and Loan Guarantees," Interim rule, 80 *Federal Register* 45397-45413, July 30, 2015, available at https://www.gpo.gov/fdsys/pkg/FR-2015-07-30/pdf/ 2015-18624.pdf.

[16] Department of Agriculture, Rural Utilities Service, "Rural Broadband Access Loan and Loan Guarantees Program," Notice of Solicitation of Applications (NOSA), 83 *Federal Register* 57406-57407, November 15, 2018, available at https://www.govinfo.gov/content/pkg/FR-2018-11-15/pdf/2018-24860.pdf.

will issue a revised regulation that implements the changes made by the 2018 farm bill. For up to one year after enactment, the Secretary shall use the previously existing rules and regulations for the broadband loan program until a final rule is issued.

Table 1. Appropriations funding for the rural broadband access loan and loan guarantee program

	Direct Appropriations (subsidy level)	Loan Levels Estimated in Annual Appropriations[a]
FY2001 (pilot)	—	$100 million
FY2002 (pilot)	—	$80 million
FY2003	b	$80 million
FY2004	$13.1 million	$602 million
FY2005	$11.715 million	$550 million
FY2006	$10.75 million	$500 million
FY2007	$10.75 million	$500 million
FY2008	$6.45 million	$300 million
FY2009	$15.619 million	$400 million
FY2010	$28.96 million	$400 million
FY2011	$22.32 million	$400 million
FY2012	$6.0 million	$212 million
FY2013	$4 million	$42 million
FY2014	$4.5 million	$34.5 million
FY2015	$4.5 million	$24.1 million
FY2016	$4.5 million	$20.6 million
FY2017	$4.5 million	$27.0 million
FY2018	$5 million	$29.0 million
FY2019	$5.8 million	$29.8 million

Source: Compiled by CRS from appropriations bills.

[a]Actual loan levels for a fiscal year can vary from what is estimated in annual appropriations bill.

[b]Program received $40 million composed of $20 million from FY2002 plus $20 million from FY2003 of mandatory funding from the Commodity Credit Corporation, as directed by P.L. 107-171. In the FY2004, FY2005, and FY2006 appropriations bills, mandatory funding from the CCC was canceled.

Community Connect Broadband Grants

The Consolidated Appropriations Act of 2004 (P.L. 108-199) appropriated $9 million "for a grant program to finance broadband transmission in rural areas eligible for Distance Learning and Telemedicine Program benefits authorized by 7 U.S.C. 950aaa." Essentially operating the same as the pilot broadband grants, the program provides grant money to applicants proposing to provide broadband on a "community-oriented connectivity" basis to currently unserved rural areas for the purpose of fostering economic growth and delivering enhanced health care, education, and public safety services. Funding for the broadband grant program is provided through annual appropriations in the Distance Learning and Telemedicine account within the Department of Agriculture appropriations bill. Table 2 shows a history of appropriations for the Community Connect Broadband Grants.

Table 2. Appropriations for the community connect broadband grants

Fiscal Year	Appropriation
FY2002	$20 million
FY2003	$10 million
FY2004	$9 million
FY2005	$9 million
FY2006	$9 million
FY2007	$9 million
FY2008	$13.4 million
FY2009	$13.4 million
FY2010	$17.9 million
FY2011	$13.4 million
FY2012	$10.4 million
FY2013	$10.4 million
FY2014	$10.4 million
FY2015	$10.4 million
FY2016	$10.4 million
FY2017	$34.5 million
FY2018	$30 million
FY2019	$30 million

Source: Compiled by CRS from appropriations bills.

Eligible applicants for broadband grants include most state and local governments, federally recognized tribes, nonprofits, and for-profit corporations.

Funded projects must serve a rural area where broadband service above a specified minimum speed does not exist, deploy free broadband service for at least two years to all community facilities, and offer broadband to residential and business customers. Up to 10% of the grant may be used for the improvement, expansion, construction, or acquisition of a community center that provides online access to the public.

On May 3, 2013, RUS issued a new final rule for Community Connect grants in the *Federal Register*.[17] The final rule changes previous requirements related to matching funds, eligible communities, and application scoring criteria. The final rule also removes the previous definition of broadband service speed (200 kbps). A new threshold for broadband service speed and broadband grant speed (the speed the grantee must deliver) will be provided in an annual Notice of Funding Availability (NOFA) in the *Federal Register*. The NOFA will also specify the deadline for applications, the total amount of funding available, and the maximum and minimum amount of funding available for each grant.

---

[17] Department of Agriculture, Rural Utilities Service, "Community Connect Broadband Grant Program," 78 *Federal Register* 25787-25795, May 3, 2013, available at http://www.gpo.gov/fdsys/pkg/FR-2013-05-03/pdf/2013-10502.pdf.

In February 2019, RUS issued a Funding Opportunity Announcement (FOA) establishing an application window for FY2019 Community Connect grants through April 15, 2019.[18] The FOA set a minimum threshold for speeds constituting broadband service at 10 Mbps download and 1 Mbps upload for both fixed and mobile broadband. The minimum broadband speed that an applicant must propose to deliver is 25 Mbps download, 3 Mbps upload for both fixed and mobile service to the customer. The minimum grant is $100,000 and the maximum is $3 million.

The 2018 farm bill (P.L. 115-334) codifies the Community Connect Grant Program and authorizes the program at $50 million for each of fiscal years 2018 through 2023.

Telecommunications Infrastructure Loans and Loan Guarantees

The Telecommunications Infrastructure Loan and Loan Guarantee Program[19] provides loans and loan guarantees for the construction, maintenance, improvement, and expansion of telephone service and broadband in rural areas. The program was first authorized in 1949 to finance rural telephone service. Since 1995, RUS has required that networks funded by this program offer broadband service as well.

Loans and loan guarantees are available only to rural areas and towns with a population of 5,000 or less. Eligible areas are those without telecommunications facilities or areas where the applicant is the recognized telecommunications provider. Funded projects cannot duplicate existing services.

The program is authorized to provide several different types of financing, including

- direct Treasury rate loans, which bear interest at the government's cost of money (or the current Treasury rate). Thus, the interest charged varies with the Treasury rate. As Treasury rates increase, so does the cost to the borrower for these loans.
- guaranteed loans, which are provided to borrowers of a nongovernment lender or from the Federal Financing Bank (FFB). The interest rate charged on FFB loans is the Treasury rate plus an administrative fee of one-eighth of 1%. The terms of these loans may vary significantly and allow borrowers more flexibility in meeting their financing needs.
- hardship direct loans, which bear interest at a fixed rate of 5% per year. These loans are intended only for borrowers with extremely high investment costs in terms of per subscriber service. These borrowers also have a very low number of subscribers for each mile of telecommunications line constructed. This low

---

[18] Department of Agriculture, Rural Utilities Service, Funding Opportunity Announcement, available at https://www.rd.usda.gov/files/2019_CC_FOAfinal.pdf.

[19] For more information, see http://www.rd.usda.gov/programs-services/telecommunications-infrastructure-loans-loan- guarantees.

> subscriber density inherently increases the cost to serve the most sparsely populated rural areas. Because of the high cost of the investment needed, these borrowers cannot typically afford higher interest rate loans.[20]

The annual loan level for the Telecommunications Infrastructure Loan and Loan Guarantee Program is $690 million. Currently, the 5% hardship loans are not offered—because of low interest rates, the Treasury and FFB loans can currently offer lower interest rates than the 5% offered by hardship loans.

Distance Learning and Telemedicine Program

The Distance Learning and Telemedicine (DLT) Program was established by the 1996 farm bill— the Federal Agriculture Improvement and Reform Act of 1996 (P.L. 104-127). Though initially providing both grants and loans, since FY2009 only DLT grants have been awarded by RUS.

DLT grants serve as initial capital assets for equipment and software that operate via telecommunications to rural end-users of telemedicine and distance learning. DLT grants do not support connectivity. Grant funds may be used for audio, video, and interactive video equipment; terminal and data terminal equipment; computer hardware, network components, and software; inside wiring and similar infrastructure; acquisition of instructional programming; broadband facilities;[21] and technical assistance. Eligible applicants include most entities in rural areas that provide education or health care through telecommunications, including most state and local governmental entities, federally recognized tribes, nonprofits, for-profit businesses, and consortia of eligible entities. The 2018 farm bill (P.L. 115-334) reauthorizes the DLT program through FY2023 at $82 million per year and sets aside 20% of DLT grant funding for applications related to substance use disorder treatment services.

ReConnect Program

An Interagency Task Force on Agriculture and Rural Prosperity was created on April 25, 2017, by Executive Order 13790 and was charged with identifying legislative,

---

[20] 2017 USDA Budget Explanatory Notes for Committee on Appropriations, p. 31 2, available at http://www.obpa.usda.gov/31rus2017notes.pdf.

[21] As of FY2018, purchasing and installing broadband facilities has been added to the approved grant purposes. This purpose is limited to a maximum of 20% of the requested grant amount and must be used for providing distance learning or telemedicine services. The awardee must own the final broadband asset in order for funding to be approved. See USDA Rural Development, *Distance Learning and Telemedicine Grant Program Application Guide Fiscal Year 2018*, p. 7, available at https://www.rd.usda.gov/files/2018-DLT_App-Guide_final.pdf.

regulatory, and policy changes to promote agriculture, economic development, job growth, infrastructure improvements, technological innovation, energy security, and quality of life in rural America. The first recommendation of the Task Force's report to the President was to expand e-connectivity in rural and tribal areas.[22]

To help implement this recommendation, the Administration requested $500 million in a discretionary add-on to the FY2018 appropriation which would fund a combination grant/loan program at USDA/RUS to deploy broadband in rural and tribal areas.

Section 779 of the Consolidated Appropriations Act, 2018 (P.L. 115-141) appropriated $600 million to RUS to "conduct a new broadband loan and grant pilot program." The law states that the funding is to "remain available until expended," and that

- at least 90% of the households to be served by a project receiving a loan or grant under the pilot program shall be in a rural area without sufficient access to broadband, defined for this pilot program as 10 Mbps downstream, and 1 Mbps upstream, which shall be reevaluated and redetermined, as necessary, on an annual basis by the Secretary of Agriculture;
- an entity to which a loan or grant is made under the pilot program shall not use the loan or grant to overbuild or duplicate broadband expansion efforts made by any entity that has received a broadband loan from RUS;
- in addition to other available funds, not more than 4% of the funds can be used for administrative costs to carry out the pilot program and up to 3% may be utilized for technical assistance and predevelopment planning activities to support the most rural communities; and
- RUS shall adhere to the notice, reporting, and service area assessment requirements previously established in the 2014 farm bill.

The Explanatory Statement that accompanied the FY2018 Consolidated Appropriations Act states

> The agreement reiterates that funding should be prioritized to areas currently lacking access to broadband service, and investments in broadband shall consider any technology that best serves the goals of broadband expansion. Lastly, the agreement restates the importance of coordination among federal agencies in expanding broadband deployment

[22] U.S. Department of Agriculture, Agriculture and Rural Prosperity Task Force, Report to the President of the United States from the Task Force on Agriculture and Rural Prosperity, October 21, 2017, pp. 17-20, available at https://www.usda.gov/sites/default/files/documents/rural-prosperity-report.pdf. The Task Force recommended that the Administration establish executive leadership to expand e-connectivity across rural America, assess the state of rural e- connectivity, reduce regulatory barriers to infrastructure deployment, assess the efficacy of current programs, and incentivize private capital investment.

and adoption and expects the Department to take caution to maximize these limited resources and not overbuild or duplicate existing broadband capable infrastructure.

The Consolidated Appropriations Act, 2019 (P.L. 116-6) provides $550 million in FY2019 for the pilot broadband loan and grant program, now called the Rural eConnectivity Pilot Program, or ReConnect Rural Broadband Program. The $550 million includes $125 million in direct appropriation, plus $425 million to be reprogrammed from the cushion of credit subaccount (7 U.S.C. 940c). Division B, Section 779 directs the Secretary of Agriculture to ensure that applicants determined to be ineligible for the ReConnect Program have a means of appealing or otherwise challenging that determination in a timely fashion. The law also directs the Secretary, in determining whether an entity may overbuild or duplicate broadband expansion efforts made by an entity that has received an RUS broadband loan, to not consider loans that were rescinded or defaulted on, or loans the terms and conditions of which were not met, if the entity under consideration has not previously defaulted on, or failed to meet the terms and conditions of, a Rural Utilities Service loan or had a Rural Utilities Service loan rescinded.

On December 14, 2018, RUS released the *Funding Opportunity Announcement* (*FOA*) and solicitation of applications for the ReConnect Program.[23] As set forth in the statute, at least 90% of the households to be served by a project receiving a loan or grant under the pilot program shall be in a rural area without sufficient access to broadband at a minimum speed of 10 Mbps/1 Mbps. RUS defines “sufficient access to broadband” as any rural area that has fixed, terrestrial broadband service delivering at least 10 Mbps downstream and 1 Mbps upstream. Mobile and satellite service will not be considered in making the determination that households in the proposed funded service area do not have sufficient access to broadband.

With the government shutdown delaying the rollout of the ReConnect Program, on February 25, 2019, RUS released an amendment and clarification to the December *FOA*, with revised application deadlines.[24]

Approximately $600 million has been set aside for funding opportunities under the FOA, with additional budget authority available for a reserve which may be used for additional loans or grants. Award recipients must complete projects within five years. Entities eligible for awards are states or local governments, U.S. territories, an Indian tribe, nonprofit entities, for-profit corporations, limited liability companies, and

---

[23] U.S. Department of Agriculture, Rural Utilities Service, “Broadband Pilot Program,” *Federal Register*, vol. 83, no. 240, December 14, 2018, pp. 64315-64325, available at https://www.govinfo.gov/content/pkg/FR-2018-12-14/pdf/ 2018-27038.pdf.

[24] U.S. Department of Agriculture, Rural Utilities Service, “Broadband Pilot Program—ReConnect Program,” *Federal Register*, vol. 84, no. 37, February 25, 2019, pp. 5981-5983, available at https://www.govinfo.gov/content/pkg/FR- 2019-02-25/pdf/2019-03163.pdf.

cooperative or mutual organizations. This includes telecommunications companies, rural electric cooperatives and utilities, internet service providers, and municipalities.

Funds will be awarded for projects that have financially sustainable business models that will bring broadband to rural homes, businesses, farms, ranches, and community facilities such as first responders, health care facilities, and schools. The ReConnect Program consists of three funding categories.

100% Loan

- Up to $200 million is available.
- The maximum amount that can be requested is $50 million.
- Interest rate is set at a fixed 2%.
- Eligible areas are where 90% of households do not have sufficient access to broadband at 10 Mbps/1 Mbps.
- Applicants must propose to build a network capable of providing service to every premise in the proposed funded service area at a minimum speed of 25 Mbps/3 Mbps.
- Applications accepted on a rolling basis through July 12, 2019.

50% Loan/50% Grant Combination

- Up to $200 million is available.
- The maximum amount that can be requested is $25 million for the loan and $25 million for the grant. Loan and grant amounts will always be equal.
- Interest rate for the loan will be set at the Treasury rate.
- Eligible areas are where 90% of households do not have sufficient access to broadband at 10 Mbps/1 Mbps.
- Applicants must propose to build a network capable of providing service to every premise in the proposed funded service area at a minimum speed of 25 Mbps/3 Mbps
- Applications accepted on a rolling basis through June 21, 2019.

100% Grant

- Up to $200 million is available.
- The maximum amount that can be requested is $25 million.
- Applicants must provide a matching contribution equal to 25% of the cost of the overall project.

- Eligible areas are where 100% of households do not have sufficient access to broadband at 10 Mbps/1 Mbps.
- Applicants must propose to build a network capable of providing service to every premise in the proposed funded service area at a minimum speed of 25 Mbps/3 Mbps.
- Applications accepted on a rolling basis through May 31, 2019.

IMPACT OF UNIVERSAL SERVICE REFORM ON RUS BROADBAND LOAN PROGRAMS

RUS has three programs that provide or have provided loans for broadband infrastructure projects: the Rural Broadband Access Loan and Loan Guarantee program (also known as the Farm Bill broadband loan program), the Broadband Initiatives Program (BIP under the ARRA), and the Telecommunications Infrastructure Loan Program (established in 1949 as the Rural Telephone Loan and Loan Guarantee program).[25]

Whereas RUS broadband loans are used as up-front capital to invest in broadband infrastructure, the Federal Communications Commission's (FCC's) Universal Service Fund (USF)— specifically, the high cost fund—has functioned as an ongoing subsidy to keep the operation of telecommunications networks in high cost areas profitable for providers. Many RUS telecommunications and broadband borrowers (loan recipients) receive high cost USF subsidies. In many cases, the subsidy received from USF helps provide the revenue necessary to keep the loan viable. The Telecommunications Infrastructure Loan Program is highly dependent on high cost USF revenues, with 99% (476 out of 480 borrowers) receiving interstate high cost USF support. This is not surprising, given that the RUS Telecommunications Infrastructure Loans are available only to the most rural and high cost areas (towns with populations less than 5,000).

Regarding broadband loans, 60% of BIP (stimulus) borrowers draw from state or interstate USF support mechanisms, while 10% of Farm Bill (Rural Broadband Access Loan and Loan Guarantee Program) broadband borrowers receive interstate high cost USF support.[26]

The FCC, in an October 2011 decision, adopted an order that calls for the USF to be transformed, in stages, over a multiyear period—from a mechanism to support voice telephone service to one that supports the deployment, adoption, and use of both fixed

---

[25] For more information on the RUS portfolio of telecommunications and broadband programs offering loans, loan guarantees, grants, and loan/grant combinations, see CRS Report R42524, *Rural Broadband: The Roles of the Rural Utilities Service and the Universal Service Fund*, by Angele A. Gilroy and Lennard G. Kruger.

[26] Jessica Zufolo, Deputy Administrator, RUS, *Overview of the RUS Telecommunications Loan and Grant Programs*, July 2011, Slide 7, http://www.narucmeetings.org/Presentations/Zufolo_7-2011.pdf.

and mobile broadband. More specifically, the high cost program is being phased out and a new fund, the Connect America Fund (CAF), which includes the targeted Mobility Fund and new Remote Areas Fund, is replacing it.[27]

During this transition, the uncertainty surrounding the FCC's proposed methodology for distributing Connect America Fund monies has led many small rural providers to postpone or cancel investment in broadband network upgrades.[28] According to RUS, "demand for RUS loans dropped to roughly 37% of the total amount of loan funds appropriated by Congress in FY2012," and "[c]urrent and prospective RUS borrowers have communicated their hesitation to increase their outstanding debt and move forward with planned construction due to the recently implemented reductions in USF support and Inter-Carrier Compensation (ICC) payments."[29]

Appropriations

Table 3. Recent and Proposed Appropriations for RUS Broadband Programs (dollars)

	FY2018 (P.L. 115-141)	FY2019 (Admin. Request)	FY2019 (P.L. 116-6)	FY2020 (Admin. Request)
Rural Broadband Loans (farm bill)	5.0 million (29.8 million loan level)	4.5 million (23.1 million loan level)	5.8 million (29.8 million loan level)	0
Telecom Infrastructure Loans and Loan Guarantees	0.863 million (690 million loan level)	0.863 million (690 million loan level)	1.725 million (690 million loan level)	1.933 million (690 million loan level)
Community Connect Grants	30 million	30 million	30 million	30 million
DLT Grants	49 milliona	23.6 million	47 millionb	44 millionc
ReConnect Program	600 million	—	550 milliond	200 million

Source: CRS, based on congressional budget documents.

[a]Includes $20 million to help address the opioid epidemic in rural America.

[b]Includes $16 million to help address the opioid epidemic in rural America.

[c]Includes $20 million to help address the opioid epidemic in rural America.

[d]Includes $125 million in appropriations plus $425 million to be reprogrammed from the cushion of credit subaccount (7 U.S.C. 940c). Includes $125 million in appropriations plus $425 million to be reprogrammed from the cushion of credit subaccount (7 U.S.C. 940c).

---

[27] For more information, see CRS Report R42524, *Rural Broadband: The Roles of the Rural Utilities Service and the Universal Service Fund*, by Angele A. Gilroy and Lennard G. Kruger.

[28] According to a January 2013 survey conducted by NTCA—The Rural Broadband Association, 69% of member companies responding to the survey had either cancelled or postponed $492.7 million in broadband investments due to the uncertainty surrounding the transition to the FCC's Connect America Fund. See NTCA—The Rural Broadband Association, *Survey: FCC USF/ICC Impacts*, January 2013, available at http://www.ntca.org/images/stories/Documents/Advocacy/SurveyReports/FCC_USF_ICC_ImpactSurvey.pdf.

[29] Letter from RUS to the FCC, February 13, 2013, available at https://prodnet.www.neca.org/publicationsdocs/wwpdf/ 21513usda.pdf.

The Rural Broadband Access Loan and Loan Guarantee Program, the Community Connect Grant Program, the Telecommunications Infrastructure Loan and Loan Guarantee program, the Rural Broadband ReConnect Program, and the Distance Learning and Telemedicine grant program are funded through the annual Agriculture, Rural Development, Food and Drug Administration, and Related Agencies Appropriations Act. The appropriations provided to the broadband loan programs are loan subsidies which support a significantly higher loan level. Table 3 shows recent and proposed appropriations for the rural broadband programs in the Rural Utilities Service.

FY2018

The Administration's FY2018 budget proposal requested the following for RUS broadband programs:

- Rural Broadband Access Loans—$4.5 million in budget authority to subsidize a broadband loan level of $27 million. According to the budget proposal, this funding level will provide for approximately 3 loans in FY2018.[30]
- Telecommunications Infrastructure Loans—$0.863 million in budget authority to subsidize a loan level of $690 million ($345 million for Treasury loans and $345 million for FFB loans). The subsidy is for Treasury loans. According to the budget proposal, this funding level will provide for approximately 40 loans in FY2018.[31]
- Community Connect and DLT grants—for FY2018, the Administration is proposing transferring Community Connect and DLT grants into a new $162 million "Rural Economic Infrastructure Program," which will also include Rural Development Community Facilities grants and Home Repair grants. Up to $80 million will be directed toward the Appalachian region. According to the Administration, the new account "combines the Rural Development grant programs into one account to provide the Administration with the flexibility to place resources where significant impact can be made for economic infrastructure development."[32]

On July 12, 2017, the House Appropriations Committee approved the Agriculture, Rural Development, Food and Drug Administration, and Related Agencies Appropriations Act, 2018 (H.R. 3268; H.Rept. 115-232). The bill provided $4.521

---

[30] 2018 USDA Budget Explanatory Notes for Committee on Appropriations, Rural Utilities Service, p. 31-34, available at https://www.obpa.usda.gov/31rusexnotes2018.pdf.

[31] Ibid., p. 31-34.

[32] 2018 USDA Budget Explanatory Notes for Committee on Appropriations, Rural Development, p. 28-18, available at https://www.obpa.usda.gov/28rdexnotes2018.pdf.

million to subsidize a loan level of $26.991 million for the broadband loan program. Funding provided for the broadband loan program was intended to promote availability in those areas where there is not otherwise a business case for private investment in a broadband network. The committee directed RUS to focus expenditures on projects that bring broadband service to underserved households and areas.

The House bill provided $122.692 million for the new Rural Economic Infrastructure Account (24% below the Administration request), which would include both Community Connect and DLT grants, along with Community Facilities grants and Home Repair grants. The bill included language requiring at least 15% of the account resources ($18.4 million) be allocated to each program area. The committee noted that tribal communities continue to struggle with gaining access to broadband service, and encouraged the Secretary to provide a report that identifies the specific challenges Indian Tribal Organizations (ITOs) have in gaining access to broadband service and provide a plan for addressing these challenges, including how the Community Connect program can assist ITOs.

Regarding telecommunications loans, the House matched the Administration proposal, providing a loan level of $690 million ($345 million in direct Treasury loans and $345 million in FFB loans) with an appropriation of $0.863 million to subsidize direct Treasury loans.

Additionally, the House Appropriations Committee report directed USDA to continue coordinating with the FCC, NTIA, and other related federal agencies to ensure that policies tied to one federal program do not undermine the objectives and functionality of another. The committee directed the department to prepare a report, in collaboration with the FCC and DOC, detailing areas of responsibility toward addressing rural broadband issues. The report shall include, but not be limited to, how the programs work complimentarily to one another; how they address broadband issues in unserved and underserved areas, including tribal lands; identify barriers to infrastructure investment in rural areas and tribal lands; data speeds which fixed, wireless, and mobile broadband users in rural areas and tribal lands experience; and cost estimates to increase speeds to 25 Mbps in unserved communities and communities currently being served by speeds less than 25 Mbps.

On July 20, 2017, the Senate Appropriations Committee approved its version of the FY2018 agriculture appropriations bill (S. 1603; S.Rept. 115-131). The bill provided $4.53 million to subsidize a loan level of $27.043 million for the broadband loan program, $30 million for the Community Connect grant program, and $26.6 million for DLT grants. Unlike the House and the Administration request, the committee did not include funding for Rural Economic Infrastructure grants. For telecommunications loans, the Senate matched the House bill and the Administration proposal, providing a loan level of $690 million ($345 million in direct Treasury loans and $345 million in FFB loans) with an appropriation of $0.863 million to subsidize direct Treasury loans.

Regarding the broadband loan program, the committee encouraged RUS to focus expenditures on projects that bring broadband service to currently unserved households, and directed RUS to report back to the committee on administrative efforts to eliminate duplicative or overbuilding of broadband technology. The committee also recommended that USDA explore a pilot grant program to demonstrate the use of multistrand fiber-optic cable that exists as part of electrical transmission infrastructure to provide state-of-the-art broadband services to currently underserved rural schools and medical centers within a mile of the existing cable.

The Consolidated Appropriations Act, 2018 (P.L. 115-141) provided $5 million to subsidize a broadband loan level of $29.851 million, $30 million to Community Connect broadband grants, and $49 million for DLT grants, which included an additional $20 million to address the opioid epidemic in rural America. P.L. 115-141 also appropriated $600 million to RUS to "conduct a new broadband loan and grant pilot program."

FY2019

The Administration's FY2019 budget proposal requested the following for RUS broadband programs:

- Rural Broadband Access Loans—$4.5 million in budget authority to subsidize a broadband loan level of $23.149 million. According to the budget proposal, this funding level will provide for approximately three loans in FY2019.[33]
- Telecommunications Infrastructure Loans and Loan Guarantees—$0.863 million in budget authority to subsidize a loan level of $690 million ($172.6 million for Treasury loans and $517.4 million for FFB loans). The subsidy is for Treasury loans. According to the budget proposal, this funding level will provide for approximately 30 loans in FY2019.[34]
- Community Connect Grants—$30 million, which will support approximately 13 broadband grants in FY2019.[35]
- Distance Learning and Telemedicine Grants—$23.6 million, which will support approximately 72 projects in FY2019.[36]

On May 16, 2018, the House Appropriations Committee approved the FY2019 Agriculture Appropriations bill (H.R. 5961; H.Rept. 115-706). The bill would provide the following:

---

[33] 2019 USDA Budget Explanatory Notes for Committee on Appropriations, Rural Utilities Service, p. 31-29, available at https://www.obpa.usda.gov/FY19explan_notes.html.

[34] Ibid., p. 31-21.

[35] Ibid., p. 31-30.

[36] Ibid., p. 31-29.

- Rural Broadband Access Loans—$5.83 million in budget authority to subsidize a broadband loan level of $29.851 million.
- Telecommunications Infrastructure Loans and Loan Guarantees—$1.125 million in budget authority to subsidize direct Treasury loans set at a level of $465 million. Along with a loan level $225 million for FFB guaranteed loans, the total loan level is $690 million.
- Community Connect Grants—$30 million.
- Distance Learning and Telemedicine Grants—$32 million.
- ReConnect Program—$550 million. This appropriation would continue the pilot broadband loan and grant program that was funded (at $600 million) in the FY2018 Consolidated Appropriations Act, 2018 (P.L. 115-141).

In the committee report, the committee expressed its view that "it is important for Departments to avoid efforts that could duplicate existing networks built by private investment or those built leveraging and utilizing other federal programs." As such, the committee "directs the Secretary of Agriculture to coordinate with the Federal Communications Commission (FCC) and the National Telecommunications Information Administration (NTIA) to ensure wherever possible that broadband loans and grants issued under the pilot program are being targeted to areas that are currently unserved." The committee directed USDA to use the NTIA's assessment of the current state of broadband access nationwide, and to explore using all broadband technologies, including, but not limited to, fiber, cable modem, fixed wireless, and television white space.

The committee also noted that tribal communities continue to struggle with gaining access to broadband service, and encouraged the Secretary to provide a report that identifies the specific challenges Indian Tribal Organizations (ITOs) have in gaining access to broadband service and provide a plan for addressing these challenges, including how the Community Connect program can assist ITOs.

On May 24, 2018, the Senate Appropriations Committee approved its FY2019 Agriculture Appropriations bill (S. 2976; S.Rept. 115-259). The bill would provide the following:

- Rural Broadband Access Loans—$5.83 million in budget authority to subsidize a broadband loan level of $29.851 million.
- Telecommunications Infrastructure Loans and Loan Guarantees—$1.725 million in budget authority to subsidize direct Treasury loans set at a level of $345 million. Along with a loan level of $345 million for FFB guaranteed loans, the total loan level is $690 million.
- Community Connect Grants—$30 million.

- Distance Learning and Telemedicine Grants—$50 million (including $20 million to help address the opioid epidemic in rural America).
- ReConnect Program—$425 million.

The committee encouraged RUS to focus expenditures on projects that bring broadband service to currently unserved households, and directed RUS to report back to the committee on administrative efforts to eliminate duplicative or overbuilding of broadband technology. The committee also recommended that USDA explore a pilot grant program to demonstrate the use of multistrand fiber-optic cable that exists as part of electrical transmission infrastructure to provide state-of-the-art broadband services to currently underserved rural schools and medical centers within a mile of the existing cable; encouraged RUS to coordinate with the FCC and other relevant federal entities when making determinations of sufficient broadband access, to ensure the most accurate and up-to-date broadband coverage data are used, while being cognizant of potential problems of overbuilding; encouraged the Secretary to utilize appropriate grant program funds to locate buried, antiquated infrastructure facilities prior to construction of new utilities infrastructure financed by RUS; and urged RUS to ensure the agency's criteria and application processes provide for fair consideration of open access projects by accounting for the unique structures and opportunities such projects present in advancing broadband deployment in unserved and underserved communities.

On February 15, 2019, the Consolidated Appropriations Act, 2019 was signed into law (P.L. 116- 6). The FY2019 appropriations and levels are as follows:

- Rural Broadband Access Loans—$5.83 million in budget authority to subsidize a broadband loan level of $29.851 million.
- Telecommunications Infrastructure Loans and Loan Guarantees—$1.725 million in budget authority to subsidize direct Treasury loans set at a level of $345 million. Along with a loan level of $345 million for FFB guaranteed loans, the total loan level is $690 million.
- Community Connect Grants—$30 million.
- Distance Learning and Telemedicine Grants—$47 million (including $16 million to address the opioid epidemic in rural America).
- ReConnect Program—$550 million ($125 million direct appropriation plus $425 million to be reprogrammed from the cushion of credit account).

P.L. 116-6 also directs USDA rural development programs, including the broadband programs, to allocate (to the maximum extent feasible) at least 10% of funds to projects in persistent poverty counties.

The conference report (H.Rept. 116-9) contains language directing USDA to avoid efforts that could duplicate existing networks built by private investment or those built leveraging and utilizing other federal programs, and directs the Secretary to coordinate with the FCC and NTIA to ensure wherever possible that broadband loans and grants are targeted to areas that are currently unserved. In particular, the conference agreement directs USDA to use the NTIA's assessment of the current state of broadband access nationwide. USDA is also directed, in implementing a strategy for broadband deployment to unserved communities, to explore using all technologies, including but not limited to, fiber, cable modem, fixed wireless, and television white space.

FY2020

The Administration's FY2020 budget proposal requested the following for RUS broadband programs:

- Rural Broadband Access Loans—Zero funding. According to the budget proposal, the elimination of funding will be offset by continued access by most eligible borrowers to the ReConnect Program (broadband pilot loan and grants).[37]
- ReConnect Program—$200 million, which, according to the budget proposal, will support approximately eight loans, grants, or loan/grant combinations in FY2020.[38]
- Telecommunications Infrastructure Loans and Loan Guarantees—$1.933 million in budget authority to subsidize a loan level of $690 million ($175.7 million for Treasury loans and $514.3 million for FFB loans). The subsidy is for the Treasury loans. According to the budget proposal, this funding level will provide for approximately 20 loans in FY2020.[39]
- Community Connect Grants—$30 million, which will support approximately 13 broadband grants in FY2020.[40]
- Distance Learning and Telemedicine Grants—$43.6 million, which will support approximately 90 projects in FY2020.[41]

PAST CRITICISMS OF RUS BROADBAND PROGRAMS

RUS broadband programs have been awarding funds to entities serving rural communities since FY2001. Since their inception, a number of criticisms have emerged.

---

[37] 2020 USDA Budget Explanatory Notes for Committee on Appropriations, Rural Utilities Service, p. 31-32, available at http://www.obpa.usda.gov/31rus2020notes.pdf.

[38] Ibid., p. 31-33.

[39] Ibid., p. 31-25.

[40] Ibid., p. 31-34.

[41] Ibid., p. 31-32.

Loan Approval and Application Process

Perhaps the major criticism of the broadband loan program was that not enough loans are approved, thereby making it difficult for rural communities to take full advantage of the program. The loan application process has been criticized as being overly complex and burdensome, requiring applicants to spend months preparing costly market research and engineering assessments. Many applications are rejected because the applicant's business plan is deemed insufficient to support a commercially viable business. The biggest reason for applications being returned has been insufficient credit support, whereby applicants do not have sufficient cash-on- hand (one year's worth is required in most cases). The requirement for cash-on-hand is viewed as particularly onerous for small start-up companies, many of whom lack sufficient capital to qualify for the loan. Such companies, critics assert, may be those entities most in need of financial assistance.

In report language to the FY2006 Department of Agriculture Appropriations Act (P.L. 109-97), the Senate Appropriations Committee (S.Rept. 109-92) directed the RUS "to reduce the burdensome application process and make the program requirements more reasonable, particularly in regard to cash-on-hand requirements." The committee also directed USDA to hire more full-time employees to remedy delays in application processing times.

At a May 17, 2006, hearing held by the Senate Committee on Agriculture, Nutrition, and Forestry, the Administrator of the RUS stated that RUS is working to make the program more user friendly, while at the same time protecting taxpayer investment:

> As good stewards of the taxpayers' money, we must make loans that are likely to be repaid. One of the challenges in determining whether a proposed project has a reasonable chance of success is validating the market analysis of the proposed service territory and ensuring that sufficient resources are available to cover operating expenses throughout the construction period until such a time that cash flow from operations become sufficient. The loan application process that we have developed ensures that the applicant addresses these areas and that appropriate resources are available for maintaining a viable operation.[42]

According to RUS, the loan program was initially overwhelmed by applications (particularly during a two-week period in August 2003), and as the program matured, application review times have dropped.[43] On May 11, 2007, RUS released a Proposed Rule which sought to revise regulations for the broadband loan program. In the

[42] Testimony of Jim Andrew, Administrator, Rural Utilities Service, U.S. Department of Agriculture, "Broadband Program Administered by USDA's Rural Utilities Service," full committee hearing before the Senate Committee on Agriculture, Nutrition, and Forestry, 109th Congress, May 17, 2006.

[43] Rural Utilities Service, private communication, January 18, 2007.

background material accompanying the Proposed Rule, RUS stated that the average application processing time in 2006 was almost half of what it was in 2003.[44]

Eligibility Criteria

Since the inception of the broadband grant and loan programs, the criteria for applicant eligibility have been criticized both for being too broad and for being too narrow. An audit report released by USDA's Office of Inspector General (IG) found that the "programs' focus has shifted away from those rural communities that would not, without Government assistance, have access to broadband technologies."[45] Specifically the IG report found that the RUS definition of rural area has been "too broad to distinguish usefully between suburban and rural communities,"[46] with the result that, as of March 10, 2005, $103.4 million in loans and grants (nearly 12% of total funding awarded) had been awarded to 64 communities located near large cities. The report cited examples of affluent suburban subdivisions qualifying as rural areas under the program guidelines and receiving broadband loans.[47]

On the other hand, eligibility requirements have also been criticized as too narrow. For example, the limitation of assistance only to communities of 20,000 or less in population excludes small rural towns that may exceed this limit, and also excludes many municipalities seeking to deploy their own networks.[48] Similarly, per capita income requirements can preclude higher income communities with higher costs of living (e.g., rural Alaska), and the limitation of grant programs only to underserved areas excludes rural communities with existing but very limited broadband access.[49]

Loans to Communities with Existing Providers

The IG report found that RUS too often has given loans to communities with existing broadband service. The IG report found that "RUS has not ensured that communities without broadband service receive first priority for loans," and that although RUS has a system in place to prioritize loans to unserved communities, the system "lacks a cutoff

---

[44] Rural Utilities Service, Department of Agriculture, "Rural Broadband Access Loans and Loan Guarantees," Proposed Rule, *Federal Register*, Vol. 72, No. 91, May 11, 2007, p. 26744.

[45] U.S. Department of Agriculture, Office of Inspector General, Southwest Region, *Audit Report: Rural Utilities Service Broadband Grant and Loan Programs*, Audit Report 09601-4-Te, September 2005, p. I, http://www.usda.gov/ oig/webdocs/09601-04-TE.pdf.

[46] Ibid., p. 6.

[47] Ibid., p. 8.

[48] Martinez, Michael, "Broadband: Loan Fund's Strict Rules Foil Small Municipalities," *National Journal's Technology Daily*, August 23, 2005.

[49] GAO, *Broadband Deployment is Extensive throughout the United States, but It Is Difficult to Assess the Extent of Deployment Gaps in Rural Areas*, pp. 33-34.

date and functions as a rolling selection process—priorities are decided based on the applicants who happen to be in the pool at any given moment."[50] The result is that a significant number of communities with some level of preexisting broadband service have received loans. According to the IG report, of 11 loans awarded in 2004, 66% of the associated communities served by those loans had existing service. According to RUS, 31% of communities served by all loans (during the period 2003 through early 2005) had preexisting competitive service (not including loans used to upgrade or expand existing service).[51] In some cases, according to the IG report, "loans were issued to companies in highly competitive business environments where multiple providers competed for relatively few customers."[52] At the May 1, 2007, hearing before the House Subcommittee on Specialty Crops, Rural Development, and Foreign Agriculture, then-RUS Administrator James Andrews testified that of the 69 broadband loans awarded since the program's inception, 40% of the communities approved for funding were unserved at the time of loan approval, and an additional 15% had only one broadband provider.[53]

Awarding loans to entities in communities with preexisting competitive service raised criticism from competitors who already offer broadband to those communities. According to the National Cable and Telecommunications Association (NCTA), "RUS loans are being used to unfairly subsidize second and third broadband providers in communities where private risk capital already has been invested to provide broadband service."[54] Critics argued that providing loans in areas with preexisting competitive broadband service creates an uneven playing field and discourages further private investment in rural broadband.[55] In response, RUS stated in the IG report that its policies are in accordance with the statute, and that they address "the need for competition to increase the quality of services and reduce the cost of those services to the consumer."[56] RUS argued that the presence of a competitor does not necessarily mean that an area is adequately served, and additionally, that in order for some borrowers to maintain a viable business in an unserved area, it may be necessary for that company to also be serving more densely populated rural areas where some level of competition already exists.[57]

---

[50] Ibid., p. 13.

[51] Ibid., p. 14.

[52] Ibid., p. 15.

[53] Testimony of James Andrew, Administrator, Rural Utilities Service, U.S. Department of Agriculture, before the Subcommittee on Specialty Crops, Rural Development, and Foreign Agriculture, House Committee on Agriculture, May 1, 2007.

[54] Letter from Kyle McSlarrow, President and CEO, National Cable & Telecommunications Association to the Honorable Mike Johanns, Secretary of the U.S. Department of Agriculture, May 16, 2006.

[55] Testimony of Tom Simmons, Vice President for Public Policy, Midcontinent Communications, before Senate Committee on Agriculture, Nutrition, and Forestry, May 17, 2006.

[56] *Audit Report: Rural Utilities Service Broadband Grant and Loan Programs*, p. 17.

[57] Rural Utilities Service, private communication, January 18, 2007.

Follow-Up Audit by USDA Office of Inspector General

In 2008, as directed by the House Appropriations Committee (H.Rept. 110-258, FY2008 Agriculture appropriations bill), the IG reexamined the RUS broadband loan and loan guarantee program to determine whether RUS had taken sufficient corrective actions in response to the issues raised in the 2005 IG report. The IG concluded "the key problems identified in our 2005 report—loans being issued to suburban and exurban communities and loans being issued where other providers already provide access—have not been resolved."[58]

Specifically, the follow-up IG report found that between 2005 and 2008, RUS broadband borrowers providing services in 148 communities were within 30 miles of cities with 200,000 inhabitants, including communities near very large urban areas such as Chicago and Las Vegas.

The IG report also found that since 2005 "RUS has continued providing loans to providers in markets where there is already competing service."[59] Of the 37 applications approved since September 2005, 34 loans were granted to applicants in areas where one or more private broadband providers already offered service. These 34 borrowers received $873 million to service 1,448 communities. The IG report found that since 2005, 77% of communities which were expected to receive service from a project financed by an approved RUS broadband loan had at least one existing broadband provider present, 59% had two or more existing providers, and 27% had three or more existing providers.[60]

In an official response to the follow-up IG report, RUS fundamentally disagreed with the IG criticisms, stating that the loans awarded between 2005 and 2008 were provided "in a way entirely consistent with the statutory requirements of the underlying legislation governing administration of the program, the regulations and guidance issued by the Department to implement the statute, and the intent of Congress."[61] Specifically, RUS argued that its May 11, 2007, Proposed Rule, and the subsequent changes to the broadband loan and loan guarantee statute made by the 2008 farm bill, both addressed concerns over loans to nonrural areas and to communities with preexisting broadband providers. However, the Final Rule based on the Proposed Rule and the 2008 farm bill had not yet been released and implemented during the 2005-2008 period examined by the IG, and RUS was compelled by law to continue awarding broadband loans under the existing law and rules.

During 2009 and 2010, the Rural Broadband Access Loan and Loan Guarantee program was in hiatus while RUS implemented the Broadband Initiatives Program (Recovery Act grants and loans) and developed new regulations implementing the 2008

---

[58] U.S. Department of Agriculture, Office of Inspector General, Southwest Region, Audit Report Rural Utilities Service Broadband Loan and Loan Guarantee Program, Report No. 09601-8-Te, March 2009, p. 9.

[59] Ibid., p. 5.

[60] Ibid., pp. 5-6.

[61] Ibid., p. 14.

farm bill. On March 14, 2011, the new rules were released. According to then-RUS Administrator Jonathan Adelstein, "this regulation and other measures taken by the agency have addressed all the concerns raised by the OIG," and on March 24, 2011, "the OIG notified RUS that it has closed its audits of the RUS broadband loan program."[62]

2014 GAO Report

In May 2014, GAO released its report, *USDA Should Evaluate the Performance of the Rural Broadband Loan Program.*[63] In the report, GAO analyzed rural broadband loans awarded between the years 2003 and 2013. GAO found that of the 100 loans awarded (worth $2 billion), 43% were no longer active due to 25 loans rescinded and 18 defaulted (RUS rejected 149 of the 249 applications received); that RUS loans can help promote limited broadband deployment and economic development, but performance goals do not fully align with the program's purpose; and that FCC reforms of the Universal Service Fund and intercarrier compensation have created temporary uncertainty that may be hindering investment in broadband.

To address its findings, GAO made two recommendations to the Secretary of Agriculture: evaluate loans made by RUS through the broadband loan program to identify characteristics of loans that may be at risk of rescission or default; and align performance goals under the "enhance rural prosperity" strategic objective in the Annual Performance Report to the broadband loan program's purpose, to the extent feasible.[64]

BROADBAND LOAN REAUTHORIZATION IN THE FARM BILL

The Rural Broadband Access Loan and Loan Guarantee program is authorized by Section 601 of the Rural Electrification Act of 1936. Since the program was established in the 2002 farm bill, it has been subsequently reauthorized and modified by the 2008 and 2014 farm bills. The 2018 farm bill seeks to again reauthorize and modify the program, as well as addressing other RUS broadband programs and issues.

---

[62] Testimony of Jonathan Adelstein, Administrator of RUS, before the House Subcommittee on Communications and Technology, Committee on Energy and Commerce, April 1, 2011, p. 8, http://republicans.energycommerce.house.gov/Media/file/Hearings/Telecom/040111/Adelstein.pdf.

[63] Government Accountability Office, *USDA Should Evaluate the Performance of the Rural Broadband Loan Program*, May 2014, 56 pp., available at http://www.gao.gov/assets/670/663578.pdf.

[64] Ibid., pp. 31-32.

2008 Farm Bill

The 110th Congress considered reauthorization of the Rural Broadband Access Loan and Loan Guarantee program as part of the 2008 farm bill. The following are some key issues which were considered during the debate over reauthorization of the RUS broadband loan and loan guarantee program.

Restricting Applicant Eligibility

The RUS broadband program was criticized for excluding too many applicants due to stringent financial requirements (e.g., the requirement that an applicant have a year's worth of cash-on- hand) and an application process—requiring detailed business plans and market surveys—that some viewed as overly expensive and burdensome to complete. During the reauthorization process, Congress considered whether the criteria for loan eligibility should be modified, and whether a more appropriate balance could be found between the need to make the program more accessible to unserved and often lower-income rural areas, and the need to protect taxpayers against bad loans.

Definition of "Rural Community"

The definition of which communities qualify as "rural" had been changed twice by statute since the broadband loan program was initiated. Under the pilot program, funds were authorized under the Distance Learning and Telemedicine Program, which defines "exceptionally rural areas" (under 5,000 inhabitants), "rural areas" (between 5,000 and 10,000), and "mid-rural areas" (between 10,000 and 20,000). RUS determined that communities of 20,000 or less would be eligible for broadband loans in cases where broadband services did not already exist.

In 2002, this definition was made narrower by the Farm Security and Rural Investment Act (P.L. 107-171), which designated eligible communities as any incorporated or unincorporated place with fewer than 20,000 inhabitants, and which was outside any standard metropolitan statistical area (MSA). The requirement that communities not be located within MSA's effectively prohibited suburban communities from receiving broadband loans. However, in 2004, the definition was again changed by the FY2004 Consolidated Appropriations Act (P.L. 108-199). The act broadened the definition, keeping the population limit at 20,000, but eliminating the MSA prohibition, thereby permitting rural communities near large cities to receive loans. Thus the current definition used for rural communities is the same as what was used for the broadband pilot program, except that loans can now be issued to communities with preexisting service.

The definition of what constitutes a "rural" community is always a difficult issue for congressional policymakers in determining how to target rural communities for broadband assistance. On the one hand, the narrower the definition the greater the possibility that deserving communities may be excluded. On the other hand, the broader the definition used, the greater the possibility that communities not traditionally considered "rural" or "underserved" may be eligible for financial assistance.

A related issue is the scope of coverage proposed by individual applications. While many of the loan applications propose broadband projects offering service to multiple rural communities, RUS identified a trend toward larger regional and national proposals, covering hundreds or even more than 1,000 communities.[65] The larger the scope of coverage, the greater the complexity of the loan application and the larger the possible benefits and risks to taxpayers.

Preexisting Broadband Service

Loans to areas with competitive preexisting service—that is, areas where existing companies already provide some level of broadband—sparked controversy because loan recipients are likely to compete with other companies already providing broadband service.

During reauthorization, Congress was asked to more sharply define whether and/or how loans should be given to companies serving rural areas with preexisting competitive service.[66] On the one hand, some argued that the federal government should not be subsidizing competitors for broadband service, particularly in sparsely populated rural markets which may be able only to support one provider. Furthermore, keeping communities with preexisting broadband service eligible may divert assistance from unserved areas that are most in need. On the other hand, many suburban and urban areas currently receive the benefits of competition between broadband providers—competition which can potentially drive down prices while improving service and performance. It is therefore appropriate, others argued, that rural areas also receive the benefits of competition, which in some areas may not be possible without federal financial assistance. It was also argued that it may not be economically feasible for borrowers to serve sparsely populated unserved communities unless they are permitted to also serve more lucrative areas which may already have existing providers.

Technological Neutrality

The 2002 farm bill (P.L. 107-171) directed RUS to use criteria that are "technologically neutral" in determining which projects to approve for loans. In other words, RUS is prohibited from typically valuing one broadband technology over another when assessing loan applications. As of November 10, 2008, 37% of approved and funded projects employed fiber-to-the-home technology, 17% employed DSL, 25% fixed

---

[65] Rural Utilities Service, private communication, January 18, 2007.

[66] The statute (7 U.S.C. 950bb) allows States and local governments to be eligible for loans only if "no other eligible entity is already offering, or has committed to offer, broadband services to the eligible rural community."

wireless, 19% hybrid fiber-coaxial (cable), and 2% broadband over powerlines (BPL).[67] No funding has been provided for projects utilizing satellite broadband.[68]

While decisions on funded projects were required to be technologically neutral, RUS (through the Secretary of Agriculture) had the latitude to determine minimum required data transmission rates for broadband projects eligible for funding. According to the statute, "the Secretary shall, from time to time as advances in technology warrant, review and recommend modifications of rate-of- data transmission criteria for purposes of the identification of broadband service technologies."

Some argued that the minimum speed thresholds should be raised to ensure that rural areas receive "next-generation" broadband technologies with faster data rates capable of more varied and sophisticated applications. On the other hand, significantly raising minimum data rates could exclude certain technologies—for example, typical data transmission rates for fiber and some wireless technologies exceed what is offered by "current generation" technologies such as DSL and cable. Proponents of keeping the minimum threshold at a low level argued that underserved rural areas are best served by any broadband technology that is economically feasible to deploy, regardless of whether it is "next" or "current" generation.

P.L. 110-246

The Food, Conservation, and Energy Act of 2008 became law on June 18, 2008 (P.L. 110-246). Section 6110, "Access to Broadband Telecommunications Services in Rural Areas," reauthorized the RUS broadband loan and loan guarantee program and addressed many of the criticisms and issues raised during the reauthorization process. The following summarizes broadband-related provisions that changed previous law.

Eligibility and Selection Criteria

- Defines rural area as any area other than (1) a city or town that has a population of greater than 20,000 and (2) an urbanized area contiguous and adjacent to a city or town with a population greater than 50,000. The Secretary may, by regulation only, consider not to be rural an area that consists of any collection of census blocks contiguous to each other with a housing density of more than 200 housing

---

[67] USDA, Rural Utilities Service, "FCC/USDA Rural Broadband Educational Workshop," power point presentation, November 20, 2008, http://www.usda.gov/rus/telecom/broadband/workshops/ FCC_USDABroadband WorkshopNov20.pdf.

[68] According to the GAO, satellite companies state that RUS's broadband loan program requirements "are not readily compatible with their business model or technology," and that "because the agency requires collateral for loans, the program is more suited for situations where the providers, rather than individual consumers, own the equipment being purchased through the loan. Yet, when consumers purchase satellite broadband, it is common for them to purchase the equipment needed to receive the satellite signal, such as the reception dish." Satellite companies argue that in some rural areas, satellite broadband might be the most feasible and cost-effective solution. See GAO, Broadband Deployment is Extensive throughout the United States, but It Is Difficult to Assess the Extent of Deployment Gaps in Rural Areas, pp. 34-35.

units per square mile and that is contiguous with or adjacent to an existing boundary of a rural area.

- Provides that the highest priority is to be given to applicants that offer to provide broadband service to the greatest proportion of households currently without broadband service. Eligible entities are required to submit a proposal to the Secretary that meets the requirements for a project to offer to provide service to a rural area and agree to complete build out of the broadband service within three years.
- Prohibits any eligible entity that provides telecommunications or broadband service to at least 20% of the households in the United States from receiving an amount of funds under this section for a fiscal year in excess of 15% of the funds authorized and appropriated for the broadband loan program.
- Directs the Secretary of Agriculture "from time to time as advances in technology warrant," to review and recommend modifications in rate-of-data-transmission criteria for the purpose of identifying eligible broadband service technologies. At the same time, the Secretary is prohibited from establishing requirements for bandwidth or speed that have the effect of precluding the use of evolving technologies appropriate for use in rural areas.

Loans to Communities with Existing Providers

- Prohibits the Secretary from making a loan in any area where there are three or more incumbent service providers unless the loan meets all of the following requirements: (1) the loan is to an incumbent service provider that is upgrading service in that provider's existing territory; (2) the loan proposes to serve an area where not less than 25% of the households are offered service by not more than 1 provider; and (3) the applicant is not eligible for funding under another provision of the Rural Electrification Act. *Incumbent service provider* is defined as an entity providing broadband service to not less than 5% of the households in the service territory proposed in the application. Also prohibits the Secretary from making a loan in any area where not less than 25% of the households are offered broadband service by not more than one provider unless a prior loan has been made in the same area.

Financial Requirements

- Directs the Secretary to consider existing recurring revenues at the time of application in determining an adequate level of credit support. Requires the Secretary to ensure that the type, amount, and method of security used to secure a loan or loan guarantee is commensurate to the risk involved with the loan or loan guarantee, particularly when the loan or loan guarantee is issued to a financially healthy, strong, and stable entity. The Secretary is also required, in determining the amount and method of security, to consider reducing the security in areas that do not have broadband service.

- Allows the Secretary to require an entity to provide a cost-share in an amount not to exceed 10% of the amount of the loan or loan guarantee.
- Retains the current law rate of interest for direct loans—which is the rate equivalent to the cost of borrowing to the Department of the Treasury for obligations of comparable maturity or 4%.
- Directs that loan or loan guarantee may have a term not to exceed 35 years if the Secretary determines that the loan security is sufficient.
- In case of substantially underserved trust areas (for example, Indian lands), where the Secretary determines a high need exists for the benefits of the program, the Secretary has the authority to provide loans with interest rates as low as 2% and may waive nonduplication restrictions, matching fund requirements, credit support requirements, or other regulations.

Loan Application Requirements

- Allows the Secretary to require an entity that proposes to have a subscriber projection of more than 20% of the broadband service market in a rural area to submit a market survey. However, the Secretary is prohibited from requiring a market survey from an entity that projects to have less than 20% of the broadband market.
- Requires public notice of each application submitted, including the identity of the applicant, the proposed area to be served, and the estimated number of households in the application without terrestrial-based broadband. Authorizes the Secretary to take steps to reduce the costs and paperwork associated with applying for a loan or loan guarantee under this section by first-time applicants, particularly those who are smaller and start-up internet providers.
- Allows the Secretary to establish a preapplication process under which a prospective applicant may seek a determination of area eligibility. Provides that an application, or a petition for reconsideration of a decision on such an application, that was pending on the date 45 days before enactment of this act and that remains pending on the date of enactment of this act is to be considered under eligibility and feasibility criteria in effect on the original date of submission of the application.

Other Provisions

- Authorizes the Rural Broadband Access Loan and Loan Guarantee program at $25 million to be appropriated for each of fiscal years 2008 through 2012.
- Requires that the Secretary annually report to Congress on the rural broadband loan and loan guarantee program. The annual report is to include information pertaining to the loans made, communities served and proposed to be served, speed of broadband service offered, types of services offered by the applicants and recipients, length of time to approve applications submitted, and outreach efforts undertaken by USDA.
- Section 6111 provides for a National Center for Rural Telecommunications Assessment. The center is to assess the effectiveness of broadband loan programs, work with existing rural development centers to identify appropriate policy initiatives, and provide an annual report that describes the activities of the center, the results of research carried out by the center, and any additional information that the Secretary may request. An appropriation of $1 million is authorized for each of the fiscal years 2008 through 2012.
- Section 6112 directs the Chairman of the Federal Communications Commission (FCC), in coordination with the Secretary, to submit to Congress a report describing a comprehensive rural broadband strategy. Requires the report to be updated during the third year after enactment.

Implementation of P.L. 110-246

During 2009 and 2010, the Farm Bill Broadband Loan Program was on hiatus as RUS implemented the Broadband Initiatives Program (BIP) established under the American Recovery and Reinvestment Act of 2009 (P.L. 111-5). At the same time, final regulations implementing the broadband loan program as reauthorized by the 2008 farm bill were on hold and were being refined to reflect, in part, RUS experience in implementing BIP. Subsequently, on March 14, 2011, an Interim Rule and Notice was published in the *Federal Register* setting forth the rules and regulations for the broadband loan program as reauthorized by P.L. 110-246.[69] While the rule was immediately effective, RUS accepted public comment before ultimately releasing a final rule.

Meanwhile, pursuant to Section 6112 of P.L. 110-246, the FCC released on May 22, 2009, its report on rural broadband strategy, entitled *Bringing Broadband to Rural America*.[70] The report made a series of recommendations including improved coordination of rural broadband efforts among federal agencies, states, and communities;

---

[69] U.S. Department of Agriculture, Rural Utilities Service, "7 CFR Part 1738, Rural Broadband Access Loans and Loan Guarantees," 76 Federal Register 13770-13796, March 14, 2011.

[70] Michael J. Copps, Acting Chairman, Federal Communications Commission, Bringing Broadband to Rural America: Report on a Rural Broadband Strategy, May 22, 2009, 83 pp.

better assessment of broadband needs, including technological considerations and broadband mapping and data; and overcoming challenges to rural broadband deployment.

2014 Farm Bill

On January 27, 2014, the conference report for the Agricultural Act of 2014 was filed (H.Rept. 113-333). The conference agreement was approved by the House on January 29, approved by the Senate on February 4, and signed into law (P.L. 113-79) by the President on February 7, 2014.

P.L. 113-79 amended Section 601 of the Rural Electrification Act of 1936 (7 U.S.C. 950bb) to reauthorize the Rural Broadband Access Loan and Loan Guarantee Program through FY2018.

P.L. 113-79 also included provisions to redefine project area eligibility with respect to existing broadband service, increase the program's transparency and reporting requirements, define a minimum level of broadband service, require a study on the gathering and use of address-level data, and establish a new Rural Gigabit Network Pilot Program. The conference agreement did not include a Senate bill proposal (S. 954) to create a new grant component to the existing broadband loan and loan guarantee program, nor did the conference agreement adopt the Senate bill's broadening of the definition for eligible rural areas.

Specifically, Section 6104 of P.L. 113-79 made the following changes to the Rural Broadband Access Loan and Loan Guarantee program:

- Project area eligibility—provides that an eligible area is one where not less than 15% of the households in the proposed service territory are unserved or have service levels below the minimum acceptable level of broadband service (which is set at 4 Mbps/1 Mbps).
- Priority—directs RUS to give the highest priority to applicants that offer to provide broadband service to the greatest proportion of unserved households or households that do not have residential broadband service that meets the minimum acceptable level of broadband service, as certified by the affected community, city, county, or designee; or demonstrated on the broadband map of the affected state if the map contains address-level data, or the National Broadband Map if address-level data are unavailable. RUS shall provide equal consideration to all qualified applicants, including those that have not previously received grants, loans, or loan guarantees. Also gives priority to applicants that offer to provide broadband service not predominantly for business service, but if at least 25% of customers in the proposed service territory are commercial interests.

- Evaluation period—directs RUS to establish not less than two evaluation periods for each fiscal year to compare loan and loan guarantee applications and to prioritize loans and loan guarantees to all or part of rural communities that do not have residential broadband service that meets the minimum acceptable level of broadband service.
- Market survey requirement—provides that survey information must be certified by the affected community, city, county, or designee; and demonstrated on the broadband map of the affected state if the map contains address-level data, or the National Broadband Map if address-level data are unavailable.
- Notice requirement—directs RUS to maintain a fully searchable database on the internet that contains a list of each entity that has applied for assistance, the status of each application, and a detailed description of each application. For each entity receiving assistance, the database shall provide the name of the entity, the type of assistance being received, the purpose for which the entity is receiving the assistance, and each semiannual report submitted.
- Reporting—requires semiannual reports from loan recipients for three years after completion of the project describing in detail the use of the assistance, and the progress toward fulfilling project objectives.
- Default and deobligation—directs RUS to establish written procedures for recovering funds from loan defaults, deobligating awards that demonstrate an insufficient level of performance or fraudulent spending, awarding those funds to new or existing applicants, and minimizing overlap among programs.
- Service area assessment—directs RUS to promptly post on its website a list of the census block groups that an applicant proposes to service. RUS will provide not less than 15 days for broadband service providers to voluntarily submit information about the broadband services that the providers offer in the groups or tracts listed so that RUS may assess whether the applications submitted meet the eligibility requirements. If no broadband service provider submits this information, RUS will consider the number of providers in the group or tract to be established by reference to the most current National Broadband Map or any other data RUS may collect or obtain through reasonable efforts.
- Definition of broadband service—establishes "the minimum acceptable level of broadband service" as at least 4 Mbps downstream and 1 Mbps upstream. At least once every two years, the Secretary shall review and may adjust this speed definition and may consider establishing different minimum speeds for fixed and mobile (wireless) broadband.
- Terms and conditions—in determining the terms and conditions of assistance, the Secretary may consider whether the recipient would be serving an area that is unserved (or has service levels below the minimum acceptable level of broadband service), and if so, can establish a limited initial deferral period or

comparable terms necessary to achieve the financial feasibility and long-term sustainability of the project.

- Report to Congress—adds requirements to the content of the annual report to Congress, including the number of residences and businesses receiving new broadband services; network improvements, including facility upgrades and equipment purchases; average broadband speeds and prices on a local and statewide basis; any changes in broadband adoption rates; and any specific activities that increase high-speed broadband access for educational institutions, health care providers, and public safety service providers.
- Reauthorization—reauthorizes the broadband loan and loan guarantee program through FY2018 at the current level of $25 million per year.
- Study on providing effective data for the National Broadband Map—directs USDA, in consultation with DOC and the FCC, to conduct a study of the ways data collected by RUS could most effectively be shared with the FCC to support the development and maintenance of the National Broadband Map. The study shall include a consideration of the circumstances under which address-level data could be collected by RUS and appropriately shared with the FCC.

In addition, Section 6105 authorized a new Rural Gigabit Network Pilot Program. Specifically, USDA was authorized to provide grants, loans, or loan guarantees for projects that would extend ultra-high-speed broadband service (defined as 1 gigabit per second downstream capacity) to rural areas where ultra-high-speed service is not provided in any part of the proposed service territory. The pilot program was authorized at $10 million per year for the years FY2014 through FY2018. However, no funding was appropriated for this pilot program over that period, and the Rural Gigabit Network Pilot Program was not implemented.

Implementation of P.L. 113-79

On July 30, 2015, the RUS published in the *Federal Register* the interim rule (7 C.F.R. part 1738) implementing the Rural Broadband Access Loan and Loan Guarantee Program as reauthorized by the February 7, 2014, enactment of the Agricultural Act of 2014 (P.L. 113-79).[71] Publication of the interim rule allowed the program to go forward, initially with two application periods per year. The interim rule was made final on June 9, 2016.

---

[71] Department of Agriculture, Rural Utilities Service, "Rural Broadband Access Loans and Loan Guarantees," Interim rule, 80 *Federal Register* 45397-45413, July 30, 2015, available at https://www.gpo.gov/fdsys/pkg/FR-2015-07-30/pdf/ 2015-18624.pdf.

2018 Farm Bill

With the 2014 farm bill expiring on September 30, 2018, the 115th Congress considered reauthorization of the RUS broadband loan and loan guarantee program and other broadband- related provisions in the 2018 farm bill.

House

On April 12, 2018, H.R. 2, the Agriculture and Nutrition Act of 2018, was introduced by Representative Conaway. Subtitle B of Title VI ("Connecting Rural Americans to High Speed Broadband") would reauthorize the Rural Broadband Access Loan and Loan Guarantee Program and make a number of changes to the RUS rural broadband programs.[72] On April 18, 2018, the House Agriculture Committee approved H.R. 2 (H.Rept. 115-661) with amendments. On June 21, 2018, the House passed H.R. 2.

Senate

On June 11, 2018, the 2018 Senate farm bill, S. 3042, was introduced by Senator Roberts. The Agriculture Improvement Act of 2018 was approved on June 13, 2018, by the Committee on Agriculture, Nutrition, and Forestry and ordered to be reported with an amendment in the nature of a substitute favorably. On June 28, 2018, the Senate passed its version of H.R. 2.

Key Differences between House and Senate Bills

The following are some key differences between the House and Senate bills with respect to the rural broadband loan and loan guarantee program.

Eligible Projects

Under current law, projects eligible for rural broadband loans and loan guarantees can only (with some exceptions) serve areas in which 15% or more of households are unserved or have service levels below the minimum acceptable level of broadband service. Additionally, under current law, an eligible service area can have no more than two incumbent broadband service providers. The House bill does not change the current service area eligibility threshold for rural broadband loans and loan guarantees.

On the other hand, the Senate bill would require that rural broadband loans, loan guarantees, and grants can only serve areas in which 90% or more of households are unserved or have service levels below the minimum acceptable level of broadband service. The Senate bill also provides that an eligible service area can have no more than one incumbent broadband service provider.

---

[72] See House Committee on Agriculture, "Section-by-Section, H.R. 2, the Agriculture and Nutrition Act of 2018," pp. 41-43, available at https://agriculture.house.gov/uploadedfiles/agriculture_and_nutrition_act_of_2018_section_by_section.pdf.

Grant Authority

Both the House and Senate bills add a grant component to the current farm bill broadband loan and loan guarantee program. In the House bill, grants are only available in combination with associated loans under the rural broadband, electric infrastructure, and telecommunications infrastructure loan and loan guarantee programs. Additionally, project areas must serve hard-to- reach communities—specifically areas with a density of less than 12 service points per road mile and where no incumbent provider delivers fixed terrestrial broadband service at or above the minimum broadband speed. The maximum federal share of a total project cost varies by the density of the project service area, ranging from a 25% to 75% federal share.

In the Senate bill, grants are subject to the same service area eligibility criteria as broadband loans and loan guarantees (no less than 90% unserved, no more than one incumbent). The maximum federal share for a grant is 50%, although USDA can adjust the federal share up to 75% if the Secretary determines that the project would serve particularly remote, unserved, and low-income areas.

Definition of Minimum Broadband Service

Both the House and Senate bills set the minimum broadband service speed at 25 Mbps (download)/3 Mbps (upload), to be reviewed by the Secretary at least once every two years. Additionally, the House bill requires USDA to establish projections of minimum acceptable standards of broadband service for 5, 10, 15, 20, and 30 years into the future. Unless cost prohibitive, projects eligible for a rural broadband loan or loan guarantee must provide broadband service at the minimum level, and must be determined capable of meeting future minimum speed standards over the life of the loan or loan guarantee.

Middle Mile Projects

The House bill authorizes RUS to make rural broadband loans or loan guarantees to middle mile infrastructure projects, which are defined as any broadband infrastructure that does not connect directly to end user locations (including anchor institutions) and may include interoffice transport, backhaul, internet connectivity, data centers, or special access transport to rural areas. The Senate bill does not contain a middle mile infrastructure provision.

Reauthorization Levels

For rural broadband loans and loan guarantees, the House bill sets an authorization level of $150 million for each of fiscal years 2019 through 2023. Additionally, the House bill provides $350 million for each of fiscal years 2019 to 2023 for grants to be available in combination with associated loans and loan guarantees.

The Senate bill sets an authorization level for broadband loans, loan guarantees, and grants of $150 million for each of fiscal years 2019 through 2023.

P.L. 115-334

On December 10, 2018, the conference report (H.Rept. 115-1072) accompanying H.R. 2, the Agriculture Improvement Act of 2018, was filed. The conference report was agreed to in the House and Senate on December 11 and December 12 respectively. On December 20, 2018, the President signed the bill (P.L. 115-334).

The following summarizes the major provisions relevant to RUS broadband programs.

Section 6201. Rural Broadband Access Grant, Loan, and Loan Guarantee Program

Adds a grant component to the existing program, which is now authorized to provide grants, loans, loan guarantees, and loan/grant combinations.

- Priority—directs the Secretary to give the highest priority to applications proposing to serve rural communities that do not have any residential broadband service of at least 10 Mbps/1 Mbps. Also receiving high priority are projects that provide the maximum level of broadband service to the greatest proportion of rural households in the proposed service area. Additional priority factors include rural communities with a high percentage of low-income residents, with populations under 10,000, that are experiencing outmigration, that are isolated from other population centers, or that propose to provide broadband for use in various applications of precision agriculture. Projects will also receive priority if they are developed or funded by two or more stakeholders (for example, public-private partnerships).
- Grant Eligibility and Cost-Sharing—projects eligible for grants (including grant/loan combinations) must be carried out in a proposed service territory in which not less than 90% of the households are unserved. Grants shall not exceed 75% of the total project cost to an area with a density fewer than 7 people per square mile, 50% to an area with a density of 7 to 12 people per square mile, and 25% to an area with a density of 12 to 20 people per square mile. However, the Secretary has the authority to adjust the federal share of a grant up to 75% for an area of rural households without any 10 Mbps/1 Mbps broadband service, or rural communities that are under 10,000 in population, with a high percentage of low-income residents, experiencing outmigration, that are isolated from other population centers, or that are proposing broadband deployment for precision agriculture applications. Additionally, the Secretary may make modifications of the density thresholds to ensure that funds are best utilized to provide broadband service in communities that are the most rural in character.

- Loan Eligibility—for broadband loans or loan guarantees, eligible proposed service areas must have not less than 50% of households unserved or below the minimum acceptable level (set at 25 Mbps/3 Mbps) of fixed broadband service, whether terrestrial or wireless. P.L. 115-334 raises the previous eligibility threshold from 15% to 50%. Left unchanged is the eligibility requirement that broadband service cannot be provided in any part of the proposed service territory by three or more incumbent service providers.
- Broadband Buildout Requirements—allows five years for applicants to complete the buildout of a project (up from three years). Requires the Secretary to set a current minimum acceptable standard of broadband service of 25 Mbps/3 Mbps, and to establish projections of minimum acceptable standards of broadband service of a project for 5 to 10 years, 11 to 15 years, 16 to 20 years, and more than 20 years into the future. The Secretary shall review and may adjust those minimum levels at least once every two years. Projects eligible for a rural broadband loan or loan guarantee must provide broadband service at the minimum level, and must be determined capable of meeting future minimum speed standards over the life of the loan or loan guarantee. However, if an applicant shows that it would be cost prohibitive to meet the minimum acceptable level of broadband service for the entirety of a proposed service territory due to its unique characteristics, the Secretary and the applicant may agree to utilize substitute standards for any unserved portion of the project.
- Technical Assistance and Training—the Secretary may provide to eligible applicants technical assistance and training to prepare applications, including required reports and surveys, and to improve financial management relating to the proposed project. Only applicants proposing to serve communities without residential broadband service of at least 10 Mbps/1 Mbps are eligible for technical assistance and training. Not less than 3% and not more than 5% of the annual appropriation for the broadband grant, loan, and loan guarantee program shall be used for technical assistance and training.
- Guaranteed Loan Fees—requires the Secretary to charge lenders of guaranteed loans a fee to offset subsidy costs. Fees shall be in such amounts as to bring down the cost of subsidies for guaranteed loans, but that do not act as a bar to participation in the program.
- Payment Assistance for Certain Loan and Grant Recipients—allows the Secretary to award grant funding—subject to agreed project milestones, objectives, and other considerations—that would allow a loan recipient to receive the benefit of a subsidized loan (with reduced interest rates) or a payment assistance loan.
- Authorization—sets an authorization level of $350 million for each of fiscal years 2019 through 2023 (up from $25 million per year), and delays the

termination of authority to make loans and loan guarantees until September 30, 2023.

Section 6202. Expansion of Middle Mile Infrastructure into Rural Areas

Authorizes $10 million for each of fiscal years 2018 through 2023 for grants, loans, and loan guarantees toward middle mile infrastructure projects. Middle mile infrastructure connects underserved rural areas to the internet backbone; it does not connect directly to end-user locations. A project is eligible if at least 75% of the interconnection points serve eligible rural areas. A grant cannot exceed 20% of the total project cost.

Section 6203. Modifications to the Rural Gigabit Program

Renames the Rural Gigabit Network Pilot Program (which was authorized in the 2014 farm bill but never funded through appropriations) as the Innovative Broadband Advancement Program, which is authorized to provide a grant, a loan, or both to an eligible entity to demonstrate innovative broadband technologies or methods of broadband deployment that significantly decrease the cost of deployment and provide substantially faster broadband speeds than are available in a rural area. The program is authorized at $10 million for each of fiscal years 2018 through 2023.

Section 6204. Community Connect Grant Program

Codifies the existing Community Connect Grant Program and authorizes the program at $50 million for each of fiscal years 2018 through 2023. Defines an eligible service area as having broadband service capacity less than speeds of 10 Mbps download and 1 Mbps upload.

Section 6205. Outdated Broadband Systems

Requires the Secretary, beginning on October 1, 2020, to consider any portion of a service territory subject to an outstanding grant agreement as unserved for the purposes of broadband loan programs if broadband service is not provided at a minimum of 10 Mbps/1 Mbps, unless the broadband provider has begun or already constructed broadband facilities in that area which would meet the minimum acceptable broadband service standard.

Section 6206. Default and Deobligation; Deferral

Requires the Secretary to establish written procedures for all broadband programs to recover funds from loan and grant defaults, deobligate awards that demonstrate an insufficient level of performance or fraudulent spending, award those funds on a competitive basis to new or existing applicants, and minimize overlap among programs. The Secretary may establish a deferral period of not shorter than the buildout period established for the project in order to support the financial feasibility and long-term sustainability of the project.

Section 6207. Public Notice, Assessments, and Reporting Requirements

- Public Notice—requires the Secretary to make available to the public a fully searchable database on the RUS website that contains information on all broadband projects provided assistance or for which assistance is sought.
- Service Area Assessment—after giving public notice for a particular project seeking assistance, the Secretary shall provide 45 days for providers to voluntarily submit information indicating their presence in a proposed service area. If no existing provider submits such information, the Secretary may collect or obtain through reasonable efforts any other data on existing providers. In the case of applications requesting funding for unserved rural areas, the Secretary shall confirm unserved rural areas by conferring with the FCC and NTIA, reviewing any other source relevant to service data validation, and performing site-specific testing to verify the unavailability of any retail broadband service.
- Reporting—the Secretary shall require entities receiving assistance to provide an annual report for three years after completion of the project that describes the use by the entity of the assistance and the progress toward fulfilling the objectives of the project. Middle mile project recipients are required to submit a semiannual report for five years after project completion. The recipient of assistance shall also provide complete, reliable, and precise geolocation information that indicates the location of new broadband service that is being provided. The Secretary is also required to submit an annual report to Congress that describes the extent of participation in the RUS broadband assistance programs for the preceding fiscal year.

Section 6208. Environmental Reviews

The Secretary may obligate, but not disperse, funds before the completion of otherwise required environmental, historical, or other types of reviews if the Secretary determines that a subsequent site-specific review shall be adequate and easily accomplished for the location of towers, poles, or other broadband facilities in the service area of the borrower without compromising the project or the required reviews.

Section 6209. Use of Loan Proceeds to Refinance Loans for Deployment of Broadband Service

The proceeds of any loan or loan guarantee may be used by the recipient for the purpose of refinancing an outstanding obligation on another telecommunications loan.

Section 6210. Smart Utility Authority for Broadband

Allows a recipient of grants, loans, or loan guarantees provided by the Office of Rural Development to use not more than 10% of the amount for rural broadband infrastructure projects, including both retail and nonretail activities, except for a recipient who is seeking to provide retail broadband service in any area where such service is available at the minimum broadband speeds. Additionally allows a recipient of electric grants, loans, or loan guarantees to set aside not more than 10% of the amount for retail broadband service, for use only in an area that is not being provided with the minimum acceptable level of broadband service. The funding cannot result in competitive harm to any existing grant, loan, or loan guarantee under the Rural Electrification Act of 1936.

Section 6211. Refinancing of Telephone Loans

Clarifies that the Secretary, through the RUS telephone loan program, may refinance loans of persons furnishing telephone service in rural areas, including indebtedness of recipients on another telecommunications loan made under the Rural Electrification Act. Also strikes the current law limitation that the refinancing may not constitute more than 40% of the loan.

Section 6212. Federal Broadband Coordination

- Consultation between USDA and NTIA—USDA shall consult with NTIA to assist in the verification of eligibility for USDA broadband programs. To this end, NTIA shall make available its broadband assessment and mapping capabilities.
- Consultation between USDA and FCC—USDA shall consult with the FCC before providing broadband assistance for a project to serve an area with respect to which another entity is receiving Connect America Fund or Mobility Fund support. The FCC shall consult with USDA before offering Connect America Fund or Mobility Fund support to serve an area with respect to which another entity has received RUS broadband assistance.
- Report to Congress—USDA, the FCC, and NTIA shall submit to Congress a report on how best to coordinate federally supported broadband programs and activities in order to achieve various objectives regarding long-term broadband service needs of rural residents.

Section 6213. Transition Rule

Provides that for one year after enactment, the Secretary shall use the previously existing rules and regulations for the broadband loan and Community Connect grant program until a final rule is issued.

Section 6214. Rural Broadband Integration Working Group

Establishes an interagency Rural Broadband Integration Working Group that shall consult with a wide spectrum of stakeholders to identify, assess, and determine possible actions relating to barriers and opportunities for broadband deployment in rural areas. Not later than 60 days after enactment, the Working Group shall publish a comprehensive survey of federal programs that currently support or could reasonably be modified to support broadband deployment and adoption; and all federal agency policies and rules with the direct or indirect effect of facilitating or regulating investment in, or deployment of, wired and wireless broadband networks. The Working Group will submit to the President a list of actions that federal agencies can take to support broadband deployment and adoption, including timelines to complete a list of priority actions and rulemakings.

Other Broadband-Related Provisions

- Section 6101 sets aside 20% of DLT grant funding for applications related to substance use disorder treatment services;
- Section 6102 reauthorizes the DLT program through FY2023 at $82 million per year;
- Section 6418 requires the Secretary to collect fees on loan guarantees in amounts that when combined with any appropriated funds equal the subsidy on such guarantees. The Secretary shall charge and collect from the lender fees in such amounts as to bring down the costs of subsidies for the guaranteed loan, except that the fees shall not act as a bar to participation in the program nor be inconsistent with current practices in the marketplace; and
- Section 12511 establishes the Task Force for Reviewing the Connectivity and Technology Needs of Precision Agriculture in the United States. The Task Force will develop policy recommendations to promote deployment of broadband on unserved agricultural land, with a goal of achieving reliable capabilities on 95% of agricultural land in the United States by 2025.

OTHER LEGISLATION IN THE 115TH CONGRESS

Aside from the 2018 farm bills and annual appropriations legislation, the following bills were introduced into the 115th Congress seeking to impact the RUS broadband programs:

- H.R. 800 (Huffman), introduced on February 1, 2017, as the New Deal Rural Broadband Act of 2017, would establish an Office of Rural Broadband within

of 2018, would amend the Rural Electrification Act of 1936 to include a satellite project or technology within the definition of broadband service. Referred to the Committee on Agriculture, Nutrition, and Forestry.

- S. 3360 (Wyden), introduced August 21, 2018, as the Broadband Internet for Small Ports Act, would establish priority for small harbors to receive RUS broadband funding. Referred to the Committee on Agriculture, Nutrition, and Forestry.

APPENDIX. RURAL DEVELOPMENT TELECOM AWARDS

Table A-1. Rural development telecom awards, FY2009-FY2016 ($ millions)

	2009-2014		2015		2016		TOTAL	
	Amount	# Awards	Amount	# Awards	Amount	# Awards	Amount	# Awards
AL	60.4	19	1.3	3	0.45	2	62.1	24
AK	149.4	40	3.8	7	1.4	4	154.6	51
AZ	46.4	25	0	0	0.21	1	46.6	26
AR	279.8	45	25.7	3	10.3	7	315.8	55
CA	46.2	27	0.46	2	2.1	7	48.7	36
CO	85.8	29	0.26	1	0.18	1	86.2	31
FL	49.4	6	0.35	2	0	0	49.8	8
GA	112.4	20	0.90	2	1.5	6	114.8	28
HI	1.7	5	0	0	0	0	1.7	5
ID	51.3	26	0.48	1	12.7	1	64.5	28
IL	155.8	33	0.15	1	0.62	2	156.6	36
IN	128.2	14	0	0	0.55	2	128.7	16
IA	290.2	53	40.9	6	15.2	2	346.3	61
KS	373.1	47	0.14	1	26.5	3	399.7	51
KY	321.8	42	0.56	3	0.97	3	323.4	48
LA	54.8	16	0	0	1.4	3	56.3	19
ME	23.4	44	2.7	7	1.6	5	27.8	56
MD	74.4	31	0.10	1	0	0	74.5	32
MA	0.50	1	0	0	0.64	2	1.1	3
MI	126.8	24	0.98	3	0.54	2	128.3	29
MN	327.4	63	30.9	4	7.8	7	366.1	74
MS	50.3	25	1.7	3	2.0	4	53.9	32
MO	214.8	56	0.67	3	0	0	215.4	59
MT	359.8	25	30.3	3	30.7	5	420.8	33
NE	153.2	34	4.2	2	0.31	2	157.7	38
NV	20.1	10	0.38	1	1.4	3	21.9	14
NH	3.9	8	0.92	2	5.5	1	10.3	11
NJ	0.32	2	0	0	0	0	0.32	2
NM	191.5	34	5.8	2	14.4	3	211.7	39
NY	64.8	42	0.48	2	0.22	1	65.5	45
NC	153.6	30	0	0	1.3	5	154.8	35
ND	376.6	39	60.3	5	55.3	2	492.3	46
OH	46.0	33	0.25	1	0.39	2	46.7	36

Table A-1. (Continued)

	2009-2014		2015		2016		TOTAL	
	Amount	# Awards	Amount	# Awards	Amount	# Awards	Amount	# Awards
OK	308.6	74	22.9	4	4.5	6	336.1	84
OR	75.3	34	0.20	1	0.99	1	75.6	36
PA	4.4	18	0	0	1.0	4	5.4	22
PR	1.7	4	0	0	0	0	1.7	4
SC	205.4	21	14.0	6	1.3	4	220.7	31
SD	133.5	25	1.1	4	17.6	5	152.2	34
TN	235.9	38	0.63	2	3.5	8	240.1	48
TX	326.5	61	13.4	1	14.7	2	354.6	64
UT	22.2	15	0.33	2	0.30	1	22.8	18
VT	124.8	13	0	0	0	0	124.8	13
VA	103.8	28	3.4	2	2.4	3	109.6	33
WA	149.4	32	0.38	1	0.46	1	150.2	34
WV	48.3	19	0	0	1.2	3	49.5	22
WI	336.9	72	9.3	7	1.1	4	347.3	83
WY	54.1	10	0	0	0	0	54.1	10
VI	0.75	1	0	0	0	0	0.75	1
Western Pacific	122.8	7	0.49	1	0	0	123.3	8
TOTAL	6,648.7	1,420	280.9	102	244.2	130	7,174.0	1,652

Source: USDA, Rural Development, *2016 Progress Report*.

Notes: Includes four RUS loan and grant programs: Rural Broadband Loan and Loan Guarantee Program, Telecommunications Infrastructure Loan and Loan Guarantee Program, Community Connect Grant Program, and Distance Learning and Telemedicine Grant Program. Data for 2009-2014 includes the 2009-2010 Broadband Initiatives Program funds appropriated by the American Recovery and Reinvestment Act (ARRA). According to the *Rural Development 2017 Performance Report*, RUS awarded 107 telecom loans and grants in FY2017, totaling $502 million. The *2017 Performance Report* does not break down awards by state.

In: Broadband
Editor: Orlando Harrison
ISBN: 978-1-53616-409-1

Chapter 7

BROADBAND INTERNET: FCC'S DATA OVERSTATE ACCESS ON TRIBAL LANDS*

United States Government Accountability Office

ABBREVIATIONS

DSL	digital subscriber line
FCC	Federal Communications Commission
GPRA	Government Performance and Results Act
GPRAMA	GPRA Modernization Act of 2010
LTE	long-term evolution
Mbps	megabits per second
NCAI	National Congress of American Indians
NTIA	National Telecommunications & Information Administration
ONAP	Office of Native Affairs and Policy
Recovery Act	American Recovery and Reinvestment Act of 2009
USDA	United States Department of Agriculture

WHY GAO DID THIS STUDY

Broadband furthers economic development, educational attainment, and public health and safety; however, residents of tribal lands have lower levels of broadband access

* This is an edited, reformatted and augmented version of United States Government Accountability Office; Report to Congressional Requesters, Publication No. GAO-18-630, dated September 2018.

relative to the U.S. population. Congress has prioritized identifying and targeting funds to unserved areas. FCC uses data from broadband providers to develop maps and reports depicting broadband availability in the United States, with specific information on tribal lands. GAO was asked to review FCC's efforts to collect broadband data for tribal lands.

This chapter examines the extent to which: (1) FCC's approach to collecting broadband data accurately captures broadband access on tribal lands and (2) FCC obtains tribal input on the data. GAO interviewed stakeholders from 25 tribal governments or tribally owned providers, and visited nine tribal lands. The selected tribes varied geographically and in levels of broadband availability, among other characteristics. GAO also reviewed FCC's rulemakings on broadband data and interviewed other tribal stakeholders, FCC officials, and 13 non-tribal broadband providers selected to include a diversity of technologies. Provider and tribal interviews were based on non- generalizable samples.

WHAT GAO RECOMMENDS

GAO is making three recommendations to FCC, including that it collect and report data that accurately measure tribal broadband access as well as develop a process to obtain tribal input on the accuracy of the data. FCC agreed with the recommendations.

WHAT GAO FOUND

The Federal Communications Commission (FCC) collects data on broadband availability from providers, but these data do not accurately or completely capture broadband access on tribal lands. Specifically, FCC collects data on broadband availability; these data capture where providers may have broadband infrastructure. However, FCC considers broadband to be "available" for an entire census block if the provider *could* serve *at least one* location in the census block. This leads to overstatements of service for specific locations like tribal lands (see figure). FCC, tribal stakeholders, and providers have noted that this approach leads to overstatements of broadband availability. Because FCC uses these data to measure broadband access, it also overstates broadband access—the ability to obtain service—on tribal lands.

Additionally, FCC does not collect information on several factors—such as affordability, quality, and denials of service—that FCC and tribal stakeholders stated can affect the extent to which Americans living on tribal lands can access broadband services. FCC provides broadband funding for unserved areas based on its broadband data. Overstatements of access limit FCC's and tribal stakeholders' abilities to target broadband funding to such areas. For example, some tribal officials stated that inaccurate data have affected their ability to plan their own broadband networks and obtain funding to address broadband gaps on their lands. By developing and implementing methods for collecting and reporting accurate and complete data on broadband access specific to tribal lands, FCC would be better able to target federal broadband funding to tribal areas that need it the most and to more accurately assess FCC's progress toward its goal of increasing all Americans' access to affordable broadband.

FCC does not have a formal process to obtain tribal input on the accuracy of provider-submitted broadband data. In the *National Broadband Plan*, FCC highlighted the need for a targeted approach to improve broadband availability data for tribal lands. As outlined in the plan, such an approach would include working with tribes to ensure that information is accurate and useful. About half of the tribal stakeholders GAO interviewed raised concerns that FCC relies solely on data from providers, and most stated FCC should work with tribes to improve the accuracy of FCC's data. Establishing a formal process to obtain input from tribal governments on the accuracy of provider-submitted broadband data could help improve the accuracy of FCC's broadband data for tribal lands.

September 7, 2018

The Honorable John Hoeven
Chairman

The Honorable Tom Udall
Vice Chairman
Committee on Indian Affairs
United States Senate

The Honorable John Barrasso

The Honorable Maria Cantwell

The Honorable Steve Daines

The Honorable Martin Heinrich

The Honorable Heidi Heitkamp

The Honorable Brian Schatz

The Honorable Jon Tester
United States Senate

Broadband infrastructure is critical for economic development, educational and job opportunities, and public health and safety. In 2016, we reported that tribal lands are generally in remote and rugged areas and that broadband access can help residents develop online businesses, access telemedicine services, and use online educational tools.[1] However, residents of tribal lands have lower levels of broadband access than residents of non-tribal lands; a reflection of what is often called the "digital divide," or disparate levels of broadband access among different socioeconomic, racial, or rural groups. According to the Federal Communications Commission (FCC), as of December 2016, 35.4 percent of Americans residing on tribal lands lacked access to fixed broadband services, compared to 7.7 percent of all Americans.[2] FCC has reported that the lack of service in tribal lands presents impediments to the efforts of tribal nations related to self-governance, economic opportunity, education, public safety, and cultural preservation.[3]

One barrier to increasing access to broadband on tribal lands is the cost of deploying infrastructure to tribal lands located in rugged, sparsely populated areas. In an attempt to address this and other issues, the federal and some state governments have administered a number of programs to incentivize companies to build broadband infrastructure in unserved and underserved areas. In addition, policy makers have noted the need for accurate information in order to target these programs to the areas lacking access, and FCC has identified the need to work with tribes to ensure such data are accurate for tribal lands. However, in 2016 we reported that tribal and federal officials had concerns that the federal map of broadband availability at the time (the National Broadband Map) did not accurately depict broadband availability on tribal lands.[4]

---

[1] GAO, *Telecommunications: Additional Coordination and Performance Measurement Needed for High-Speed Internet Access Programs on Tribal Lands*, GAO-16-222 (Washington, D.C.: Jan. 29, 2016).

[2] For the purposes of this chapter, to determine which high-speed internet services qualify as "broadband," we are using the threshold for "advanced telecommunications capability" as used by FCC in its 2018 *In the Matter of Inquiry Concerning the Deployment of Advanced Telecommunications Capability to All Americans in a Reasonable and Timely Fashion,* 2018 Broadband Deployment Report, 33 FCC Rcd 1660 (2018) (Broadband Deployment Report). Therefore, the estimate of broadband access above refers to services capable of providing speeds of 25 megabits per second (Mbps) download and 3 Mbps upload. Megabits per second is a measure of the network's data transfer rate (speed) and refers to the number of bits per second that travel to a user's device (the download speed) and from a user's device (the upload speed). The estimate also does not include satellite services, as FCC currently reports on these services separately and until recently, satellite providers were not capable of providing broadband speeds, according to FCC officials. For the purposes of this analysis, mobile broadband refers to long-term evolution (LTE) services. LTE is an industry standard that is part of the fourth generation of wireless telecommunications technology, which is currently in common use.

[3] *Improving Communications Services for Native Nations,* Notice of Inquiry, 26 FCC Rcd 2672, 2673 (2011).

[4] GAO-16-222.

The federal government has not updated the National Broadband Map since April 2015, with the last update containing data as of June 30, 2014.[5] Currently, the primary source of information regarding where broadband is and is not available is the FCC, which collects this information from broadband providers. You asked us to review FCC's efforts to collect broadband data for tribal lands. This chapter examines:

- the extent to which FCC's approach to collecting broadband availability data accurately captures the ability of Americans living on tribal lands to access broadband Internet services, and
- the extent to which FCC obtains tribal input on the accuracy of provider-submitted broadband data for tribal lands.

To address both objectives, we analyzed FCC's December 2016—the most recent data at the time of our review—fixed and mobile broadband- availability data for federally recognized tribal lands.[6] Providers currently report this information to FCC by filing a Form 477, twice a year. We also used 2010 U.S. Census data to identify census blocks on tribal lands. To assess the reliability of FCC and Census data, we reviewed a previous GAO reliability assessment and performed additional work, such as electronic testing of the data and interviews with agency officials.[7] Based on the results of our analysis, we determined the data to be reliable for our purposes, which were to (1) inform our selection of tribal governments and providers for interviews and visits and (2) develop maps of fixed and mobile broadband availability for the 9 tribal lands we selected for visits, in order to obtain tribal representatives' feedback on the accuracy of the data. For both objectives, we also reviewed FCC documents regarding the Form 477 process and interviewed FCC officials as well as stakeholders representing tribal governments and broadband providers. These interviews included representatives from 25 tribal governments or tribally owned providers, including visits to 9 tribal lands. When we selected these tribes, we considered variation in location, level of broadband deployment according to FCC, land mass, and population size and density. The results of

---

[5] However, Congress recently provided $7.5 million to the National Telecommunications and Information Administration (NTIA) to update the National Broadband Map in conjunction with FCC and the states. Consolidated Appropriations Act, 2018, Pub. L. No. 115-141, 132 Stat. 348 (2018). In addition, as discussed later, FCC began collecting and creating maps of its own broadband data.

[6] We defined federally recognized tribal lands consistent with FCC's definition in its 2018 Broadband Deployment Report. Specifically, we considered tribal lands to be: (1) Joint Use Areas; (2) legal federally recognized American Indian area consisting of reservation and associated off-reservation trust land; (3) legal federally recognized American Indian area consisting of reservation only; (4) legal federally recognized American Indian area consisting of off-reservation trust land only; (5) Statistical American Indian area defined for a federally recognized tribe that does not have reservation or off-reservation trust land, specifically a Tribal Designated Statistical Area (TDSA) or Oklahoma Tribal Statistical Area (OTSA); (6) Alaskan Native village statistical area; and (7) Hawaiian Home Lands established by the Hawaiian Homes Commission Act of 1921. *See* 33 FCC Rcd 1660 (2018).

[7] We reviewed the data reliability assessment from GAO, *Broadband: Additional Stakeholder Input Could Inform FCC Actions to Promote Competition*, GAO-17-742 (Washington, D.C.: Sept.19, 2017).

our interviews are not generalizable to all tribal governments or tribally owned broadband providers. In addition to tribal governments and tribally owned providers, we interviewed six organizations that include tribal entities and four stakeholders who work with tribes on broadband issues. For reporting purposes, we developed the following series of indefinite quantifiers to describe the tribal responses from the 35 entities representing tribal stakeholders we interviewed.

- 3 to 7 is defined as "a few."
- 8 to 15 is described as "some,"
- 16 to 20 is described as "about half,"
- 21 to 27 is described as "most"; and
- 28 to 34 is described as "almost all."

Further, to obtain industry perspectives, we reviewed public comments submitted by providers and industry associations in relevant FCC rulemaking proceedings and interviewed 10 non-tribally owned fixed and mobile broadband providers and three industry associations. We selected providers to reflect a range of carrier size, as well as the technologies used to provide broadband service. In addition, we interviewed representatives from other government entities, as well as private companies that collect and report broadband data. The results of these interviews are not generalizable. A full list of the stakeholders we interviewed can be found in appendix I.

In addition, to identify the extent to which FCC's approach to collecting broadband data accurately captures Americans' ability to access broadband Internet services on tribal lands, we identified factors that affect broadband access by interviewing tribal stakeholders, as described above, and reviewing FCC documents and previous GAO work.[8] We also reviewed relevant statutes and FCC's proceedings, plans, and broadband deployment and progress reports.[9] We compared the Form 477 process to FCC's strategic goals and to factors affecting broadband access to determine the extent to which the Form 477 collects information on those factors and aligns with FCC's goals. We further evaluated this information against the *Government Performance and Results Act (GPRA),*

[8] GAO-16-222.

[9] *Modernizing the FCC Form 477 Data Program*, Further Notice of Proposed Rulemaking, 32 FCC Rcd 6329 (2017); *Connect America Fund Universal Service Reform – Mobility Fund*, Order on Reconsideration and Second Report and Order, 32 FCC Rcd 6282 (2017); *Instructions for Filing 4G LTE Coverage Data to Determine Areas Presumptively Eligible for Mobility Fund II Support*, Public Notice, 32 FCC Rcd 7023 (2017). FCC, *Strategic Plan 2018-2022* (Washington, D.C.); FCC, *Connecting America: The National Broadband Plan*; 33 FCC Rcd 1660 (2018); *In the Matter of Inquiry Concerning the Deployment of Advanced Telecommunications Capability to All Americans in a Reasonable and Timely Fashion*, 2016 Broadband Progress Report, 31 FCC Rcd 699 (2016) "Broadband Progress Report."

as enhanced by the GPRA Modernization Act of 2010 (GPRAMA) and *Standards for Internal Control in the Federal Government*.[10]

To determine the extent to which FCC obtains tribal input on the accuracy of provider-submitted broadband data for tribal lands, we reviewed FCC's policies for working with tribal governments and interviewed tribal stakeholders, among other entities.[11] We compared this information to recommendations from FCC's *National Broadband Plan*, and *Standards for Internal Control in the Federal Government*.[12] For additional details on our scope and methodology, see appendix II.

We conducted this performance audit from June 2017 to September 2018 in accordance with generally accepted government auditing standards. Those standards require that we plan and perform the audit to obtain sufficient, appropriate evidence to provide a reasonable basis for our findings and conclusions based on our audit objectives. We believe that the evidence obtained provides a reasonable basis for our findings and conclusions based on our audit objectives.

BACKGROUND

The federal government has recognized 573 Indian tribes as distinct, independent political communities with tribal sovereignty. There are different categories of tribal lands, with differing implications with respect to ownership and administration. Reservations are defined geographic areas with established boundaries recognized by the United States.[13] Tribal lands vary in size, demographics, and location. For example, those lands smallest in size are less than one square mile, and the largest, the Navajo Nation, is more than 24,000 square miles (the size of West Virginia). Tribal land locations can

---

[10] Government Performance and Results Act, Pub. L. No. 103-62, 107 Stat. 285 (1993), as enhanced by GPRA Modernization Act of 2010, Pub. L. No. 111-352, 124 Stat. 3866 (2011); GAO, *Standards for Internal Control in the Federal Government*, GAO-14-704G (Washington, D.C.: September 2014).

[11] *Statement of Policy on Establishing a Government-to-Government Relationship with Indian Tribes*, Policy Statement, 16 FCC Rcd 4078 (2000); 47 C.F.R. § 54.313 (a)(5); *Further Guidance on Tribal Government Engagement Obligation Provisions of the Connect America Fund*, Public Notice, 27 FCC Rcd 8176 (2012).

[12] *Connecting America: The National Broadband Plan*, Notice of Inquiry and Notice of Proposed Rulemaking, 25 FCC Rcd 6657 (2010) and GAO, *Standards for Internal Control in the Federal Government*, GAO-14-704G (Washington, D.C.: September 2014).

[13] The land within the reservation's boundaries may include a mixture (or checkerboard) of tribal, individual Indian, and non-Indian land. Tribal and individual Indian land may be held in trust, restricted, or fee status. The allotment and assimilation period, which began with The General Allotment Act in 1887 (also known as the Dawes Act) included a number of federal efforts to divide tribal lands into individual parcels, give each tribal member a parcel, and sell the "surplus' parcels to non-Indians. In some cases, the United States government still holds individual allotments in trust, while others have transferred to private (Indian and non-Indian) ownership. In addition, restricted status, or restricted fee lands, are lands for which the title to the land is held by an individual Indian person or a tribe and "which can only be alienated or encumbered by the owner with the approval of the Secretary of the Interior because of limitations contained in the conveyance instrument pursuant to federal law." In addition, some tribes have purchased land within and outside of their reservation's boundaries.

range from extremely remote, rural locations to urban areas. Figure 1 shows tribal lands in the United States according to the 2010 Census.

The term "broadband" commonly refers to Internet access that is high speed and provides an "always-on" connection, so users do not have to reestablish a connection each time they access the Internet. Broadband service may be "fixed"—that is, providing service to a single location, such as a customer's home—or "mobile," that is, providing service wherever a customer has access to a mobile wireless network, including while on the move, through a mobile device, such as a smartphone. Fixed and mobile broadband providers deploy and maintain infrastructure to connect consumers to the Internet.

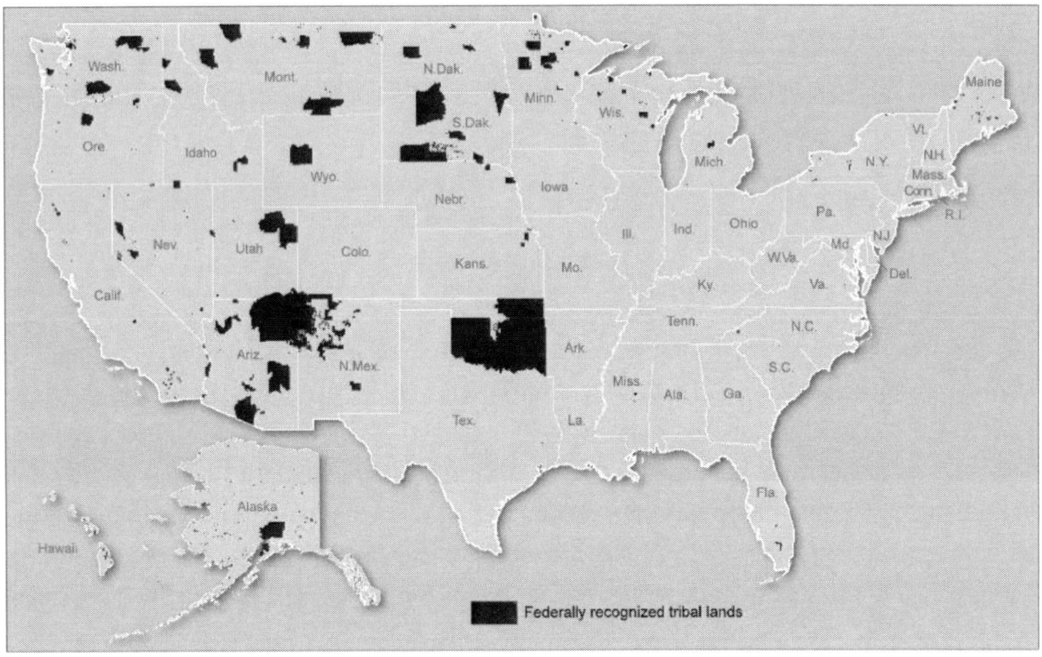

Source: GAO based on U.S. Census Bureau data and the Federal Communications Commission's definition of tribal lands. | GAO-18-630

Figure 1. Map of Federally Recognized Tribal Lands.

Providers offer fixed Internet service through a number of technologies, such as copper phone lines, fiber-optic lines, coaxial cables, wireless antennas, satellites,[14] or a mix of technologies (see Figure 2). To install fixed or wireless infrastructure, providers must obtain permits from government entities with jurisdiction over the land or permission from public utilities to deploy infrastructure on existing utility poles.

[14] We did not include satellite broadband in our assessment of broadband availability on tribal lands because FCC concluded in its 2016 Broadband Progress Report that this type of service had not yet reached FCC's speed benchmark of 25 Mbps/3 Mbps. In the 2018 Broadband Deployment Report, FCC noted that some satellite services were reporting providing speeds of 25 Mbps/3 Mbps, but did not include these services in its data tables depicting broadband deployment on tribal lands.

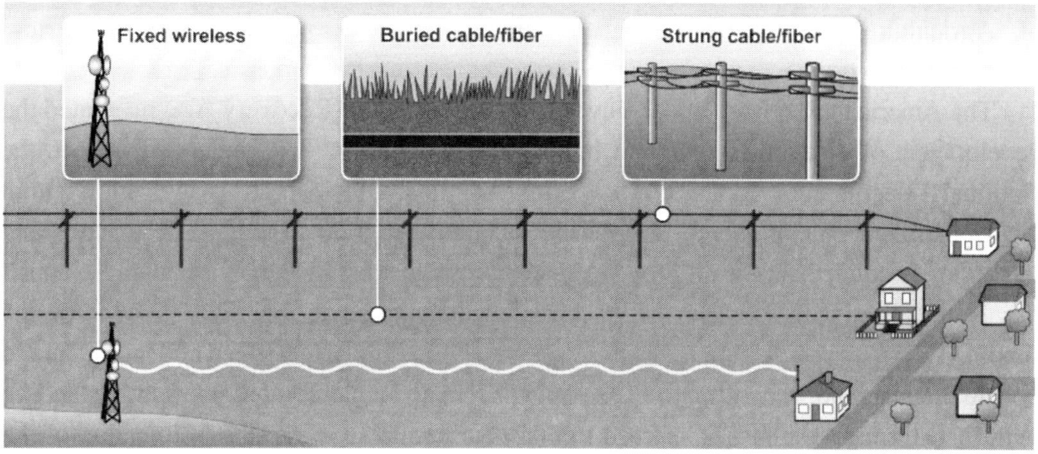

Source: GAO. | GAO-18-630

Figure 2. Examples of Fixed Broadband Technologies.

The federal government has emphasized the importance of ensuring Americans have access to broadband, and a number of agencies, including FCC, currently provide funding to subsidize broadband deployment in areas in which the return on investment has not attracted private investment. The Communications Act of 1934, as amended by the Telecommunications Act of 1996, specifies that consumers in "rural, insular, and high-cost areas" should have access to telecommunication services and rates that are "reasonably comparable" to consumers in urban areas.[15] To achieve this goal, FCC administers the High-Cost program, which provides subsidies to providers of phone service in rural, insular, and other remote areas.

In 2011, FCC launched a series of reforms to its High-Cost program, including adding support for broadband services, and created the Connect America Fund, which provides subsidies to fixed and mobile providers of telecommunications and broadband services in rural, insular, and other remote areas where the costs of providing service is high. To be eligible for Universal Service Fund support from FCC, a provider must be designated an Eligible Telecommunications Carrier by the appropriate state or by FCC and must meet certain service obligations.[16] The Connect America Fund has distributed approximately $4.5 billion per year, and has separate funding mechanisms targeted to specific goals. For example, there are funds for fixed-phone and broadband service and funds for mobile service, including a Tribal Mobility Fund (Phase 1) that awarded nearly $50 million in 2014 for the provision of 3G and 4G service to unserved tribal areas.

In addition to FCC, a number of other agencies provide funding for broadband deployment in unserved or underserved areas. For example, the United States Department

[15] 47 U.S.C. § 254(b)(3).

[16] 47 C.F.R. § 54.201(a)(1). The Universal Service Fund is paid for by contributions from providers of telecommunications based of an assessment on their interstate and international end-user revenues.

In February 2018, FCC launched an updated map of fixed broadband deployment (https://broadbandmap.fcc.gov/#/). This map allows users to search for broadband deployment by address and provides summary-level statistics regarding broadband deployment in specific tribal lands (see Figure 3). According to FCC officials, this new map format will support more frequent data updates.

FCC also provides national maps of mobile LTE coverage; these maps do not allow users to access data at the same level of granularity as the maps of fixed broadband (see Figure 4).[22]

FCC Collects Data on Broadband Availability but Lacks Accurate and Complete Data on Broadband Access on Tribal Lands

FCC collects and uses data that capture broadband availability to measure broadband access on tribal lands, leading to overstatements of broadband access on tribal lands.[23] Specifically, FCC's method of collecting mobile and fixed broadband data from providers (the Form 477) does not accurately or completely capture broadband access on tribal lands because it (1) captures nationwide broadband *availability* data— areas where providers may have broadband infrastructure—but does so in a way that leads to overstatements of availability, and (2) does not capture information on factors that FCC and tribal stakeholders have stated can affect broadband *access* on tribal lands, such as affordability, service quality, and denials of service.[24] Nonetheless, FCC uses its Form 477 broadband availability data in annual broadband deployment reports to measure the percentage of Americans living on tribal lands with or without access to broadband, and to measure progress toward FCC's strategic goal of increasing all Americans' access to affordable broadband.[25] By using broadband *availability* data to measure broadband *access* on tribal lands, FCC overstates broadband access on tribal lands.

---

[22] See, for example: FCC, *Nationwide LTE Coverage – YE 2016,* (accessed May 30, 2018), https://www.fcc.gov/reports-research/maps/nationwide-lte-coverage-ye-2016/; and FCC, *LTE Coverage by Number of Providers – YE* 2016, (accessed May 30, 2018), https://www.fcc.gov/reports-research/maps/lte-coverage-number-providers-ye-2016/.

[23] We use the term broadband availability to refer to broadband deployment. FCC officials noted that the data collected by the Form 477 reflects broadband deployment. We use the term broadband availability because FCC's Form 477 instructs fixed broadband providers to report fixed broadband deployment by submitting a list of census blocks in which the filer makes broadband connections available.

[24] FCC officials we interviewed stated that FCC has not defined the term "broadband access," and noted that the use of the term may vary across FCC documents. However, FCC and tribal stakeholders have noted that broadband access can be affected by factors such as the affordability and quality of the broadband services being offered and the extent to which providers deny service to those who request it. For example, see 2016 Broadband Progress Report 31 FCC Rcd 699 ¶ 62 (2016); FCC, *National Broadband Plan*; FCC, *Strategic Plan 2018-2022*. This is discussed in further detail below. FCC officials also identified the cost of deployment and regulatory barriers as important factors when determining whether an area has access to broadband.

[25] *See* 33 FCC Rcd 1660 ¶ 2 (2018). See also, for example, Appendix G – Americans (Thousands) Living on Tribal Lands with Access to Fixed Terrestrial 25 Mbps/3 Mbps Services and/or Mobile LTE with a Minimum

FCC Collects Broadband Availability Data, but Its Collection Method Leads to Overstatement of Availability on Tribal Lands

FCC's Form 477, its primary method of collecting nationwide broadband data, collects information on broadband availability, which identifies where providers have broadband infrastructure and could potentially provide broadband services but not where consumers can actually access those services. Moreover, the Form 477's mobile broadband data-collection methods are not standardized, and its fixed broadband data-collection methods are not sufficiently granular to provide information about broadband availability on tribal lands.

Mobile Broadband Data Collection

FCC's Form 477 requires mobile broadband providers to report their coverage areas by submitting geospatial data depicting the areas in which consumers could expect to receive the minimum advertised speed.[26] FCC has previously noted the importance of collecting nationally standardized, uniform broadband data from providers to assess broadband availability and allow for easy comparison across providers. However, the Form 477 does not require that providers use a standardized method with defined technical parameters (such as signal strength, or amount of interference) when determining their coverage area, resulting in data that cannot be meaningfully compared across providers, according to FCC. To map their coverage areas, providers may use predictive models based on different measurement methods and a variety of factors known to affect mobile broadband service such as topography, tree cover, and buildings, among other factors.

Providers and tribal stakeholders have expressed concern with the accuracy of FCC's mobile broadband data, and FCC has acknowledged concerns that the lack of a standardized method resulted in data that were unreliable for the purposes of determining mobile broadband coverage for specific geographic areas, such as tribal lands. About half of the tribal government representatives we interviewed told us that they believe FCC's data overstate mobile LTE broadband availability on their lands. For example, a few representatives expressed concerns with the accuracy of the mobile data in areas with varied terrain, such as mountains and valleys. In comments to FCC, broadband providers have also raised concerns regarding the accuracy of the mobile coverage data generated by the Form 477 for the purposes of identifying areas eligible for funding through FCC's

Advertised Speed of 5 Mbps/1 Mbps by State. Prior to 2018, FCC referred to these reports as "Broadband Progress Reports." *Connecting America: The National Broadband Plan*, Notice of Inquiry and Notice of Proposed Rulemaking, 25 FCC Rcd 6657 (2010).

[26] Mobile providers also submit a list of all census tracts in which the providers' service is advertised and available to actual and potential subscribers. FCC's requirement only applies to facilities-based mobile broadband providers, which are providers that provide services using their own network facilities and spectrum for which they hold a license, manage, or have obtained the right to use via a spectrum leasing arrangement. This would not include mobile voice service resellers.

Mobility Fund Phase II program, which provides federal funding to increase mobile broadband services in unserved areas.[27] In 2017, in response to such concerns, FCC reversed its prior decision to use the Form 477 data to identify specific areas eligible for federal funding through the Mobility Fund Phase II program.[28] Instead, FCC undertook a one-time special data collection, for which it required providers to measure their coverage based on a common set of standards, in order to better identify unserved areas that would be presumptively eligible for funding.[29] FCC plans to allow parties, including tribal governments, to challenge the data where they believe the data overstate mobile broadband coverage through August, 2018.[30] Additionally, in an August 2017 Notice of Proposed Rulemaking, FCC requested comment on potential changes to modernize its Form 477 data collection, including whether it should require all providers to use a standardized method when submitting mobile coverage data on the form. FCC officials told us that they do not have a timeline for the development of a final rule, and as of August 2018, FCC had not yet issued a final rule on modernizing the Form 477.

Fixed Broadband Data Collection

The Form 477 collects fixed broadband data that are not sufficiently granular to accurately depict broadband availability on tribal lands. Specifically, FCC directs fixed broadband providers to submit a list of census blocks where service is available on the Form 477. FCC defines "available" as whether the provider does—*or could, within a typical service interval or without an extraordinary commitment of resources*— provide service to *at least one* end-user premises in a census block.[31] Thus, in its annual reports and maps of fixed broadband service, FCC considers an entire block to be served if a provider reports that it does, or *could* offer, service to *at least one* household in the census block. FCC does not define a typical service interval or an extraordinary commitment of resources in its Form 477 instructions. However, FCC officials stated that providers should not report service in areas in which major construction would be required to provide service. A few providers told us that the lack of clear guidance from FCC regarding how to determine where broadband is available has led different providers

---

[27] *See* 32 FCC Rcd 6282 (2017).

[28] As part of its universal service reforms, FCC established the Mobility Fund to target support for mobile service. For the Mobility Fund Phase 1, FCC identified unserved areas and held a reverse auction in 2012. In the reverse auction, eligible telecommunications carriers submitted "bids," outlining how much support they needed to serve specific unserved areas, and FCC awarded support based on the lowest bid as well as the number of road miles covered by the bids. In March 2017, FCC announced that it would be conducting another reverse auction to distribute up to $4.53 billion to providers that will deploy service to areas lacking LTE service (Mobility Fund Phase 2). FCC has not announced a date for the auction.

[29] Only providers that previously reported 4G LTE services on the Form 477 were required to submit data for FCC's Mobility Fund data collection. According to FCC, limiting the scope of the special data collection reduced the burden on providers, especially smaller providers. 32 FCC Rcd 6282 ¶ 11 (2017).

[30] In February 2018, FCC completed its initial analysis of the special data collection and released a map of areas it initially deemed eligible for the Mobility Fund Phase II auction based on the data. According to FCC, 64 entities have access to the challenge process as of May 2018, including 11 tribal governments.

[31] A "typical service interval" refers to the amount of time between when a customer requests service, and when a provider is able to begin providing service.

to interpret the Form 477 directions in different ways, which can affect the accuracy and consistency of reporting from provider to provider. For example, in a filing with FCC, one provider stated that it had misapplied the definition of "available" and, as a result, overstated the availability of its services by almost 3,000 census blocks.[32] As shown in Figure 5, FCC's definition of availability leads to overstatements of fixed broadband availability on tribal lands by: (1) counting an entire census block as served if only one location has broadband, and (2) allowing providers to report availability in blocks where they do not have any infrastructure connecting homes to their networks if the providers determine they could offer service to at least one household. Almost all the providers and private companies, and most of the representatives of tribal governments and organizations we spoke with told us that due to these issues, FCC's definition of availability results in data that overstate broadband availability.

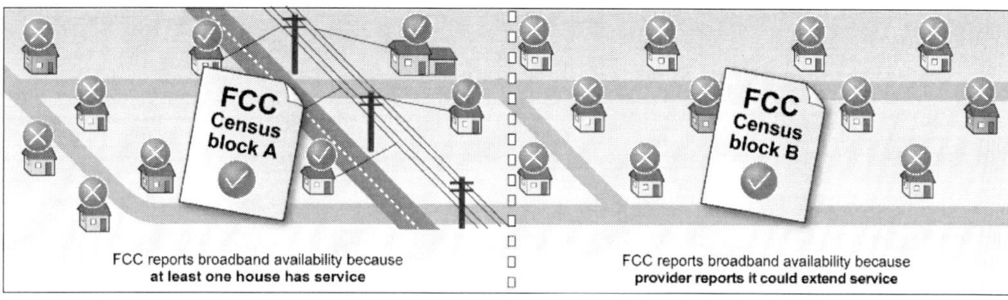

Source: GAO analysis of Federal Communications Commission (FCC) documents. | GAO-18-630

Figure 5. Overstatement of Broadband Availability in the Federal Communications Commission's Form 477 Data.

According to FCC officials, FCC requires providers to report fixed broadband availability where they could provide service within a "typical service interval" and without "an extraordinary commitment of resources" in order to: (1) ensure that it captures instances in which a provider has a network nearby but has not installed the last connection to the homes, and (2) identify where service is connected to homes, but homes have not subscribed. FCC officials also told us that FCC measures availability at the census block level because sub-census block data may be costly to collect. In 2013, FCC considered collecting more granular nationwide data on broadband deployment but decided against collecting these data because it determined that the burden would outweigh the benefit.[33]

However, FCC, tribal stakeholders, and providers have noted that FCC's approach leads to overstatements of availability. For example, in its 2017 Notice of Proposed Rulemaking on modernizing the Form 477 data collection, FCC acknowledged that by

---

[32] 31 FCC Rcd 7790 (2016). FCC officials noted that there are more than 11 million total census blocks nationwide.

[33] *In the Matter of Modernizing the FCC Form 477 Data Program*, Report and Order, 28 FCC Rcd 9887 ¶ 35 (2013).

requiring a provider to report where it *could* provide service, it is impossible to tell whether the provider would be unable or unwilling to take on additional subscribers in a census block it lists as served. According to FCC, this limits the value of the data to inform FCC policies. In addition, several providers and tribal stakeholders we interviewed said that some "digital subscriber line" (DSL) and fixed wireless providers may overstate their service areas on the Form 477 because they may not take into account technological or terrain limitations that would affect their ability to actually provide service.[34] FCC has also recognized that by measuring availability at the census block level, not every person will have access to broadband in a block that the data show as served, and FCC has noted that in rural areas, such as tribal lands, census blocks can be large and providers may only deploy service to a portion of the census block.[35] A few representatives for tribal governments and organizations noted that the use of census blocks may uniquely overstate broadband availability on tribal lands when census blocks contain both tribal and non-tribal areas, because availability in the non-tribal portion of the block can result in the tribal area of the census block also being counted as served.

FCC is considering requiring providers to report whether they are willing and able to serve additional customers in a census block and collecting sub-census block data in its 2017 proposed rulemaking on modernizing the Form 477.[36] About one-third of the parties that commented on FCC's proposals were not in favor of FCC collecting these more granular data on the Form 477, stating that the data would be less accurate and more burdensome for providers to collect and report, among other reasons, and questioned whether more detailed information on nationwide broadband availability is necessary.[37] We heard similar concerns from a few of the providers and trade associations we interviewed. However, about one- third of the parties that commented on FCC's proposals were in favor of collecting more granular data, stating that such data would be more useful for policymakers and more accurate. Additionally, a few tribally owned and non-tribal providers we interviewed told us that providers already maintain data for business purposes that would allow them to report more granular information on broadband availability. One stakeholder we spoke with pointed out that, as the federal government and states work to ensure the last remaining unserved areas—rural, low-

---

[34] "Digital subscriber line" (DSL) service typically refers to internet services delivered over traditional copper phone lines.

[35] *Modernizing the FCC Form 477 Data Program*, Further Notice of Proposed Rulemaking, 32 FCC Rcd 6329 (2017).

[36] FCC requested comment on the cost and burden of requesting more detailed data from providers, namely, whether to require fixed broadband providers to submit information identifying areas where: (1) there are existing customers and a provider could add new customers within a standard time interval upon request; (2) existing customers are served but providers cannot add new customers; and (3) there are no existing customers but new customers could be added within a standard time interval upon request. FCC also requested comment on whether to collect more granular data, such as data by street address. 32 FCC Rcd 6329 (2017).

[37] For example, commenters raised concerns that the lack of addresses in rural areas, such as tribal lands, would impose a burden on providers that are required to file a Form 477 and that the use of inconsistent geolocation methodologies would result in inaccurate data.

population density areas including tribal lands—have service, sub- census-block-level data are needed to ensure that governments are making wise and accurate investments.

FCC Does Not Collect Data on Several Factors that Affect Broadband Access on Tribal Lands

FCC does not collect information on several factors that FCC and tribal stakeholders have stated can affect broadband access. FCC and tribal stakeholders have noted that broadband access can be affected by factors such as the affordability and quality of the broadband services being offered, and the extent to which providers deny service to those who request it. By collecting and using data on factors that can affect broadband access, FCC would have more complete information on the extent to which Americans living on tribal lands have access to broadband Internet services.

- *Affordability*: FCC has noted that affordability of broadband services can affect broadband access but does not collect information on the cost of broadband service on tribal lands on the Form 477. For example, in the *National Broadband Plan*, FCC cited affordable access to robust broadband service as a long-term goal, and in its *Strategic Plan 2018–2022*, FCC acknowledged that affordability is an important factor affecting broadband access and a key driver of the digital divide.[38] Moreover, most of the representatives of tribal governments and organizations we spoke to told us that the affordability of broadband services is an important factor for understanding whether or not people on tribal lands could realistically access broadband services.[39] Tribal government officials from one tribe we spoke with told us that residents on their lands cannot access broadband because it is too costly. For example, a provider that advertises services on the tribe's land charges $130 per month for broadband services, approximately one-and-a-half times the average rate providers charge for comparable services in urban areas, according to FCC (see Figure 6).[40]

[38] FCC officials also noted that affordability is a key factor that affects whether people choose to subscribe to broadband services—known as "broadband adoption." As a result, according to officials, availability alone may be an incomplete indicator of broadband adoption.

[39] We have previously reported that tribal officials and providers identified affordability as a barrier to broadening the availability of services on tribal lands. GAO-16-222.

[40] Tribal government officials told us that this was the cost for services with maximum speeds of 10 megabits per second (Mbps)/4 Mbps. FCC's Urban Rate Survey collects information on the prices providers charge for fixed services in urban areas, in order to determine the benchmark rate Universal Service Fund recipients can charge customers. In its 2018 Urban Rate Survey, FCC surveyed providers to identify fixed broadband rates in urban areas and determined that the average rate plus two standard deviations for 10 Mbps/1 Mbps services ranged from $87.68 to $88.13. FCC uses the Urban Rate Survey to set rate benchmarks, and requires recipients of high-cost and/or Connect America Fund support to offer broadband services at rates that are at or below the relevant reasonable comparability benchmark.

In the 2018 Broadband Deployment Report, FCC acknowledged that affordability can influence a consumer's decision on whether to purchase broadband, but FCC did not consider cost in its assessment of broadband access on tribal lands, stating that pricing does not go to the congressional requirement to assess deployment and availability in conducting its inquiry as required by Congress under section 706 of the Telecommunications Act and also citing a lack of reliable comprehensive data on this issue. In addition, FCC officials we interviewed acknowledged that while broadband service may be technically available, it may be prohibitively expensive for some, which may make availability alone an incomplete indicator of broadband access.

- *Quality of Service*: In the *Telecommunications Act of 1996* Congress recognized the importance of service quality by defining advanced telecommunications capability as any technology that enables users to originate and receive *high-quality* voice, data, graphics, and video telecommunications. In keeping with this legislation, FCC has consistently set thresholds for speeds that qualify as broadband services and has stated that "latency" and consistency of service figure prominently into whether a broadband service is able to provide advanced capabilities and thus whether users can access high-quality telecommunications.[41] Likewise, almost all of the representatives for tribal governments or organizations we interviewed told us that quality of service is a key component of access to broadband and that routine outages, slow speeds, and high latency keep people on tribal lands from consistently accessing the Internet. Most tribal stakeholders and a few providers we interviewed told us that factors such as terrain, weather, and type of technology can all affect the quality of service an end user receives and, ultimately, the subscribers' ability to access the Internet (see Figure 7). For example, some representatives of tribal governments and organizations told us issues like oversubscription— when a provider signs up more customers than its equipment can handle—and outdated or limited infrastructure result in low-quality services that cannot support advanced and, in some cases, basic functions.[42]

[41] "Latency" refers to the amount of time it takes for data to travel from a computer to a server and back again. A high-latency network connection experiences long delay times, which can affect the performance of videoconferencing, phone, and streaming media services. In the 2016 Broadband Progress Report, FCC noted that latency may make a variety of applications unusable, regardless of the download/upload speeds being offered. 31 FCC Rcd 699 ¶ 62 (2016); FCC has made similar statements in other contexts as well. *See also In the Matter of Inquiry Concerning the Deployment of Advanced Telecommunications Capability to All Americans in a Reasonable and Timely Fashion*, 2015 Broadband Progress Report, 30 FCC Rcd 1375 ¶ 24 (2015); FCC, *National Broadband Plan*; FCC, 2016 *Measuring Broadband America Fixed Broadband Report: A Report on Consumer Fixed Broadband Performance in the United States* (Washington, D.C.: 2016).

[42] Advanced functions include, for example, Voice over Internet Protocol (VOIP), videoconferencing, and video streaming. Basic functions would include, for example, e- mail.

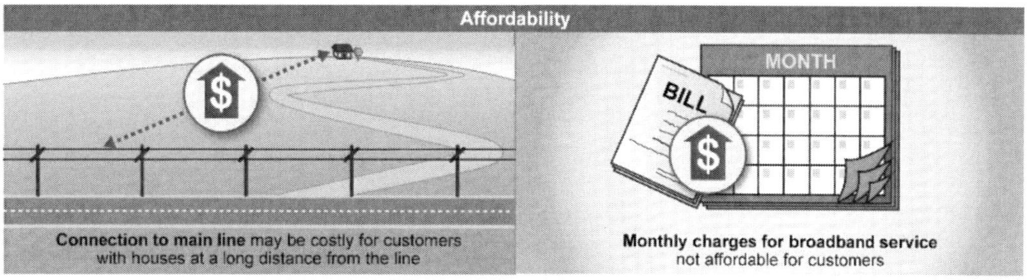

Source: GAO. | GAO-18-630

Figure 6. Examples of Scenarios That Affect Broadband Access: Affordability.

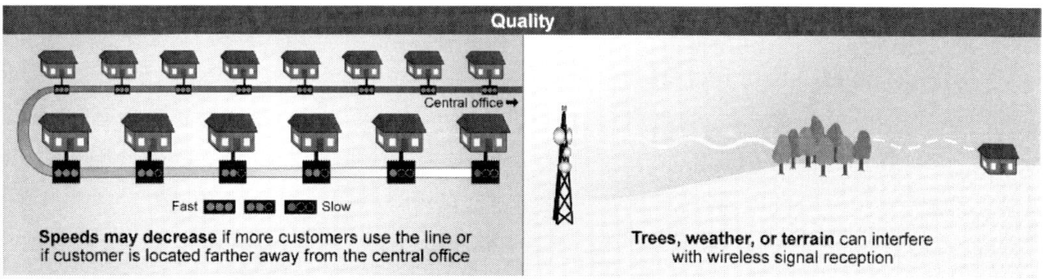

Source: GAO. | GAO-18-630

Figure 7. Examples of Scenarios That Affect Broadband Access: Quality of Service.

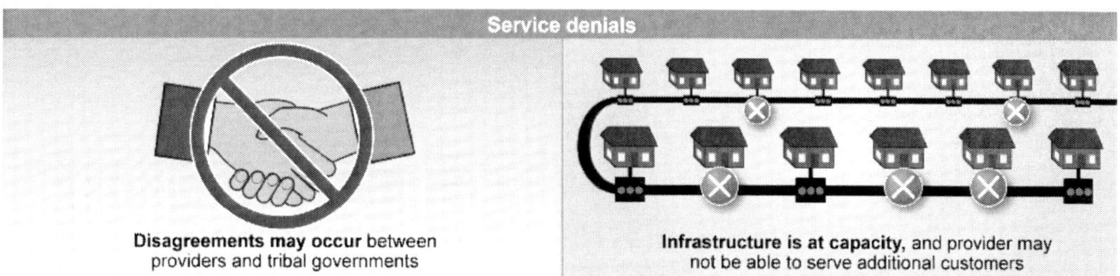

Source: GAO. | GAO-18-630

Figure 8. Examples of Scenarios That Affect Broadband Access: Service Denials.

Though FCC uses the Form 477 to collect some data on advertised speeds from providers, FCC does not collect data on actual speeds, service outages, and latency on the form.[43]

[43] On the Form 477, FCC collects data from providers on maximum advertised speeds for fixed broadband, and minimum advertised speeds for mobile broadband. In a 2011 Notice of Proposed Rulemaking on modernizing the Form 477 data program, FCC requested comment on whether to include measures of service quality, including service outages and latency, on the Form 477. In its subsequent 2013 Rule on the Form 477 program, FCC did not address the collection of service quality data, noting that this issue remained open for consideration. FCC did not address this issue in the 2017 Notice of Proposed Rulemaking on the 477 program. *Modernizing the FCC Form 477 Data Program,* Notice of Proposed Rulemaking, 26 FCC Rcd 1508 ¶¶ 91, 97-98 (2011); and *In the Matter of Modernizing the FCC Form 477 Data Program*, Report and Order, 28 FCC Rcd 9887 FN 29 (2013).

In its 2018 Broadband Deployment Report, FCC stated that it did not consider FCC data on actual speed, latency, or consistency of service when evaluating broadband access due to the lack of appropriate data. FCC noted that the lack of Form 477 data on actual speeds in particular constrained evaluation of mobile broadband access.[44]

- *Service Denials*: FCC has recognized that information on denials of service is pertinent to understanding actual broadband access but does not collect data on service denials in the Form 477. Specifically, in the *National Broadband Plan*, FCC recommended that FCC collect data to determine whether broadband service is being denied to potential residential customers based on the income of the residents in a particular geographic area. Some representatives of the tribal governments or organizations told us that that they were aware of a provider denying service to residents of tribal lands, despite the provider reporting broadband availability on at least a portion of those lands, according to our analysis of the Form 477 data. These representatives told us that they believed service was denied because of disputes with the tribal government, low demand for service, or the high costs of extending services to the home on tribal lands. Some representatives of tribal governments or organizations we spoke with also told us that providers may have denied service because their equipment was at capacity and could not accommodate new users (see Figure 8).

For example, on three of the tribal lands we visited, we observed fiber optic cable located close to government and residential structures that did not have broadband access via fiber. According to tribal government officials, despite the physical proximity of the fiber optic cable, the tribal government and residents could not access it because the provider was not offering service or was unwilling or unable to build to the structures. A few providers we interviewed stated that they may not provide services to individuals who request them because of high-costs, administrative barriers, or technical limitations. However, FCC does not collect data on service denials on the Form 477.[45]

---

[44] 33 FCC Rcd 1660 ¶¶ 31-32 (2018). FCC also considered third-party data on actual speeds in its 2018 Report, but noted that these data have some limitations because they were not collected pursuant to statistical sampling techniques.

[45] Separate from the Form 477 process, FCC used to collect information on "unfulfilled service requests," as part of an effort to determine whether certain broadband providers receiving funding were meeting obligations to offer broadband service upon a customers' reasonable request. However, FCC stopped requiring that providers submit data on unfulfilled requests after modifying the obligations to outline specific deployment thresholds, rather than requiring that service be available at a customer's "reasonable request." *In the Matter of Connect America Fund, ETC Annual Reports and Certifications*, Report and Order, 32 FCC Rcd 5944 ¶ 6 (2017).

FCC Uses Broadband Availability Data to Measure Broadband Access on Tribal Lands, Overstating Access on Tribal Lands

In its *Strategic Plan 2018–2022* and the *National Broadband Plan*, FCC identified increasing all Americans' access to affordable broadband as a long-term, strategic goal.[46] Congress has similarly directed FCC to develop policies and programs aimed at increasing access to affordable broadband in all regions of the United States, including tribal lands, and required FCC to report annually on its progress.[47] According to the *Government Performance and Results Act (GPRA), as enhanced by the GPRA Modernization Act of 2010 (GPRAMA)*, agencies should use accurate and reliable data to measure progress toward achieving their goals. Additionally, *Standards for Internal Control in the Federal Government* state that agencies should use quality information—information that is complete, appropriate, and reliable—to inform decision-making processes and evaluate the agency's performance in achieving goals. According to these standards, agencies should also communicate quality information externally to achieve the agency's goals.

However, FCC has used its Form 477 data, which do not accurately or completely measure broadband access on tribal lands, as its primary source to evaluate progress toward FCC's strategic goal of increasing broadband access and to develop maps and reports intended to depict broadband access on tribal lands. For example, in its 2018 Broadband Deployment Report, FCC found that 64.6 percent of Americans residing on tribal lands have access to fixed broadband services. By using these data, FCC has overstated the extent to which Americans living on tribal lands can actually access broadband Internet services and FCC's progress toward increasing broadband access. As a result, the digital divide may appear less significant as a national challenge, and FCC and tribal stakeholders working to target broadband funding to unserved or underserved tribal lands will be limited in their ability to make informed decisions. This increases the risk that residents living on tribal lands will continue to lack broadband access. Some tribal officials stated that inaccurate data have affected their ability to plan their own broadband networks and obtain federal broadband funding, and most of the tribal stakeholders we interviewed identified a pressing need for accurate data on the gaps in broadband access on tribal lands in order to ensure that tribes can qualify for federal funding and to effectively target the areas that need it most. For example, representatives for one tribal government that is providing broadband services said the government will not be able to use a federal grant to build broadband infrastructure in areas of their reservation that lack access, because the Form 477 data overstate actual access on the tribe's land. As more than three quarters of the tribal governments we spoke to are

---

[46] *See* FCC, *Strategic Plan 2018—2022*, Strategic Goal 1, and; FCC, *Connecting America: The National Broadband Plan*, Long-Term Goal No. 3.

[47] 47 U.S.C. § 1302(b); 47 U.S.C. §§ 254(b)(1), (b)(2).

working to provide broadband services on their lands in some capacity, overstating broadband access on tribal lands could affect the ability of a number of tribes to access federal funding to increase broadband access on their lands.

As previously discussed, FCC is considering proposals to modify its Form 477 data collection as part of a 2017 Notice of Proposed Rulemaking, but FCC officials told us that the Commission does not have a timeline for issuance of a final rule. While some of FCC's proposals could help address some of the limitations identified above by, for example, collecting more granular nationwide broadband availability data, FCC has not addressed specifically the collection of more accurate and complete data on broadband access for tribal lands in this proceeding. FCC has identified the need to improve broadband data for tribal lands in particular, and as previously noted, in 2018 Congress directed FCC to develop a report evaluating broadband coverage in certain tribal lands and initiate a proceeding to address the unserved areas identified in the report.[48] FCC officials told us that FCC has not determined how it will address this requirement, but it is currently considering its options, including potentially addressing the requirement as part of its ongoing proposed rulemaking on modernizing the Form 477 data collection. An evaluation of broadband coverage on tribal lands that relies on the current Form 477 data would be subject to the limitations described above, including the overstatement of broadband access on tribal lands.

Additionally, FCC has demonstrated that it is possible in some circumstances to collect more granular data when such data collection is targeted to a specific need or area. For example, in 2017 FCC began requiring certain providers that receive funding through the Connect America Fund to report the latitude and longitude of locations where broadband is available, and FCC has noted that these more granular data are extremely useful to the Commission, especially for rural areas where census blocks can be quite large.[49] A few large providers and trade associations similarly stated in public comments on FCC's proposed rulemaking to modernize the Form 477 process that FCC should target its collection of more granular broadband data to areas where the data are most likely to be overstated—specifically, large, rural census blocks with low population densities, such as those on tribal lands. Additionally, as discussed above, FCC undertook a one-time special data collection for Mobility Fund II to ensure that the mobile broadband data it collected would be reliable for the intended use. By developing and implementing methods for collecting and reporting accurate and complete data on broadband access specific to tribal lands, FCC would be able to better identify tribal areas

---

[48] FCC, *Connecting America: The National Broadband Plan;* Consolidated Appropriations Act, 2018, div. P, §§ 508(a)(1), (a)(2)(B), (b).

[49] This requirement applies only to locations that were deployed or upgraded with Connect America Funds after May 25, 2016. Under this data collection, FCC requires providers to report broadband as available at locations where (1) there is a current subscriber, or (2) a provider could offer service within 10 days upon request, and to report maximum available—not actual—speeds. FCC began collecting geolocation data from some carriers in 2016 and is expanding this requirement to remaining recipients of Universal Service Funds on a rolling basis. 32 FCC Rcd 6329 (2017).

without access to broadband and to target federal broadband funding to the tribal areas most in need.

FCC Does Not Have a Formal Process to Obtain Tribal Input on Its Broadband Data, and Tribal Stakeholders Reported a Lack of Provider Engagement

FCC Does Not Have a Formal Process to Obtain Tribal Input on Its Broadband Data

FCC uses data submitted by broadband providers via the Form 477 process to develop maps and datasets depicting broadband services nationwide, and in specific locations, such as tribal lands, but does not have a formal process to obtain input from tribes on the accuracy of the broadband data. FCC's 2010 *National Broadband Plan* noted the need for the federal government to improve the quality of data regarding broadband on tribal lands and recommended that FCC work with tribes to ensure that any information collected is accurate and useful. It also noted that tribal representatives should have the opportunity to review mapping data about tribal lands and offer supplemental data or corrections. Similarly, federal internal control standards note the need for federal agencies to communicate with external entities, such as tribal governments, and to enable these entities to provide quality information to the agency that will help it achieve its objectives.[50] FCC officials told us that they address questions and concerns regarding provider coverage claims submitted to the Office of Native Affairs and Policy, which will work with tribal governments to help them identify inaccurate broadband data for tribal lands, and share tribal questions and concerns with the appropriate FCC bureaus.[51] However, FCC does not have a formal process for tribes (or other governmental entities) to provide input to ensure that the broadband data FCC collects through the 477 process, or the resulting maps that FCC creates to depict broadband on tribal lands, are accurate. Similarly, FCC does not use other methods to verify provider-submitted Form 477 data on tribal lands against other sources of information, such as on-site tests or data collected by other agencies.[52]

---

[50] GAO-14-704G.

[51] FCC officials also stated that they conducted a more proactive outreach campaign to inform tribes of the ability to participate in the Mobility Fund Phase II challenge process, which is discussed earlier and later in this chapter.

[52] For example, USDA's Rural Utilities Service sends field representatives to verify the presence or absence of broadband infrastructure before funding broadband grant projects. The field representatives may meet with local representatives as well as local providers, in addition to identifying any existing broadband infrastructure and testing the performance of the services provided. However, FCC officials said that they do not have the resources to conduct field tests of the data.

When discussing the lack of a formal process for tribal representatives or other governmental entities to provide feedback on the accuracy of the 477 broadband data, FCC officials noted that if consumers and local officials have information on individual locations that lack broadband service, such information does not indicate that the entire census block lacks broadband service. Additionally, FCC officials noted that providers attest to the accuracy of the data and that FCC staff validate the data by conducting internal checks to identify possible errors, such as unlikely changes in a providers' coverage area, and may follow-up with a provider to discuss such changes. However, these checks do not include soliciting input from tribes.

About half of the tribal stakeholders we spoke to raised concerns that FCC's broadband deployment data rely solely on unverified information submitted by providers. Additionally, most tribal stakeholders we interviewed told us that consistent with the recommendations in the *National Broadband Plan*, FCC should work directly with tribes to obtain information from them to improve the accuracy of its broadband deployment data for tribal lands. These stakeholders identified several ways in which FCC could work with tribes on this issue, including:

- conducting on-site visits with tribal stakeholders to observe the extent to which broadband infrastructure and services are present;[53]
- conducting outreach and technical assistance for tribal stakeholders to raise awareness and use of FCC's broadband data; and
- providing opportunities for the tribes to collect their own data or submit feedback regarding the accuracy of FCC's data.

FCC's *National Broadband Plan* notes the importance of supporting tribal efforts to build technical expertise with respect to broadband issues, and federal internal control standards state that federal agencies should obtain quality information from external entities.[54] Officials we interviewed in FCC's Office of Native Affairs and Policy told us that they provide some outreach and technical assistance to tribal officials at regional and national workshops, and FCC officials stated that they conducted specific outreach to tribal entities regarding the Mobility Fund Phase II challenge process,[55] while, about half of the tribal representatives we spoke to stated that they were not aware of the Form 477 data or corresponding maps, or raised concerns about a lack of outreach from FCC to

---

[53] FCC has done such site visits in the past and reported on discrepancies between their observations of broadband infrastructure on tribal lands and the National Broadband Map, noting, "[w]e walked along a route where a carrier had reported broadband service via fiber on the National Broadband Map, yet saw none." Federal Communications Commission, Office of Native Affairs and Policy, *2012 Annual Report* (2012).

[54] GAO-14-704G.

[55] In commenting on a draft of this chapter, FCC described its outreach to tribal entities regarding the Mobility Fund Phase II challenge process as sending e-mails to the leaders and information technology managers of all 573 federally recognized tribes, conference calls and webinars open to all tribes, formal presentations at multiple inter-tribal conferences around the country, and a session at a tribal workshop conducted at the Lac du Flambeau Reservation in Wisconsin that was open to all tribes.

inform tribes about the data. Some tribal stakeholders stated that if FCC were to solicit tribal input as part of its verification of the broadband data and maps, technical training and assistance could help tribes use and provide feedback on the data, or improve the collection and submission of their own data. A few of the stakeholders we interviewed noted that tribes can face difficulties when they attempt to challenge FCC's broadband availability data. For example, in 2013, prior to the auction that distributed Tribal Mobility Fund Phase 1 support, FCC allowed interested parties to challenge FCC's preliminary determinations regarding which census blocks lacked 3G or better service and would be eligible for support in the auctions.[56] However, all of the tribal entities that challenged the accuracy of FCC's data were unsuccessful in increasing the number of eligible areas. According to FCC officials, the tribal entities did not provide sufficient or sufficiently verifiable information to support their challenges. A few tribal stakeholders provided varying reasons for this, one of which was the need for more technical expertise to help the tribe meet FCC's requirements.

Because FCC lacks a formal process to obtain tribal input on its broadband data, FCC is missing an important source of information regarding areas in which the data may overstate broadband service on tribal lands. Tribal stakeholders are able to provide a first-hand perspective on the extent to which service is available within their lands and the extent to which factors like affordability, service quality, and service denials affect residents' ability to access broadband. FCC plans to award nearly $2 billion in support from the Connect America Fund to areas that it has identified as lacking broadband, including tribal lands. Any inaccuracies in its broadband data could affect FCC's funding decisions and the ability of tribal lands to access broadband in the future.[57] Additionally, in its 2017 report on tribal infrastructure, the National Congress of American Indians stressed the importance of including tribal governments in a leadership role with respect to collecting data on local infrastructure needs.[58] Specifically, it stressed the need for the federal government to invest in tribal data systems and researchers to generate useful, locally specific data that can inform the development and implementation of

---

[56] In 2011, as part of its reform to the Universal Service Fund programs and the establishment of new funding mechanisms, FCC decided to use data from a third-party source to identify census blocks without wireless coverage, stating that it could not use Form 477 data due to a lack of census block-level data for wireless service or data from the National Broadband Map due to concerns regarding inconsistencies in how wireless services were reported.

[57] As part of Connect America Fund Phase II, FCC is conducting a reverse auction by which providers submit bids for support to provide fixed broadband service to specific unserved areas. The auction is scheduled to begin July 24, 2018. FCC is not conducting a challenge process to determine whether any census blocks are incorrectly listed as having fixed broadband service, although it has conducted challenge processes in the past before awarding funds. In explaining this decision, FCC stated that a prior challenge process was time consuming and administratively burdensome, and it was difficult for challengers to prove that a company was not serving an area it claimed to serve. FCC also stated that the 477 data was reliable because providers are required to file the data and attest to its accuracy. *In the Matter of Connect America Fund, ETC Annual Reports and Certifications, Rural Broadband Experiments*, Report and Order and Further Notice of Proposed Rulemaking, 31 FCC Rcd 5949 ¶¶ 58-59 (2016).

[58] National Congress of American Indians (NCAI), *Tribal Infrastructure: Investing in Indian Country for a Stronger America,* An initial report by NCAI to the Administration and Congress, 2017.

infrastructure development projects and assess the effectiveness of those projects over time. By establishing a process to obtain input from tribal governments on the accuracy of provider- submitted broadband data that includes outreach and technical assistance, as recommended in the *National Broadband Plan*, FCC could help tribes develop and share locally specific information on broadband access, which would in turn improve the accuracy of FCC's broadband data for tribal lands. The success of such an effort may rely on the tribes' knowledge of, and technical ability to participate in, the process.

Half of the Tribal Stakeholders We Interviewed Reported a Lack of Provider Engagement

When discussing the need to improve data regarding broadband on tribal lands, FCC's 2010 *National Broadband Plan* recommended that FCC develop a process for tribes to receive information from providers about broadband services on tribal lands. In 2011, FCC required that Eligible Telecommunications Carriers (providers receiving Universal Service Funds from FCC) serving tribal lands meaningfully engage with tribes regarding communications services (including broadband).[59] Specifically, the providers must file an annual report documenting that this engagement included a discussion of, among other things, a needs assessment and deployment planning for communications services, including broadband. FCC's 2012 guidance on fulfilling the engagement obligations, which FCC officials confirmed is still in effect, noted that the stated goal of the engagement requirement was to benefit tribal government leaders, providers, and consumers by fostering a dialogue between tribal governments and providers that would lead to improved services on tribal lands. The guidance further noted that the tribal engagement process "cannot be viewed as simply another 'check the box' requirement by either party," and states that a provider should "demonstrate repeated good faith efforts to meaningfully engage with the tribal government."[60] Finally, FCC noted in its 2012 guidance that the guidance would evolve over time based on the feedback of both tribal governments and broadband providers and that FCC would develop further guidance and best practices.[61] This approach is consistent with federal internal control standards, which call for agencies to communicate with, and obtain quality information from, external parties.[62]

---

[59] *In the Matter of Connect America Fund*, Report and Order and Further Notice of Proposed Rulemaking, 26 FCC Rcd 17663 (2011).

[60] *Office of Native Affairs and Policy, Wireless Telecommunications Bureau, and Wireline Competition Bureau Issue Further Guidance on Tribal Government Engagement Obligation Provisions of the Connect America Fund*, Public Notice, 27 FCC Rcd 8176 (2012).

[61] 27 FCC Rcd 8176 (2012).

[62] GAO-14-704G.

About half of the tribal stakeholders we interviewed raised concerns about difficulties accessing information from providers regarding broadband deployment on their tribe's lands, a key part of the provider engagement process, according to FCC's guidance. For example, a representative from one tribe stated that a provider declined his requests to meet more than once a year to discuss the provider's deployment of broadband services on the tribe's land. A representative from another tribal government stated that some providers are very focused and transparent about their broadband plans and work with the tribe, while other providers treat tribal engagement as a "box to check" and send the tribe broadband deployment information that is not useful because it is redacted. Similarly, some tribal stakeholders stated that providers heavily redacted deployment information (which providers may consider proprietary) or required the tribe sign non-disclosure agreements to access deployment data. According to one tribal stakeholder, these non-disclosure agreements could possibly require tribes to waive tribal sovereign immunity in order to view the data.

Some of the industry stakeholders we interviewed stated that they attempt to engage with tribes but the level of responsiveness from tribes varies. For example, some stakeholders stated that they send letters and do not hear back from tribes. One stakeholder stated that they make repeated attempts to contact tribes when they do not hear back after their initial contact, while another stated that a provider meets regularly with some tribes.

Although FCC stated in its 2012 guidance that it would update the tribal engagement guidance and develop best practices based on feedback from tribal governments and broadband providers, it has taken limited steps to obtain such feedback from providers and tribal governments to determine whether its guidance is enabling meaningful tribal engagement. Additionally, FCC has not updated the guidance or issued best practices.[63] Thus, FCC has limited information regarding whether its tribal engagement requirement is fulfilling its intended purpose. FCC officials we interviewed said that the Office of Native Affairs and Policy (ONAP) provided information and, in some cases, held training sessions about the tribal engagement obligation during workshops with tribal representatives, and encouraged representatives to contact ONAP with any concerns. ONAP officials also noted that they handle complaints from tribes regarding a lack of provider engagement and reach out to providers to address tribal concerns. ONAP officials stated that they have had internal discussions about whether the guidance is clear or needs revision, but this has not gone beyond internal discussions.[64] A few of the tribal stakeholders provided examples of the benefits of providers engaging with tribes to

[63] *Office of Native Affairs and Policy, Wireless Telecommunications Bureau, and Wireline Competition Bureau Issue Further Guidance on Tribal Government Engagement Obligation Provisions of the Connect America Fund*, Public Notice, 27 FCC Rcd 8176 (2012).

[64] FCC officials also noted that industry stakeholders filed petitions for reconsideration of the tribal engagement obligation and FCC's 2012 guidance (these petitions cited concerns with FCC's process for developing the requirements, among others), but noted that these petitions remain pending.

ensure tribal representatives have access to information regarding broadband availability on their lands. For example, one representative stated that this information could help the tribes plan deployments by focusing on areas that they know the provider does not plan to serve. Another representative stated that tribal engagement could help improve the accuracy of FCC's broadband maps. By obtaining feedback from both tribal stakeholders and providers on the effectiveness of FCC's tribal engagement guidance to determine whether changes are needed, FCC would be better positioned to ensure that tribal governments and providers are sharing information in a manner that will lead to improved services on tribal lands.

CONCLUSION

FCC has collected data and developed maps and reports depicting broadband on tribal lands and has noted the lower levels of broadband access on tribal lands, in comparison to other areas. However, limitations in FCC's existing process for collecting and reporting broadband data have led FCC to overstate broadband access on tribal lands. By taking steps to address these limitations and to collect data that more accurately and completely depict broadband access on tribal lands, FCC would have greater assurance that it is making progress on reducing the digital divide on tribal lands and targeting broadband funding to tribal lands most in need. Without taking these steps, FCC increases the risk that residents living on tribal lands will continue to lack broadband access.

Compounding the limitations in FCC's data collection process is FCC's lack of a formal process to obtain tribal input on the accuracy of provider- submitted broadband data for tribal lands. By developing a process to solicit tribal input and ensuring that tribes know about the process and are equipped with the technical skills and abilities necessary to provide this information, FCC would be better able to ensure the accuracy of its broadband data for tribal lands. Moreover, FCC would be able to obtain firsthand, locally specific information on broadband access that could inform FCC's policies and funding decisions and help FCC achieve its goal of increasing broadband access for all Americans, including those living on tribal lands. Finally, by obtaining feedback from providers and tribal stakeholders on the effectiveness of FCC's tribal engagement guidance, FCC would be better positioned to assess whether its guidance is helping providers meet requirements and ultimately whether providers' engagement is fulling its intended purpose of fostering a dialogue between tribal governments and providers that would lead to improved services on tribal lands.

Recommendations

We are making the following three recommendations to the Chairman of the Federal Communications Commission.

- The Chairman of the Federal Communications Commission should develop and implement methods—such as a targeted data collection—for collecting and reporting accurate and complete data on broadband access specific to tribal lands. (Recommendation 1)
- The Chairman of the Federal Communications Commission should develop a formal process to obtain tribal input on the accuracy of provider-submitted broadband data that includes outreach and technical assistance to help tribes participate in the process. (Recommendation 2)
- The Chairman of the Federal Communications Commission should obtain feedback from tribal stakeholders and providers on the effectiveness of FCC's 2012 statement to providers on how to fulfill their tribal engagement requirements to determine whether FCC needs to clarify the agency's tribal engagement statement. (Recommendation 3)

Agency Comments

We provided a draft of this chapter to FCC for review and comment. In written comments provided by FCC (reproduced in appendix III), FCC agreed with our findings and recommendations. In its written comments, FCC described efforts, some of which are already under way, that it felt would address each recommendation and stated its intent to build upon those efforts. For example, FCC explained that it is exploring methods to collect more granular broadband deployment data and noted the need to balance the burden on Form 477 filers. FCC also noted that it is starting work to address a statutorily-required evaluation of broadband coverage on certain tribal lands. We agree that increasing the granularity of deployment data is helpful in addressing data accuracy issues, but we also note that it is important to collect data related to factors that affect broadband access on tribal lands.

FCC also described informal efforts to collect tribal feedback on providers' broadband data and stated it would explore options for a formal process to collect feedback. Regarding our recommendation related to providers' engagement efforts, FCC outlined its existing methods by which tribal stakeholders can provide feedback on providers' engagement efforts and agreed that seeking additional feedback from tribal stakeholders and providers would be desirable. We agree that improving feedback in these ways could help FCC determine whether it needs to clarify its tribal engagement statement. FCC also provided technical comments, which we incorporated as appropriate.

We are sending copies of this chapter to the appropriate congressional committees, the Chairman of the Federal Communications Commission, and other interested parties.

Mark L. Goldstein
Director, Physical Infrastructure Issues

APPENDIX I: LIST OF INTERVIEWEES

Representatives from tribal governments or tribally owned broadband providers
Choctaw Nation of Oklahoma (OK)
Confederated Tribes of the Colville Reservation (WA)
Fond du Lac Band of Lake Superior Chippewa (MN)
Fort Belknap Indian Community (MT)
Gila River Telecommunications, Inc. (AZ)
Hopi Telecommunications, Inc. (AZ)
Jamestown S'Klallam Tribe (WA)
Karuk Tribe (CA)
Leech Lake Band of Ojibwe (MN)
Makah Tribe (WA)
Navajo Tribal Utility Authority (AZ, NM, UT)
Nez Perce Tribe (ID)
Osage Nation (OK)
Pueblo of Acoma (NM)
Pueblo of Pojoaque (NM)
Pueblo of San Ildefonso (NM)
Taos Pueblo (NM)
Red Spectrum Communications (Coeur d'Alene Tribe (ID))
Saint Regis Mohawk Tribe and Mohawk Networks, LLC (NY)
San Carlos Apache Telecommunications Utility, Inc. (AZ)
Southern California Tribal Chairmen's Association - Tribal Digital Village Network (CA)
Spokane Tribe of Indians and Spokane Tribe Telecom Exchange (WA)
Standing Rock Telecommunications, Inc. (ND, SD)
Warm Springs Telecommunications Co. (OR)
Yurok Tribe and Yurok Connect (CA)
Representatives from tribal associations/consortiums that include tribes
Affiliated Tribes of Northwest Indians
Middle Rio Grande Pueblo Consortium
National Congress of American Indians
National Tribal Telecommunications Association
Native American Finance Officers Association (NAFOA)
REDINet
Representatives from companies/academic groups that work with tribes
AMERIND Risk
Arizona State University, American Indian Policy Institute and School of Public Affairs
Mobius Legal Group PLLC
Turtle Island Communications

Representatives from providers/trade associations (non-tribally owned)
AT&T
CenturyLink
CTIA
Commnet
Frontier
Inland Cellular
King Street Wireless
Kit Carson Electric Cooperative
NTCA
Pine Telephone Company
Rural Wireless Association
Verizon
[a]
Representatives from companies that collect broadband data
Alexicon
Connected Nation
Government Agencies (non-tribal)
Census Bureau
U.S. Department of Agriculture's Rural Utilities Service
Federal Communications Commission
Department of Interior's Bureau of Indian Affairs
National Telecommunications and Information Administration
Housing and Urban Development
Indian Health Service
Minnesota Office of Broadband Development

Source: GAO. | GAO-18-630

[a] One broadband provider we interviewed did not want to be included in this appendix.

APPENDIX II: OBJECTIVES, SCOPE, AND METHODOLOGY

This chapter discusses the extent to which: (1) the Federal Communications Commission's (FCC) approach to collecting broadband availability data accurately captures the ability of Americans living on tribal lands to access broadband Internet services and (2) FCC obtains tribal input on the accuracy of provider-submitted broadband data for tribal lands.

To address both objectives, we analyzed FCC's December 2016 fixed and mobile broadband availability data—the most recent data at the time of our review—to identify the speeds, technologies, and availability providers reported for federally recognized tribal lands.[65] Providers currently report this information to FCC by filing a "Form 477,"

[65] We defined federally recognized tribal lands consistent with FCC's definition in its 2018 Broadband Deployment Report. Specifically, we considered tribal lands to be: (1) Join Use Areas; (2) legal federally recognized American Indian area consisting of reservation and associated off-reservation trust land; (3) legal federally recognized American Indian area consisting of reservation only; (4) legal federally recognized American Indian area consisting of off-reservation trust land only; (5) Statistical American Indian area defined for a federally recognized tribe that does not have reservation or off-reservation trust land, specifically a Tribal Designated Statistical Area (TDSA) or Oklahoma Tribal Statistical Area (OTSA); (6) Alaskan Native village

twice a year. We also used 2010 U.S. Census data to identify census blocks completely or partially on tribal lands. To assess the reliability of FCC's data and 2010 U.S. Census data, we reviewed a previous GAO reliability assessment, and for FCC's data we conducted electronic testing and analysis of the data, reviewed FCC guidance and documentation, and interviewed FCC officials.[66] Based on the results of our analysis, we determined the data to be reliable for our purposes, which were: (1) to inform our selection of tribal governments and providers for interviews and visits, as described below, and (2) to develop maps depicting fixed and mobile broadband availability for the nine tribal lands we selected for visits, in order to obtain tribal representatives' feedback on the data. Specifically, we mapped;

- fixed broadband data according to speed and technology, and
- mobile data for long-term evolution (LTE) services by provider for each tribal land.[67]

We used those maps during our visits to discuss the accuracy of the data with representatives for each tribal government or tribally owned provider. Though we analyzed all up and download speeds that providers reported in the Form 477, for the purposes of this chapter we defined "broadband" as fixed Internet service reaching at least 25 megabits per second (Mbps) download and 3 Mbps upload speeds, in accordance with FCC's advanced telecommunications capability benchmark in its 2018 Broadband Deployment Report.[68] We also report on the availability of mobile broadband, which, for the purposes of this chapter, does not have a speed threshold and refers to long-term evolution (LTE) services.

To address both objectives and obtain tribal government representatives' feedback on the accuracy of FCC's broadband data for their lands, we interviewed representatives from 25 tribal governments or tribally owned providers, including visits to 9 tribal lands. We considered a range of factors when we selected tribal governments and tribally owned providers for interviews, including our analysis of Form 477 data, recommendations from tribal, industry, or government stakeholders regarding tribal and non- tribal representatives familiar with broadband data issues, and demographic and geographic characteristics, among others. For example, we considered demographic characteristics such as unemployment rate from the 2011– 2015 American Community Survey data, and geographic characteristics such as rurality from the United States

statistical area; and (7) Hawaiian Home Lands established by the Hawaiian Homes Commission Act of 1921. *See* 33 FCC Rcd 1660 ¶¶ 31-32 (2018).

[66] We reviewed the data reliability assessment from GAO, *Broadband: Additional Stakeholder Input Could Inform FCC Actions to Promote Competition*, GAO-17-742 (Washington, D.C.: Sept.19, 2017).

[67] LTE is an industry standard that is part of the fourth generation of wireless telecommunications technology, which is currently in common use.

[68] 33 FCC Rcd 1660 ¶¶ 31-32 (2018).

Department of Agriculture (USDA) Rural-Urban Commuting Area Codes data. The tribes included in our review vary with respect to location, level of broadband availability according to FCC, land mass, and population size and density. The results of our interviews are not generalizable to all tribal governments or tribally owned broadband providers. In addition to tribal governments and tribally owned providers, we interviewed six tribal organizations and four stakeholders who work with tribes on broadband issues. For reporting purposes, we developed the following series of indefinite quantifiers to describe the tribal responses from the 35 entities representing tribal stakeholders we interviewed:

- 3 to 7 is defined as "a few;"
- 8 to 15 is described as "some;"
- 16 to 20 is described as "about half;"
- 21 to 27 is described as "most;" and
- 28 to 34 is described as "almost all."

A full list of the tribal stakeholders we interviewed can be found in appendix I.

Further, to obtain industry perspectives, we reviewed public comments submitted by providers and industry associations in FCC's ongoing 2017 *Notice of Proposed Rulemaking on Modernizing the Form 477 Data Program*.[69] We also interviewed 10 non-tribally owned fixed and mobile broadband providers and three industry associations to understand providers' views on the Form 477 and how providers interact with tribal governments. When selecting providers for interviews, we included providers that reported serving the lands of tribal governments we interviewed and selected providers that varied in the percentage of tribal lands they reported serving. The providers we interviewed represent large, nationwide carriers as well as small, local carriers, and offer broadband via a variety of technologies, including fiber optics, digital subscriber line (DSL), fixed wireless, and mobile LTE.[70] The results of our interviews with providers are not generalizable to all broadband providers. In addition, to address both objectives, we interviewed representatives from other government entities, as well as private companies that collect and report broadband data. A full list of the industry stakeholders we interviewed can be found in appendix I.

To identify the extent to which FCC's approach to collecting broadband availability data reflects the ability of Americans living on tribal lands to actually access broadband Internet services, we reviewed documentation of the Form 477 process, including submission guidance, and FCC's proposals and public comments in its 2017 *Notice of*

---

[69] *Modernizing the FCC Form 477 Data Program*, Further Notice of Proposed Rulemaking, 32 FCC Rcd 6329 (2017).

[70] Digital subscriber line (DSL) service typically refers to Internet services delivered over traditional copper phone lines.

Proposed Rulemaking on Modernizing the Form 477 Data Program and Mobility Fund Phase II proceedings.[71] We also interviewed FCC officials, industry stakeholders, and tribally owned broadband providers to understand FCC's current process for collecting broadband data. To understand the purpose of the Form 477 data collection process and FCC's strategic goals, we reviewed relevant statutes, and FCC documents, including FCC's *Strategic Plan 2018—2022*, the *National Broadband Plan*, and FCC's broadband deployment and progress reports.[72] Given the importance placed on broadband access in these documents, we interviewed tribal stakeholders, as described above and reviewed FCC documents to identify factors affecting the ability of Americans living on tribal lands to access broadband Internet services. We also reviewed previous GAO work that identified barriers to broadband access on tribal lands.[73] We compared the Form 477 process to FCC's strategic goals and to factors affecting broadband access to determine the extent to which the Form 477 was designed to collect information on those factors and to meet FCC's goals. We further evaluated this information against the *Government Performance and Results Act, as enhanced by the GPRA Modernization Act of 2010* and *Standards for Internal Control in the Federal Government*.[74] We also reviewed documentation for other FCC data collection programs, including the *Measuring Broadband America* program and the Urban Rate Survey, to determine the extent to which FCC collected data on factors affecting broadband access outside of the Form 477 process.

To determine the extent to which FCC obtains tribal input on the accuracy of provider-submitted broadband data for tribal lands, we interviewed FCC officials and analyzed FCC documents regarding the collection procedures for the Form 477 data and FCC's policies for working with tribal governments, as well as Connect America Fund documents regarding requirements for providers to share information with tribal governments.[75] We also reviewed documents on past FCC Universal Service Fund processes to challenge broadband data and identified prior instances in which tribal governments or tribally owned providers challenged FCC's broadband data and the outcomes of those challenges. Additionally, we interviewed tribal stakeholders, as

---

[71] 32 FCC Rcd 6329 (2017); *Connect America Fund Universal Service Reform—Mobility Fund*, Order on Reconsideration and Second Report and Order, 32 FCC Rcd 6282 (2017); *Instructions for Filing 4G LTE Coverage Data to Determine Areas Presumptively Eligible for Mobility Fund II Support*, Public Notice, 32 FCC Rcd 7023 (2017).

[72] Communications Act of 1934, Pub. L. No. 73-416, 48 Stat. 1064 (1934), *as amended by Telecommunications Act of 1996*, Pub. L. No. 104-104, § 706, 110 Stat. 56, 153 (1996) (codified as amended at 47 U.S.C. § 151); FCC, *Strategic Plan 2018-2022* (Washington, D.C.); 33 FCC Rcd 1660 ¶¶ 31-32 (2018); 31 FCC Rcd 699 (2016).

[73] GAO, *Telecommunications: Additional Coordination and Performance Measurement Needed for High-Speed Internet Access Programs on Tribal Lands*, GAO-16-222 (Washington, D.C.: Jan. 29, 2016)

[74] Government Performance and Results Act, Pub. L. No. 103-62, 107 Stat. 285 (1993), as enhanced by GPRA Modernization Act of 2010, Pub. L. No. 111-352, 124 Stat. 3866, 3867 (2011). (codified at 31 U.S.C. § 1115(b)(2), (6)); GAO, *Standards for Internal Control in the Federal Government*, GAO-14-704G (Washington, D.C.: Sept. 2014).

[75] *Statement of Policy on Establishing a Government-to-Government Relationship with Indian Tribes*, Policy Statement, 16 FCC Rcd 4078 (2000). 47 C.F.R. § 54.313 (a)(5); *Further Guidance on Tribal Government Engagement Obligation Provisions of the Connect America Fund*, Public Notice, 27 FCC Rcd 8176 (2012).

described above, to understand the extent to which: (1) FCC involves tribal governments and other stakeholders in the validation of Form 477 broadband data, (2) tribal governments can access broadband data from FCC or providers, and (3) FCC's Form 477 data accurately reflected broadband access on their lands. For the nine tribal lands we visited, we asked tribal governments or tribally owned providers to identify where the data do or do not accurately reflect broadband access on maps of FCC's data. Further, to identify how providers complied with FCC's tribal engagement requirement and obtain their perspectives, we interviewed providers and industry associations. We compared FCC's data validation procedures and tribal stakeholders' feedback on the process to FCC's policies for working with tribal governments, FCC recommendations from the *National Broadband Plan* and *Standards for Internal Control in the Federal Government*.[76] We also interviewed and received written comments from officials from other federal agencies that have broadband programs, including USDA Rural Utilities Service, the National Telecommunications and Information Administration (NTIA), and others, in addition to a state agency and three private companies that collect and report broadband data to understand how other entities collect and validate broadband data.[77]

We conducted this performance audit from June 2017 to September 2018 in accordance with generally accepted government auditing standards. Those standards require that we plan and perform the audit to obtain sufficient, appropriate evidence to provide a reasonable basis for our findings and conclusions based on our audit objectives. We believe that the evidence obtained provides a reasonable basis for our findings and conclusions based on our audit objectives.

APPENDIX III: COMMENTS FROM THE FEDERAL COMMUNICATIONS COMMISSION

Federal Communications Commission

Washington, D.C. 20554

August 22, 2018

[76] *Connecting America: The National Broadband Plan*, Notice of Inquiry and Notice of Proposed Rulemaking, 25 FCC Rcd 6657 (2010); GAO-14-704G.

[77] We received written comments from one company that collects broadband data.

Mark L. Goldstein, Ph.D.
Director, Physical Infrastructure Issues
Government Accountability Office
441 G Street, NW
Washington, D.C. 20548

Dear Director Goldstein:

Thank you for the opportunity to review GAO's draft report, "FCC's Data Overstate Access on Tribal Lands." The Commission has long recognized the particular challenges associated with providing telecommunications services on Tribal lands, and we agree with GAO that accurate, comprehensive data are vital to the Commission's efforts to bridge the digital divide, including on Tribal lands. As described below, we have efforts underway to ensure that we collect the best possible data and, indeed, we believe that we already have work in progress to address each of the three recommendations GAO advances. We will continue with the work underway and re-double our efforts to close the digital divide for all Americans, including those on Tribal lands.

Methods to Collect and Report Data on Broadband Access to Specific Tribal Lands. GAO's first recommendation is that the FCC Chairman "should develop and implement methods—such as targeted data collection—for collecting and reporting accurate and complete data on broadband access to specific tribal lands." The Commission agrees with the importance of having access to quality data. For this reason, the Commission has initiated a rulemaking proceeding to explore ways in which it could improve aspects of the Form 477 collection.[1] In that proceeding, the Commission sought comment on a wide variety of issues related to making the Form 477 collection as efficient and effective as possible. Among these issues is whether the Commission should revise the Form 477 to collect deployment data on a more granular level than it does currently. Given a better understanding of the Form 477 instructions, the issue of granularity appears to underlie many of GAO's concerns in the draft report about Form 477 deployment data—the issue that individual locations might not have network coverage in a Census block with some deployment. We are cognizant that increasing granularity in the collection would require the Commission to resolve significant technical issues and would

[1] *Modernizing the FCC Form 477 Data Program*; Further Notice of Proposed Rulemaking; WC Docket 11-10 (2017).

likely increase the burden on Form 477 filers. If an appropriate method for such a collection can be identified, however, this may address many of the concerns GAO raises in the draft report by providing the Commission with a more precise picture of broadband deployment. We are continuing to work on these issues in the context of WC Docket 11-10.

Also relevant to GAO's first recommendation is the requirement in the *Consolidated Appropriations Act of 2018* that the Commission conduct an assessment regarding the availability of broadband services in Indian country and to report on the results by March 23, 2019.[2] Based on the results of that assessment, the legislation directs the Commission to conduct a rulemaking proceeding to address the unserved areas identified in the report. We have initiated work on that effort and will work with Tribal and other stakeholders to develop a clear picture of broadband deployment on Tribal lands and address unserved areas.

The Commission recognizes that the digital divide is all too real, especially in Indian country. That's why the Commission has primarily relied on Form 477 data for a limited purpose—identifying the too-many census blocks where "no" Internet service provider has deployed broadband infrastructure, and thus the areas that unambiguously need federal funding through the Connect America Fund to get broadband. This divide is particularly stark on Tribal lands, as GAO recognizes, with more than 35% of Tribal residents lacking "any" chance to access broadband infrastructure.

And the Commission has recognized that more granular data will be needed in the future. As our policies bring broadband deployment into wholly unserved blocks, it will be more important to understand availability in partially served blocks. That's why the Commission opened a proceeding into this issue last year, and the Commission remains dedicated to moving forward with a proceeding that explores ways to collect more granular data without unnecessarily burdening those who are deploying on Tribal lands and often with few resources to spare.

Process to Obtain Tribal Input on Provider-Submitted Broadband Data. The draft report's second recommendation is that the Chairman of the FCC "develop a process to obtain tribal input on the accuracy of provider-submitted broadband data that includes outreach and technical assistance to help tribes participate in the process." The Commission agrees that tribal input on the accuracy of provider-submitted broadband data is important. Indeed, the FCC currently has in place a number of informal means by which Tribal and other stakeholders can raise any concerns. For example, Tribal stakeholders can, and do, raise concerns and questions about the data to the Commission's Office of

[2] *Consolidated Appropriations Act, 2018*, H.R. 1625, 115th Cong., Division P, *Ray Baum's Act*, § 508. (2018).

Native Affairs and Policy (ONAP), which shares them with the relevant bureaus. In addition, the Commission has given Tribes a direct role in evaluating and challenging providers' claims of service coverage in the ongoing Mobility Fund Phase II (MF-II) proceeding. ONAP and the Commission's Rural Broadband Auctions Task Force have cooperated on a number of initiatives to make Tribal leaders and others aware of the MF-II challenge process for the Mobility Fund II auction eligible areas and the importance of participating in that process. These efforts have included sending information in emails to the leaders and IT managers of all 573 federally recognized Tribes; conducting outreach, including conference calls and webinars open to all Tribes; formal presentations at multiple inter-tribal conferences around the country; and a session at a July 31 Tribal workshop conducted at the Lac du Flambeau Reservation in Wisconsin that was open to all Tribes. We agree that, in addition to these mechanisms, implementing a formal process for continuing Tribal engagement could have significant value in helping the FCC understand both the extent of, and the specific issues that drive or hinder, broadband deployment on Tribal lands. We will work with stakeholders to explore options for implementing such a formal process.

Feedback from Tribal Stakeholders and Providers on Providers' Tribal Engagement Requirements. Finally, the draft report recommends that the FCC Chairman "obtain feedback from tribal stakeholders and providers on the effectiveness of the FCC's 2012 statement to providers on how to fulfill their tribal engagement requirements to determine whether the Commission needs to clarify its tribal engagement statement." We agree that seeking additional feedback on the overall effectiveness of the Commission's Further Guidance Public Notice is desirable. We note that the Commission's ONAP solicits and receives feedback from Tribes on whether and how providers are fulfilling the requirements of the rule, the effectiveness of the Commission's guidance, and any problems encountered in the engagement process. ONAP regularly includes presentations on the Tribal engagement obligation at its Tribal workshops, which it conducts at different locations around the country throughout the year. Additionally, ONAP solicits and receives feedback on the engagement requirements from Tribes and other participants at inter-Tribal conferences and similar events. As a result of feedback concerning the availability of compliance reporting, the Commission has made changes to its filing requirements and Tribal Nations will soon be able to obtain providers' reports on their Tribal engagement efforts directly through a Universal Service Administrative Company online portal. We will continue to seek additional feedback from Tribal stakeholders, as well as feedback from providers, regarding the effectiveness of the guidance provided by the Commission thus far on how providers may fulfill their Tribal engagement requirements.

Thank you for the opportunity to review GAO's recommendations. We look forward to working with GAO in the future.

Sincerely,

Kris A Monteith

Kris Anne Monteith
Chief, Wireline Competition Bureau

Patrick Webre
Chief, Consumer and Governmental Affairs Bureau

In: Broadband
Editor: Orlando Harrison

ISBN: 978-1-53616-409-1

Chapter 8

TRIBAL BROADBAND: FEW PARTNERSHIPS EXIST AND THE RURAL UTILITIES SERVICE NEEDS TO IDENTIFY AND ADDRESS ANY FUNDING BARRIERS TRIBES FACE*

United States Government Accountability Office

ABBREVIATIONS

BIP	Broadband Initiatives Program
BTOP	Broadband Technology Opportunities Program
CAF	Connect America Fund
EDA	Economic Development Administration
ETC	eligible telecommunications carrier
FCC	Federal Communications Commission
KCEC	Kit Carson Electric Cooperative
NoaNet	Northwest Open Access Network
NTIA	National Telecommunications and Information Administration
NTUA	Navajo Tribal Utility Authority
RUS	Rural Utilities Service

* This is an edited, reformatted and augmented version of United States Government Accountability Office; Report to Congressional Requesters, Publication No. GAO-18-682, dated September 2018.

Why GAO Did This Study

In 2018, FCC estimated that 35 percent of Americans living on tribal lands lack broadband service compared to 8 percent of Americans overall. Various federal programs support increasing broadband deployment in unserved areas, including tribal lands. Tribes can form partnerships with private sector companies and others to deploy broadband infrastructure on tribal lands. GAO was asked to provide information on these partnerships.

This chapter discusses (1) examples and outcomes of tribal partnership arrangements, (2) the amount of federal funding provided to tribal entities for broadband deployment, and (3) stakeholder-identified barriers that tribes face in obtaining federal funding and the extent to which federal agencies have addressed those barriers. GAO identified partnerships by reviewing federally funded broadband projects that included a partnership component; analyzed federal funding dedicated to broadband deployment; interviewed agency and tribal government officials, tribal associations, tribally owned broadband providers, and industry stakeholders; and assessed RUS's efforts to address the regulatory funding barriers tribes may face. The information presented is illustrative and is not generalizable to all tribes or all partnerships.

What GAO Recommends

GAO recommends that RUS identify and address regulatory barriers that impede tribal entities from obtaining RUS funding for broadband deployment. RUS neither agreed nor disagreed with this recommendation.

What GAO Found

GAO identified some partnership arrangements between tribes and other entities to increase broadband access on tribal lands. Among the seven examples GAO identified, tribes partnered with different types of entities including private broadband providers, a community access network provider, an electric cooperative, a regional consortium, and tribally owned broadband providers. According to the tribes participating in the partnerships, almost all of the partnerships improved broadband service on tribal lands.

The Federal Communications Commission (FCC) and the Rural Utilities Service (RUS) are the primary sources of federal funding to deploy broadband infrastructure where the cost of providing service is high, including on tribal lands. GAO reviewed funding for four programs, three in FCC and one grant program in RUS, and found that in total, less than 1 percent has gone directly to tribes or to tribally owned broadband providers to expand broadband service. GAO found that 14 tribal entities received federal funding from FCC and RUS to increase broadband deployment for 2010–2017.

The tribal officials, tribal associations, and tribally owned broadband providers GAO contacted cited several barriers tribes face in obtaining federal funding to deploy broadband service on tribal lands. For example, they said tribes face regulatory barriers in applying for RUS's grant funding, including (1) preparing existing and proposed network design, (2) demonstrating financial sustainability of the broadband project within 5 years, and (3) obtaining matching funds. An interagency council recently recommended that federal agencies identify and address regulatory barriers that may unduly impede broadband deployment. Although RUS conducts some outreach with tribes, it has not undertaken a formal assessment to identify and address the regulatory barriers that tribes may face in obtaining RUS's funding for broadband deployment. RUS officials told GAO that they do not have the resources to do so. Nevertheless, lacking such an assessment, tribes may continue to face regulatory barriers in obtaining RUS funding for broadband deployment on their lands. By identifying and addressing any regulatory barriers that impede tribal entities' access to RUS funding, RUS could help tribes obtain funding to expand broadband deployment on tribal lands.

September 28, 2018

The Honorable John Hoeven
Chairman

The Honorable Tom Udall
Vice Chairman
Committee on Indian Affairs
United States Senate

The Honorable John Barrasso

The Honorable Maria Cantwell

The Honorable Steve Daines

The Honorable Martin Heinrich

The Honorable Heidi Heitkamp

The Honorable Brian Schatz

The Honorable Jon Tester
United States Senate

Broadband service provides users with many opportunities to improve communications, including enhancements in e-commerce, telemedicine, and education tools.[1] In 2018, the Federal Communications Commission (FCC) estimated that 35 percent of Americans living on tribal lands lack broadband service, compared to 8 percent of Americans overall.[2] According to FCC, the gap in broadband access between rural areas and rural tribal lands is even larger. Further, when including mobile broadband service, FCC reported that nearly 60 percent of Americans living in rural tribal lands nationwide lack broadband access, compared to about 31 percent of rural Americans overall.[3] This large difference is due, in part, to the relatively high proportion of tribal lands that are located in rural areas with rugged terrain and low population density, which decreases business incentives to invest in broadband infrastructure in these areas. According to FCC, this lack of service on tribal lands could impede efforts by Indian tribes to achieve self-governance and promote economic opportunity, education, public safety, and cultural preservation.

Currently, the federal government has programs that support increasing broadband deployment in rural and unserved areas (including tribal lands). Most notably, Congress has tasked FCC with encouraging the deployment of advanced telecommunications capability in a reasonable and timely fashion to all Americans, including those in rural areas and on tribal lands.[4] To accomplish this requirement, FCC is responsible for implementing Universal Service Fund programs, one of which, the Connect America Fund, provides approximately $4.5 billion annually to support broadband service in underserved and unserved areas.[5] The Rural Utilities Service (RUS), a component agency of the U.S. Department of Agriculture, also has the Community Connect Program that

---

[1] The term "broadband" commonly refers to Internet access that is high speed and provides an "always-on" connection, so users do not have to reestablish a connection each time they access the Internet.

[2] See FCC, *Inquiry Concerning Deployment of Advanced Telecommunications Capability to All Americans in a Reasonable and Timely Fashion,* 33 FCC Rcd 1660 (2018) (2018 Broadband Deployment Report). For the purposes of this chapter, we use the definition of "tribal lands" from this chapter. That report defines tribal lands as: (1) Joint Use Areas; (2) legal federally recognized American Indian area consisting of reservation and associated off-reservation trust land; (3) legal federally recognized American Indian area consisting of reservation only; (4) legal federally recognized American Indian area consisting of off- reservation trust land only; (5) statistical American Indian area defined for a federally recognized tribe that does not have reservation or off-reservation trust land, specifically a tribal designated statistical area or Oklahoma tribal statistical area; (6) Alaskan Native village statistical area, and; (7) Hawaiian Home Lands established by the Hawaiian Homes Commission Act of 1921. See Federal Communications Commission (FCC), *2018 Broadband Deployment Report*, FCC 18-10 (Washington, D.C.: Feb. 2, 2018).

[3] Levels of broadband access may vary between specific tribal lands, which may not be reflected in these nationwide figures. We previously reported that FCC's broadband data overstate broadband access on tribal lands. As such, it is possible that the percentage of Americans living on tribal lands that lack broadband access is higher than reported. See GAO, *Broadband Internet: FCC's Data Overstate Access on Tribal Lands,* GAO-18-630 (Washington, D.C.: Sept. 7, 2018).

[4] 47 U.S.C. § 1302(a). Advanced telecommunications capability enables users to originate and receive high-quality voice, data, graphics, and video telecommunications using any technology.

[5] 47 U.S.C. § 254.

provides grant funding to improve broadband service.[6] Tribes as well as partnerships between tribes and other entities may apply for such funding in order to expand service on tribal lands.

You asked us to examine several issues related to broadband access on tribal lands, including the use of partnership arrangements between tribal entities—that is, tribal governments and telecommunications providers owned by tribes—and other entities. This is the second report we are issuing in response to your request.[7] This chapter discusses (1) examples of partnership arrangements that tribal entities have used to increase broadband deployment on tribal lands and the outcomes of those partnerships, (2) the amount of funding provided to tribal entities for broadband deployment from key federal programs, and (3) stakeholder- identified barriers that tribal entities face in obtaining federal funding for broadband deployment and the extent to which federal agencies have taken action to address those barriers. In this chapter, we use the term "partnerships" to refer to when an Indian tribe works with another entity to design, build, or operate infrastructure assets to improve or enhance broadband access.

To address these objectives, we reviewed the Communications Act of 1934, as amended,[8] the *National Broadband Plan*,[9] and relevant literature identified through a literature search of tribal partnerships for broadband deployment. In addition, we reviewed documentation and interviewed officials from FCC, RUS, and the Department of Commerce's National Telecommunications and Information Administration (NTIA).[10]

To identify examples of tribal broadband partnerships for our review, we first interviewed agency officials, tribes, private providers, and tribal associations to determine if they were aware of any partnerships focused on broadband deployment on tribal lands. We then identified any broadband projects with a tribal partnership component by reviewing financial information from 2010 to 2017 for the following federal programs: (1) FCC's Universal Service Fund high-cost program and Connect America Fund (including the Mobility Fund Phase I (Auction 901) and Tribal Mobility Fund Phase I (Auction 902)); (2) RUS's Community Connect Grant Program; (3) RUS's Broadband Initiatives Program; and (4) NTIA's Broadband Technology Opportunities Program. While there may be other tribal partnerships that exist, through these efforts we found seven tribal partnership examples that we discuss in this chapter. The identified tribal broadband partnerships are intended to be illustrative examples and, accordingly, are not generalizable to others that may exist.

---

[6] Other federal programs can also be used to fund broadband deployment, including additional RUS programs. A list of funding resources is available at: https:// broadbandusa.ntia.doc.gov/funding-list.

[7] GAO-18-630.

[8] Pub. L. No. 73-416, 48 Stat. 1064 (1934) codified as amended at 47 USC §§ 151 et seq.

[9] In March 2010, an FCC task force issued the *National Broadband Plan* that included a centralized vision for achieving affordability and maximizing use of high-speed Internet. See, FCC, *Connecting America: The National Broadband Plan* (Washington, D.C.: 2010).

[10] NTIA advises the President on telecommunications and information policy issues, including broadband access.

To determine the amount of funding from key federal programs provided to tribal entities for broadband deployment, we first identified the federal programs that provide broadband funding from NTIA's *Guide to Federal Funding of Broadband Projects*.[11] The guide lists 17 federal programs that fund broadband infrastructure. Of those federal programs, we focused our review on four programs—three in FCC and one grant program in RUS— selected because they provide the most directly relevant funding for broadband deployment in unserved areas, which includes tribal lands.[12] We compiled total funding data for these four federal programs and the amount of funding provided to tribes and tribal entities for broadband deployment projects for 2010 to 2017. We took steps to assess the reliability of the data, such as cross-checking the data, following up with agency officials, and reviewing documentation, and found the data were sufficiently reliable for the purposes of summarizing total funding and the amount provided to tribes and tribal entities. Because we relied upon titles or names of grant recipients to identify those grants awarded to tribes and tribally owned providers, our analysis may not include some grants awarded to broadband providers that deploy infrastructure to larger service areas that may also include tribal lands. We obtained stakeholder views on barriers that tribal entities face in obtaining federal funding and federal government efforts to address those barriers by interviewing FCC and RUS officials and 17 tribal government officials, 9 tribally owned broadband providers, and 5 tribal associations. We selected tribal governments to interview to include variation in geographic location, level of broadband deployment, and population density. We visited six tribes in Idaho, New Mexico, and Washington State. We also interviewed other stakeholders selected to represent a range of views and those with experience working with Indian tribes and broadband service. These stakeholders included 10 private broadband providers and associations; 2 regional consortium; 2 community access providers; an academic institution; and a cooperative. The views obtained from the interviews are not generalizable to all tribes, all broadband providers, or all industry stakeholders. We reviewed a report from the Broadband Opportunity Council recommending that agencies identify and address regulatory barriers that may unduly impede broadband deployment and assessed RUS's efforts to address the regulatory barriers tribes may face in attempting to obtain RUS funding for broadband deployment.[13] Appendix I describes our scope and methodology in greater detail.

---

[11] Department of Commerce, National Telecommunications and Information Administration, *Broadband USA: Guide to Federal Funding of Broadband Projects* (June 2017).

[12] We found that broadband deployment and infrastructure projects are often included in federal funding programs as auxiliary components of the project and that the primary purpose is not telecommunications related, but rather related to healthcare, telemedicine, or economic growth. Therefore, we excluded those types of programs from our review.

[13] The Broadband Opportunity Council was created by the March 2015 Presidential Memorandum "Expanding Broadband Deployment and Adoption by Addressing Regulatory Barriers and Encouraging Investment and Training." The Broadband Opportunity Council includes 25 federal agencies and departments with missions or programs with the potential to drive broadband infrastructure investment and adoption. The memorandum asked the Council to produce specific recommendations to increase broadband deployment, competition, and adoption through executive actions within the scope of agency programs, mission, and budgets.

We conducted this performance audit from September 2017 to September 2018 in accordance with generally accepted government auditing standards. Those standards require that we plan and perform the audit to obtain sufficient, appropriate evidence to provide a reasonable basis for our findings and conclusions based on our audit objectives. We believe that the evidence obtained provides a reasonable basis for our findings and conclusions based on our audit objectives.

BACKGROUND

The federal government has recognized 573 Indian tribes as distinct, independent political communities with inherent powers of a limited sovereignty, which has never been extinguished. These tribes can vary significantly in regard to tribal size, population, and ownership status of land. For instance, some tribal lands include reservations—land set aside by treaty, federal law, or executive order for the residence or use of an Indian tribe.[14] Some tribal lands include parcels with different ownership; parcels may be held in trust by the federal government for the benefit of a tribe or an individual tribal citizen. Trust and restricted lands can affect a tribe's ability to use their land as collateral to obtain a loan. In addition, the size of a tribe's land base can range from less than one square mile to more than 24,000 square miles (the size of West Virginia). Some tribes are located in extremely remote, rural locations and others are located in urban areas.

The term "broadband" commonly refers to Internet access that is high speed and provides an "always-on" connection, so users do not have to reestablish a connection each time they access the Internet. Broadband providers deploy and maintain infrastructure to connect consumers to the Internet and provide Internet service through a number of technologies. Broadband infrastructure may include burying fiber-optic or copper cables, stringing cable on existing poles, or erecting towers for wireless microwave links, which relay wireless Internet connections from tower to tower. Figure 1 illustrates some of the options for broadband deployment infrastructure. To install this infrastructure, providers must obtain permits from government entities with jurisdiction over the land or permission from public utilities to deploy infrastructure on existing utility poles. The federal government has emphasized the importance of ensuring Americans have access to broadband, and a number of agencies, including FCC and RUS, currently provide funding to subsidize broadband deployment in areas in which the return on investment has not attracted private investment. FCC funds a number of programs through the Universal Service Fund, which may increase broadband deployment on tribal lands.[15] One program is the high-cost program (renamed the

[14] The land within the reservation may include a mixture (or checkerboard) of tribal trust land, individual Indian trust land, and non-Indian land.

[15] The Universal Service Fund also includes the Schools and Libraries Support Program (commonly referred to as E-rate). We did not include this program in our review because it primarily provides assistance to eligible

Connect America Fund (CAF) in 2011), which provides financial support to both wireline and wireless telecommunications carriers that provide telecommunications services (referred to as providers in this chapter) to supplement their operating costs to serve consumers in rural or remote areas, where the cost of providing service is high.[16] From 2010 to 2017, a total of $34.5 billion in annual and standalone Universal Service Fund high-cost support was disbursed to providers, as follows:

- In total, the high-cost program and Connect America Fund provided $34.1 billion from 2010 to 2017 in financial support to providers, consisting of annual disbursements between $3.7 and $5 billion.
- The Mobility Fund Phase I provided $300 million in 2012 in one-time support to providers to expand broadband service in areas where service was not available, including tribal lands.[17]
- The Tribal Mobility Fund Phase I provided $49.8 million in 2014 in one-time support to providers to deploy broadband service to unserved tribal lands.[18]

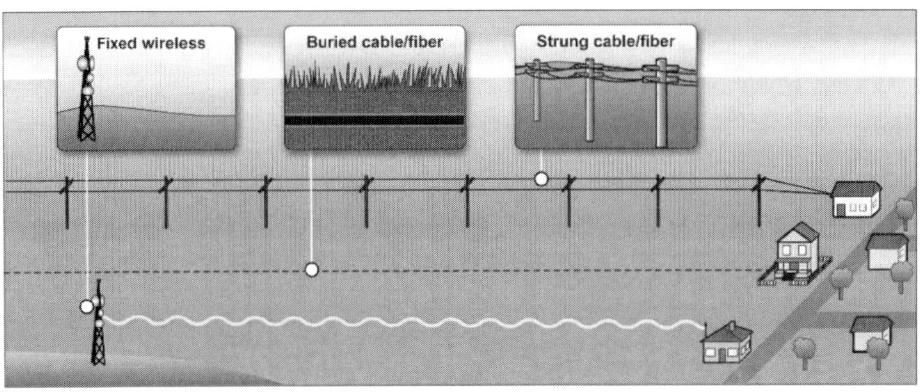

Source: GAO. | GAO-18-682

Figure 1: Examples of Broadband Infrastructure.

---

schools and libraries through discounted telecommunications and information services, rather than funding broadband deployment.

[16] In November 2011, FCC fundamentally reformed the high-cost program so that it could support both telephone and broadband service. See FCC, *Connect America Fund* et al., Report and Order and Further Notice of Proposed Rulemaking, 26 FCC Rcd 17663 (2011).

[17] Those providers wishing to receive this support were asked to indicate the amount of one-time support they needed to provide service to a particular area in a process FCC refers to as a "reverse auction." Winning bids were based on the lowest per-unit bid amounts and one provider per area received support to deploy the broadband-capable network in the areas winning bids. FCC sometimes refers to this competitive bidding process as a reverse auction since the lower bids win. Mobility Fund Phase II will make up to $4.53 billion in support available over 10 years to primarily rural areas that lack unsubsidized 4G Long Term Evolution (LTE) service.

[18] According to FCC's March 2017 Report and Order and Further Notice of Proposed Rulemaking, funds reserved for tribal lands will be included as part of the Mobility Fund Phase II auction. See FCC, *In the Matter of Connect America Fund Universal Service Reform – Mobility Fund*, Report and Order and Further Notice of Proposed Rulemaking, FCC 17-11, 32 FFC Rcd 2152 (2017)).

To be eligible to receive for Universal Service Fund program support, a provider must be designated an eligible telecommunications carrier (ETC) by the appropriate state or by FCC. Under FCC rules, which many state programs mirror, ETCs must meet certain service obligations as described below:[19]

- provide a 5-year plan showing how program support will be used to improve its coverage, service quality, or capacity in each service area where it seeks designation;
- demonstrate its ability to remain functional in emergency situations;
- demonstrate that it will satisfy consumer protection and service quality standards;
- offer local usage plans comparable to those offered by the incumbent carrier in the areas for which it seeks designation; and
- acknowledge that it may be required to provide equal access to other providers within the service area if all other ETCs in the designated service area relinquish their designations.

In addition to FCC's funding, RUS has a current program and had a prior program and NTIA had a prior program that provided funding to improve broadband service in unserved or underserved areas.[20] The RUS and NTIA prior programs were authorized by the American Recovery and Reinvestment Act of 2009 (Recovery Act) to expand high-speed Internet service in unserved areas, and there is no current funding for these programs.[21]

- RUS's Community Connect program currently provides grants to rural communities to provide high-speed Internet service to unserved areas.[22] The Community Connect program is significantly smaller than FCC's programs, with $95.2 million awarded to 36 recipients from 2010 to 2017.[23] The purpose of the RUS Community Connect program is to provide financial assistance to eligible applicants that will provide broadband service that fosters economic growth and delivers enhanced educational, healthcare, and public-safety benefits. In addition, RUS previously administered the Broadband Initiatives Program (BIP), authorized by the Recovery Act to expand high-speed Internet service in

[19] 47 C.F.R. § 54.202

[20] RUS has other programs to finance broadband deployment in rural areas for which tribal entities are eligible, but those programs are not the subject of this chapter.

[21] Pub. L. No. 111-5, 123 Stat. 115, 118-119 (2009).

[22] The fiscal year 2018 omnibus spending bill includes $600 million for RUS for a new broadband loan and grant pilot program in which at least 90 percent of the households served by projects receiving a loan or grant do not have sufficient access to broadband.

[23] RUS has a more robust broadband loan program for broadband infrastructure projects, including the Telecommunications Infrastructure Loans and Guarantees program and the Rural Broadband Access Loan and Loan Guarantee program. However, the scope of this review does not include federal loan programs.

unserved areas. BIP funding included $2.2 billion dedicated to deploy broadband infrastructure. Through the program, RUS funded a total of 247 infrastructure projects with the requirement that all projects be fully completed by June 30, 2015. In addition to the infrastructure awards, 12 technical assistance grants went to tribal communities to develop regional plans to provide broadband service in rural areas that remain critically unserved.[24]

- NTIA administered a prior program also authorized by the Recovery Act called the Broadband Technology Opportunities Program (BTOP). NTIA made available $3.1 billion in BTOP funding to deploy broadband infrastructure. Through the program, NTIA awarded a total of 116 infrastructure grants with the requirement that all projects be fully complete within 5 years of the award date.

Selected Tribes Partnered with Various Entities to Increase Broadband Deployment and Outcomes Varied

Although we identified some partnership arrangements between tribes and other entities to increase broadband deployment on tribal lands using prior authorized funding, based on our review, these arrangements are not being used under currently available programs.[25] As previously noted, there are greater costs associated with deploying broadband on unserved tribal lands because the unserved areas are generally rural, with possibly rugged terrain, and have low population densities. Because of these greater costs, there may be little to no private sector incentive to deploy broadband or enter into a partnership arrangement to do so.

During our review, we did not find any partnership arrangements that leveraged currently available federal funding from FCC's CAF or RUS's Community Connect Program. The seven partnership examples we identified were ones that obtained federal funding under past programs, namely BIP and BTOP that were funded by the Recovery Act. Among these examples, tribes partnered with several different types of entities that were eligible to receive federal grants, including (1) private providers; (2) a community access network provider; (3) an electric cooperative; (4) a regional consortium; and (5) tribally owned telecommunications companies (which we will refer to as tribally owned providers). These types of arrangements are explained below. Outcomes of these

[24] Through BIP, RUS funded 19 technical assistance programs, the majority of which went to tribal communities.

[25] Specifically, we looked for broadband projects with a tribal partnership component by reviewing financial information from 2010 to 2017 for (1) FCC's high-cost program and Connect America Fund (including the Mobility Fund Phase I (Auction 901), and Tribal Mobility Fund Phase I (Auction 902)); (2) RUS's Community Connect Program; (3) RUS's BIP; and (4) NTIA's BTOP.

partnership arrangements varied, as reported by tribal officials and other stakeholders we interviewed, but these stakeholders did not always agree on the outcomes.[26]

Private Providers

Private providers can partner with a tribe to deploy broadband infrastructure on tribal lands. We found two instances in which a tribe partnered with private providers to improve broadband service.

- *Pine Telephone Company and Choctaw Nation.* With the land the tribe has jurisdiction over covering over 10 counties across 12,000 square miles in Oklahoma, the Choctaw Nation's lands encompasses about 15 percent of the State of Oklahoma's total area—an area larger than the entire state of Maryland. According to the Oklahoma Department of Transportation, the Choctaw Nation is the largest employer in the southeastern Oklahoma region and its businesses are key contributors to the state's economy. However, a tribal official told us that the tribal government has struggled to meet the tribe's broadband needs. According to the Choctaw Nation official, Pine Telephone Company (Pine), a privately owned company, has a history of partnering with the Choctaw Nation. In 2010, Pine received a BTOP grant of $9.5 million to deploy broadband infrastructure to underserved areas of Southeastern Oklahoma, including Choctaw Nation lands.

  According to tribal officials and representatives from Pine, the partnership enabled tribal government agencies and buildings— including public schools, public safety agencies, fire and police departments, and a health clinic—to get broadband service. The partnership also improved broadband service for the Choctaw Nation. Additionally, Pine had been proactive in partnering with the Choctaw Nation to secure federal grants and assist with land use and rights-of-way issues, according to a tribal official.[27] Pine representatives told us that partnering with the Choctaw Nation had been beneficial based on their common interest to increase broadband service to the area.
- *Inland Cellular, First Step, and Nez Perce.* The Nez Perce Tribe's reservation consists of 750,000 acres located in north central Idaho. Tribal officials told us that the terrain on the reservation makes broadband deployment challenging

[26] We did not independently verify or evaluate the accuracy of project information as reported by the tribal officials and other stakeholders involved in the partnership arrangements.

[27] Securing rights-of-way across Indian lands is an important component of providing tribal lands with the critical infrastructure needed to support economic activity. Obtaining rights- of-ways to deploy telecommunications equipment across tribal lands could involve individual landholders, tribal governments, service providers, and the federal government.

because it has very large hills and deep valleys; additionally the reservation is sparsely populated. Tribal officials told us that prior to 2010, there was no broadband service available on the Nez Perce reservation. In 2010, the tribe received a BTOP grant of $1.6 million for the Nez Perce Broadband Enhancement Project; the project was completed in 2013. The tribe used that federal grant to deploy 216 miles of broadband (wireless) infrastructure across its reservation to provide broadband service in four northern Idaho counties. As part of the project, the tribe partnered with two private providers, Inland Cellular and First Step, to expand broadband service on the reservation. The tribe used BTOP funding for infrastructure buildout in areas in need of connectivity, while Inland Cellular and First Step focused their efforts on infrastructure buildout in more populated areas.

According to Nez Perce and Inland Cellular officials, the partnership resulted in broadband service being provided to previously underserved rural communities and 17 community institutions, including schools and public safety organizations. Because the partners each own towers on the reservation, the officials told us they could collocate equipment on each other's towers, an approach that resulted in more reliable service. Further, Nez Perce officials and Inland Cellular representatives told us that their partnership was complementary, in that Inland Cellular offered voice services and the tribe's enterprise offered data services.

Community Access Network Provider

Community access network providers are typically owned and operated by public entities rather than by a private corporation. All profits are reinvested to operate, maintain, and expand the community network. Community access networks focus on building broadband infrastructure that allows multiple Internet service providers to offer their services to customers. For example, rather than having one choice for Internet service, community access network providers will allow several service providers to compete for customers.

- *Northwest Open Access Network (NoaNet).* NoaNet, a utility network that offers communities access to broadband infrastructure, has deployed infrastructure in rural areas of Washington State, including on tribal lands. NoaNet received two BTOP grants in 2010—one grant for $84 million and the other for $54 million—to enhance existing infrastructure and improve broadband service in unserved areas. NoaNet representatives told us that over the course of several years, NoaNet deployed 2,300 miles of fiber-optic cable across tribal lands in Washington State and partnered with several Indian tribes and nations, including the Kalispel Indian Community of the Kalispel Reservation, Lower Elwha Tribal Community, and Yakama Nation, to deploy broadband infrastructure. For example, NoaNet representatives told us they partnered with Yakama Nation and exchanged a NoaNet-owned asset for access to a power source and the right to install fiber-optic lines on Yakama tribal land.

According to NoaNet representatives, NoaNet's infrastructure buildout improved broadband services and created new economic development opportunities for several tribes in Washington State. For example, they said NoaNet collaborated with Yakama Nation Networks—a wireless network and tribal enterprise serving the tribe— to provide faster broadband service to the reservation. Further, the NoaNet representatives said the availability of broadband service created new technical jobs with professional growth opportunities on the reservation. NoaNet representatives added that NoaNet enabled high-speed Internet service to the Makah Tribe's health clinic, government offices, school, and library, where they previously had no Internet service at all. Moreover, they told us that partnerships are beneficial in helping tribes gain telecommunications experience. Similarly, according to a tribal representative from Jamestown S'Klallam Tribe, NoaNet's infrastructure buildout helped the tribe obtain broadband services for its library and also helped create economic opportunities for the tribe.

Electric Cooperative

Rural electric cooperative networks typically serve areas that have low population density where traditional providers do not want to serve because of limited opportunities for financial return on investment.

- *Kit Carson Electric Cooperative and Taos Pueblo.* Kit Carson Electric Cooperative (KCEC) is a member-owned, nonprofit electric distribution cooperative that operates a fiber-optic broadband network. In 2010, KCEC received $64 million in grant funding from RUS's BIP to create a 2,400-mile broadband network in northern New Mexico and provide broadband service to businesses and homes, including those on the Taos Pueblo and Picuris Pueblo.

  In an August 2016 presentation to the New Mexico state legislature, KCEC stated that that the project connected tribal members and community institutions, created job opportunities, and improved public safety by improving emergency communications services.[28] According to Taos Pueblo officials, the impetus to work with KCEC was to improve broadband service to meet immediate

[28] Kit Carson Electric Cooperative presentation to the New Mexico Committee on Science, Technology, and Telecommunications presented by Luis A. Reyes, Chief Executive Office on August 22, 2016.

economic, education, health service, and public safety needs of the tribe. However, based on our meetings with both KCEC representatives and Taos Pueblo officials they have different perspectives about the success of this partnership at delivering broadband service to the tribe. For example, KCEC representatives told us that the cooperative constructed the fiber-optic network and connected the government buildings and homes of Taos Pueblo and Picuris Pueblo members as promised, and that KCEC has responded to service interruptions when they occurred on tribal lands. On the other hand Taos Pueblo officials told us, that KCEC did not deploy broadband infrastructure to enable service to all homes and buildings on tribal lands as the tribe had expected. Similarly, KCEC representatives told us that they worked regularly with the Taos and Picuris tribal governments and had good relationships with them; they noted that they meet with tribal leadership every quarter to maintain effective communications and address any issues. In contrast, according to Taos Pueblo officials, KCEC did not solicit tribal input when building out the fiber-optic network, and only met with Taos Pueblo officials about once a year and did not follow up on the issues the tribe raised. Further, according to KCEC representatives, in its federal funding application, KCEC made a commitment that the Taos and Picuris tribal lands would be the first areas targeted for building out the network, and the representatives said that KCEC completed 100 percent of the construction and connected the tribal governments as promised. Taos Pueblo officials, however, said that their tribe was the last to receive service and that KCEC did not complete the broadband construction, including service to the homes of some of its members, because KCEC exhausted its BIP funding.

Regional Consortium

We have previously reported that regional consortium, which are typically formed by groups to undertake an enterprise beyond the resources of any one member, can sponsor regional networks that focus on building broadband networks and providing broadband services to schools, medical providers, public safety agencies, and other community institutions.[29]

- *North Central New Mexico Economic Development District.* Located in northern New Mexico, the Pueblo of Pojoaque, Santa Clara Pueblo, Tesuque Pueblo, and Ohkay Owingeh partnered with local governments to establish the North Central New Mexico Economic Development District (the District), a regional

[29] GAO, *Telecommunications: Projects and Policies Related to Deploying Broadband in Unserved and Underserved Areas,* GAO-14-409 (Washington, D.C.: Apr. 23, 2014).

consortium, to address the socio-economic needs of its members. In 2008, regional planners and government officials identified broadband as the region's number- one infrastructure priority because rural north central New Mexico relied significantly on dial-up Internet service and lacked affordable service to small businesses, libraries, schools, and other community institutions. In 2010, the District received a BTOP grant of $10.6 million to build a community-owned broadband network, known as REDI Net. The District sought the BTOP grant to improve rural healthcare services, make public and higher education more accessible, and improve local government services, like public safety. REDI Net's construction included upgrading existing infrastructure and deploying 136 miles of new fiber-optic cable across the region and on pueblo lands to replace low-performing dial-up service with faster, more affordable broadband service.
According to the project's progress report submitted to NTIA, the partnership enabled broadband infrastructure to be deployed across the four participating lands and connect 110 community institutions. The project's description stated that REDI Net was being used to deliver telemedicine services, distance-learning applications, and critical communications for emergency first-responders. According to REDI Net representatives, in 2017, REDI Net became a standalone organization, separate from the District, and currently charges a monthly fee for the pueblos to use the broadband network. A REDI Net representative told us that the biggest outcome of the partnership has been the improved relationships and collaboration among the Pueblo of Pojoaque, Santa Clara Pueblo, Tesuque Pueblo, and Ohkay Owingeh and other local municipalities.

Tribally Owned Providers

Some tribes have created their own telecommunications companies to provide broadband access to their communities. Based on the examples we identified, a tribe may create its own telecommunications or broadband company or a tribe may partner with an existing tribal enterprise such as an electrical utility to provide broadband services.

- *Navajo Nation and Navajo Tribal Utility Authority.* The Navajo Nation—which spans across Arizona, New Mexico, and Utah— partners with a tribally owned entity, the Navajo Tribal Utility Authority (NTUA), to provide broadband service to residents and households. According to a NTUA representative, the Navajo Nation has diverse, challenging terrain—which includes canyons, valleys, timber forest, desert, and mountains—making it difficult to provide broadband service to tribal residents. In 2010, NTUA received a BTOP grant of $32 million to deploy broadband infrastructure covering 15,000 square miles across the three states.

According to the project's progress report submitted to NTIA, by 2013, NTUA leveraged BTOP funding to deploy 570 miles of fiber-optic cable and 775 miles of wireless infrastructure resulting in a total of 1,345 new network miles.

According to NTUA representatives, the partnership between the Navajo Nation and NTUA increased broadband deployment on the nation and created new opportunities for NTUA to partner with other private providers to further expand broadband services. For example, NTUA representatives said NTUA partnered with a private broadband provider, Commnet, to deploy wireless broadband infrastructure that enabled tribal citizens to receive 4G LTE service. NTUA and Commnet representatives told us NTUA's relationship with Navajo Nation represented an attractive business opportunity for Commnet because of NTUA's established rights-of-ways on the Navajo Nation's tribal lands.

- *Saint Regis Mohawk Tribe and Mohawk Networks.* The Saint Regis Mohawk Tribe, located in the northern region of New York, received a $10.5 million BIP grant in 2010 to complete a large broadband project expanding access to unserved areas. According to tribal officials and Mohawk Networks representatives, the tribe completed a $15 million broadband infrastructure project laying 68 miles of fiber and connecting 1,500 tribal households and community institutions. Upon completion of the BIP broadband project, the tribal officials said the tribe launched its tribally owned broadband provider, Mohawk Networks, LLC in 2015, to respond to tribal residents' need for reliable, cost-effective broadband service. Tribal officials said Mohawk Networks currently provides high-speed Internet to tribal homes and businesses.

  According to tribal officials, in addition to providing broadband service to tribal residents for the first time, the partnership between Saint Regis Mohawk Tribe and its tribally owned broadband provider created new jobs and opportunities to expand broadband services. For example, the officials said the partnership resulted in the creation of a tribal subsidiary, North Country Broadband Services, Inc., to deploy wireless infrastructure to neighboring counties, thus generating new revenue for Mohawk Networks.

Few Federal Funds were Provided to Tribal Entities to Increase Broadband Deployment from 2010 to 2017

FCC and RUS are the primary sources of federal funding to deploy broadband infrastructure in rural and remote areas where the cost of providing service is high, including tribal lands. Based on our review of the funding provided by four federal programs targeted to increase deployment in unserved areas, very little has gone directly to tribes or to tribally owned broadband providers. Specifically, from 2010 to 2017, we found that less than 1 percent of FCC funding and about 14 percent of RUS funding went directly to tribes and tribally owned providers. Combined, FCC and RUS funding totaled $34.6 billion during that time period and tribes and tribally owned providers received $235 million, or about 0.7 percent.[30]

While the majority of the funding from the four programs we reviewed from both agencies is provided to deploy broadband to rural, unserved, or underserved areas, only one source of funding, FCC's Tribal Mobility Fund Phase I, is dedicated specifically to deploying broadband on tribal lands.[31]

The National Broadband Plan stated in 2010 that tribes needed substantially greater financial support than was available to them at the time and that accelerating tribal broadband deployment would require increased funding.[32] Furthermore, the National Congress of American Indians expressed concerns that the needs for federally funded broadband projects are greater on tribal lands but tribes do not receive the appropriate share of federal funding aimed at increasing broadband deployment.[33]

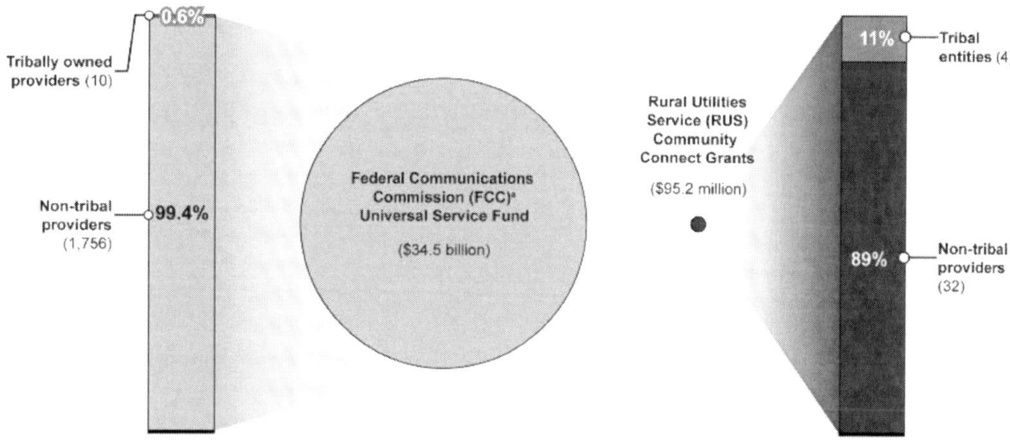

Sources: GAO analysis of FCC and RUS data. | GAO-18-682
ª FCC funding includes Mobility Fund Phase I, Tribal Mobility Fund Phase I, and Connect America Fund.

Figure 2. Percentage of Tribal Entities and Non-Tribal Broadband Providers Receiving Funds from Ongoing FCC and RUS Programs to Increase Broadband Deployment, 2010-2017.

Through our analysis we found that 14 tribal entities received federal funding from FCC and RUS to increase broadband deployment from 2010-2017 (see Figure 2). Of the

---

[30] According to RUS officials, Community Connect grant funding has also been awarded to non-tribal providers that serve tribal lands.
[31] The Mobility Fund Phase II will include a tribal reserve to ensure some of the support is directed to tribal lands.
[32] FCC, *Connecting America: The National Broadband Plan* (Washington, D.C., 2010).
[33] According to its website, the National Congress of American Indians is the oldest, largest, and most representative American Indian and Alaska Native organization serving the broad interests of tribal governments and communities.

four main programs we reviewed, tribes and tribally owned providers received the following funds:

- Connect America Fund: Nine tribally owned providers received high- cost support funding totaling $218.1 million.
- Mobility Fund Phase I: One tribally owned provider received support totaling $3.3 million.
- Tribal Mobility Fund Phase I: No tribal providers received funding.
- RUS Community Connect Grants: Four tribal entities received $13.5 million.

STAKEHOLDERS CITED BARRIERS FOR TRIBES TO OBTAIN FEDERAL FUNDING AND FEDERAL AGENCIES HAVE TAKEN FEW ACTIONS TO ADDRESS THE BARRIERS

The tribal officials, tribal associations, and tribally owned broadband providers we interviewed cited several barriers that tribes may face when seeking federal funding for broadband deployment. The two primary barriers these interviewees cited were (1) the statutory requirement for ETC designation and (2) grant application requirements.

Statutory Requirement for ETC Designation

FCC's Connect America Fund (CAF) is the largest source of federal funding for broadband deployment in unserved and underserved areas; however, very few tribes are currently eligible for this source of funding. At the time of our review, FCC officials told us there were 11 tribes that have providers that are designated as ETCs and therefore would be eligible to receive CAF funding. Although FCC adopted rules in 2011 to create CAF and modernize the program so that it could support broadband capable networks, FCC officials told us that most ETCs are the telephone companies that were in existence when Congress passed the Telecommunications Act of 1996.[34] According to FCC officials, FCC has explored whether it has authority to allow non-ETC providers to receive CAF support payments but determined that the statute is clear that only ETCs can receive program support. Between 2012 and 2017, FCC officials said FCC received nine ETC applications, four of which were from tribally owned providers. Of those four, only one tribally owned provider was designated an ETC. Three tribes we contacted said they would like the opportunity to receive CAF support to deploy broadband on tribal lands,

[34] Pub. L. No. 104-104, 110 Stat. 56 (1996).

but they realize they are not eligible to receive funding unless they have the ETC designation. Moreover, officials from two tribes and a tribal association stated that while they want to provide broadband services in their communities, they did not seek the ETC designation because of the ETC service obligations described above.

The Leech Lake Band of Ojibwe applied for ETC status in 2013. We met with tribal officials who told us that the tribe was providing broadband service in its community through its own, tribally chartered telecommunications company and at the time of our visit, they had been waiting several years for a decision from FCC on their ETC application. The tribal officials told us that if FCC did not make a decision soon, the tribal government would need to shut down the broadband network, as the tribe's original decision to fund the network assumed there would be a CAF subsidy to help defray the costs. The Leech Lake reservation is rural with low population density and is surrounded by the Chippewa National Forest. Subsequent to our meeting with the tribe in November 2017, the tribe withdrew its application in March 2018, noting that it was ceasing its attempt to run its telecommunications company specifically "due to inaction" by FCC.

According to representatives from a tribal association we contacted, FCC has provided ETCs with billions of dollars to deploy service to unserved areas through the Universal Service Fund programs, but FCC's efforts have not always been successful in the hardest to reach areas, particularly tribal lands. The representatives noted that FCC's competitive market approach does not work where competition cannot be supported and that there needs to be a different approach. Similarly, tribal officials from Idaho told us that rural service providers are able to operate due to CAF support, but the tribe is not eligible to receive those subsidies. Officials said although the provider in their area has received millions of dollars in CAF subsidies, it has not deployed broadband on the tribal lands. Other tribal officials from Washington State told us that although private providers received CAF subsidies to deploy broadband service to their reservation, the private providers told the tribe it would be years before they offer service on tribal lands.

In 2014, FCC conducted its Rural Broadband Experiment to open up eligibility for CAF funding to non-ETC providers. FCC made $100 million available for the experiment and applicants included a diverse group of entities, including competitive providers, electric utilities, wireless Internet service providers, and others. However, while this experiment opened the application process to non-ETC providers, it did not remove the ETC requirement. CAF support awarded through this experiment was provisional pending the broadband providers' obtaining ETC status. According to FCC documentation, there were 181 applicants for the experiment, but only 16 ended up meeting all the requirements to receive funding. None of those 16 entities was tribal.

Grant Application Requirements

Stakeholders we interviewed said tribes may face barriers completing federal grant applications to obtain funding for broadband deployment. In particular, two community access providers, five tribally owned providers, and one regional consortium we contacted said that meeting the application requirements was difficult. Representatives from eight of the tribes we contacted told us that in general, the language included in the federal grant applications is difficult to understand or the administrative requirements of federal grants are burdensome. Another tribal representative told us he would only recommend applying for RUS's Community Connect program if the tribe has an entire team of dedicated people to manage the grant process. Some of the tribal officials we contacted cited difficulties preparing required application materials between the time a grant announcement was made and the submission deadline. For example, tribal officials we contacted from New Mexico and Oklahoma stated that the constrained time frames prevented them from effectively preparing a comprehensive application package. In some cases, the narrow application windows prevented the tribes from applying at all.[35] Furthermore, tribal officials, tribal associations, and tribally owned broadband providers told us that complying with the following regulatory requirements for RUS Community Connect grants could be challenging for tribes:

- *Preparing existing and proposed network design:* RUS's Community Connect program requires applicants to submit information on the network's design that contains all the technical information on the applicant's existing (if applicable) and proposed network. The network design is typically completed by a licensed engineer. Tribal officials in Washington State told us that conducting analyses of existing infrastructure and what improvements are needed can be cost-prohibitive for some tribes because it requires financial resources that the tribe may not have before applying for the grant. Many of these costs are related to the expense of bringing in outside experts or consultants who are needed to perform the technical studies. Another tribal representative told us since the tribe has no way of knowing if the grant will be approved, spending money to complete the application is a large risk. According to RUS officials, the Community Connect program is not authorized to fund pre-planning activities.
- *Demonstrating financial sustainability within 5 years:* The RUS Community Connect grant application requires a "financial forecast" that includes the applicant's existing operations and the proposed project and must be supported by a detailed narrative that explains the methodology and assumptions used to develop the projections, including the number of subscribers projected to take the

[35] For example, the Community Connect grants for 2018 were announced in March 2018, and completed applications were due in May 2018.

applicant's service. The financial forecast must cover at least 5 years, and it is used by RUS to determine whether the proposed project is financially sustainable. However, tribal officials from Idaho told us that it is not feasible for tribes to show financial sustainability (a return on investment) in 5 years in high cost areas. They noted that a period of 15 years may be needed to produce a return on investment in those areas, and this requirement prevents tribes from qualifying for Community Connect grants.

- *Obtaining matching funds required to apply for federal grants:* RUS's Community Connect program requires grant applicants to provide matching funds of at least 15 percent from non-federal sources and does not accept in-kind contributions of goods or services.[36] The matching fund requirement can be difficult for some tribes to obtain. For example, officials from RUS and the tribal entities we contacted told us that tribes often times do not have the upfront cash to meet the matching requirement. According to a tribal association we contacted, obtaining credit is a serious problem for some tribes. In general, tribes cannot collateralize tribal property, and therefore often times are unable to get bank loans for infrastructure projects.

The *National Broadband Plan* recommended that federal agencies facilitate tribal access to broadband funding opportunities. Furthermore, recognizing the need to reduce barriers to expand broadband deployment, the Broadband Opportunity Council, established in March 2015, issued a report stating that federal agencies should use all available and appropriate authorities to identify and address regulatory barriers that may unduly impede either broadband deployment or the infrastructure to augment broadband deployment.[37] RUS officials said they have held a number of external training and outreach events, such as workshops and seminars, with tribes over the past 5 years to provide information about RUS's broadband programs.[38] For example, in April 2018, before the 2018 Community Connect grant's application deadline, RUS hosted a webinar on various requirements for grant applications. RUS officials told us that RUS's outreach efforts generally focus on specific programs and instructing potential applicants on program requirements and how to complete application packages.

However, beyond these outreach efforts, RUS officials said they have not undertaken a formal assessment to identify and address the regulatory barriers that tribes may face in

---

[36] Examples of in-kind contributions include the donation of personal service, material, equipment, buildings, land and other non-cash goods or services.

[37] The Broadband Opportunity Council was tasked with producing specific recommendations to increase broadband deployment, competition, and adoption through executive actions within the scope of agency programs, mission, and budgets.

[38] Additionally, according to the Department of Agriculture's Office of Tribal Relations 2017 Communication and Outreach Plan, that office intends to identify and expand current modes of outreach to prospective tribal applicants to ensure that relevant programs and policies are efficient, easy to understand, accessible, and developed in consultation with the American Indian and Alaska Native constituents they impact.

obtaining RUS funding for broadband deployment. When we asked RUS officials about the feasibility of doing so, they said that they have limited resources and multiple competing priorities for those resources. RUS officials also noted that BIP authorized and provided funding for technical assistance for applicants, funding that enabled RUS to address some of the barriers tribes face. Nevertheless, lacking such an assessment, tribes may continue to face the regulatory barriers described above in obtaining RUS funding for broadband deployment on their lands.[39] According to the *National Broadband Plan*, local entities (including tribal, state, regional, and local governments) decide to offer broadband services when no providers exist that meet local needs, and local entities do so after trying to work with established carriers to meet local needs. Several of the tribes we visited told us they were trying to deploy broadband infrastructure or offer service because the private providers were not building out on their lands. For example, one tribe stressed that unlike private providers, they would prioritize tribal areas needing broadband service, but they need federal funding to do so.

CONCLUSION

An estimated 35 percent of Americans living on tribal lands lack broadband service, which could hinder tribal efforts to promote self- governance, economic opportunity, education, public safety, and cultural preservation. However, little federal funding aimed at increasing broadband service actually goes to tribal entities, even though the *National Broadband Plan* stressed that tribes needed substantially greater financial support and recommended that federal agencies facilitate tribal access to broadband funding opportunities. Tribes may face barriers in obtaining federal funds to deploy broadband, and the Broadband Opportunity Council recognized the need for federal agencies to reduce the barriers that are impeding broadband deployment. However, RUS has not taken steps to identify or address the barriers tribes face when applying for RUS grant funding. By identifying and addressing any regulatory barriers that impede tribal entities' access to RUS funding, RUS could help tribes obtain funding to expand broadband deployment on tribal lands.

RECOMMENDATION FOR EXECUTIVE ACTION

The Secretary of Agriculture should direct the Administrator of RUS to undertake an assessment to identify any regulatory barriers that may unduly impede efforts by tribes to

---

[39] Because the information we present is not generalizable to all tribes, tribes may face other barriers in obtaining RUS funding.

obtain RUS federal grant funds for broadband deployment on tribal lands and implement any steps necessary to address the identified barriers. (Recommendation 1)

AGENCY COMMENTS

We provided a draft of this chapter to FCC, RUS, and NTIA for comment. FCC and RUS provided technical comments, which we incorporated as appropriate; NTIA did not have any comments. A Department of Agriculture official indicated in an e-mail message that RUS neither agreed nor disagreed with the recommendation. RUS's technical comments noted that RUS has and will continue to work with tribes to facilitate broadband deployment, whether tribes have the desire and capacity to provide the service or whether another provider is able to bring that service to tribal areas.

We are sending copies of this chapter to the appropriate congressional committees, the Chairman of FCC, the Secretary of Agriculture, the Secretary of Commerce, and other interested parties.

Mark L. Goldstein
Director, Physical Infrastructure

APPENDIX I: OBJECTIVES, SCOPE, AND METHODOLOGY

This chapter discusses (1) examples of partnership arrangements that tribal entities have used to increase broadband deployment on tribal lands and the outcomes of those partnerships, (2) the amount of funding provided to tribal entities for broadband deployment from key federal programs, and (3) stakeholder-identified barriers that tribal entities face in obtaining federal funding for broadband deployment and the extent to which federal agencies have taken action to address those barriers.

To address these objectives, we reviewed relevant federal statutes, including the Communications Act of 1934, as amended,[40] and Federal Communications Commission's (FCC) regulations, orders, and policy statements including FCC's *Statement of Policy on Establishing a Government-to-Government Relationship with Indian Tribes*.[41] In addition, we reviewed documentation and interviewed officials from FCC, including officials from the Office of Native Affairs and Policy; U.S. Department of Agriculture's Rural Utilities

---

[40] The Communications Act of 1934, as amended, codified at 47 U.S.C. § 151, et seq.

[41] FCC, *In the Matter of Wireless Accelerating Wireless Broadband Deployment by Removing Barriers to Infrastructure Investment*, Second Report and Order (FCC 18-30) (2018) and In *the Matter of Statement of Policy on Establishing a Government to Government Relationship with Indian Tribes*, 16 FCC Rcd 4078 (2000).

Service (RUS); U.S. Department of Commerce's National Telecommunications and Information Administration (NTIA); and U.S. Department of Housing and Urban Development's Office of Native American Programs.

To gather information on partnership arrangements that tribes have entered to increase broadband deployment on tribal lands and their outcomes, we conducted a review of relevant published literature that included government reports, industry articles, and publications from associations, non-profits, and public policy research organizations.[42] Although we were not able to identify an industry-accepted definition of partnerships, we used the term partnerships to refer to instances in which a tribal nation or tribal government works with another entity to design, build, or operate infrastructure assets, or other capital assets to improve or enhance broadband service. This also included partnerships between a tribe and its tribally owned broadband provider. To identify examples of tribal broadband partnerships for our review, we first interviewed agency officials, tribes, private providers, and other stakeholders such as tribal associations. We also identified broadband projects with a tribal partnership component by reviewing reports from 2010 to 2017 for the following federal programs: (1) FCC's Universal Service Fund high-cost program and Connect America Fund (including the Mobility Fund Phase I (Auction 901) and Tribal Mobility Fund Phase I (Auction 902)); (2) RUS's Community Connect Program; (3) RUS's Broadband Initiatives Program; and (4) NTIA's Broadband Technology Opportunities Program. While there may be other tribal partnership examples that exist, through these efforts we identified seven broadband projects with a tribal partnership component completed within the last 5 years (2013 to 2017). We interviewed tribal leaders and officials from the seven tribes that were involved in the selected partnerships, and visited six tribes in Idaho, New Mexico, and Washington State. When meeting with tribal leaders and officials, we used the same semi-structured interview questions for all tribes; however, tribal officials may not have answered all questions. Because we limited our review to these seven selected partnership examples, our findings are not generalizable.

To determine the amount of funding from key federal programs provided to tribal entities for broadband deployment, we first identified the federal programs that provide broadband funding from NTIA's *Guide to Federal Funding of Broadband Projects*.[43] The

---

[42] For the purposes of this chapter, we use the definition of "tribal lands" from FCC's 2018 Broadband Deployment Report. That report defines tribal lands as: (1) Joint Use Areas; (2) legal federally recognized American Indian area consisting of reservation and associated off-reservation trust land; (3) legal federally recognized American Indian area consisting of reservation only; (4) legal federally recognized American Indian area consisting of off-reservation trust land only; (5) statistical American Indian area defined for a federally recognized tribe that does not have reservation or off-reservation trust land, specifically a tribal designated statistical area or Oklahoma tribal statistical area; (6) Alaskan Native village statistical area, and; (7) Hawaiian Home Lands established by the Hawaiian Homes Commission Act of 1921. See FCC, *Inquiry Concerning Deployment of Advanced Telecommunications Capability to All Americans in a Reasonable and Timely Fashion,* 33 FCC Rcd 1660 (2018) (2018 Broadband Deployment Report).

[43] Department of Commerce, National Telecommunications and Information Administration, *Broadband USA: Guide to Federal Funding of Broadband Projects* (June 2017).

guide lists 17 federal programs that fund broadband infrastructure. Of those federal programs, we focused our review on four programs, three in FCC and one grant program in RUS, selected because they provide the most directly relevant funding for broadband deployment in unserved areas, which includes tribal lands. We first identified federal agencies and programs that provide grants or loans to tribal and non-tribal entities to buildout broadband infrastructure on tribal lands including: FCC, RUS, U.S. Department of Commerce Economic Development Administration (EDA), and NTIA. We interviewed federal agency officials to identify any additional federal programs that provided funding in the last 7 years. We excluded federal loan programs because they may require letters of credit and or assets as collateral, which is often not a feasible option for tribes given land ownership issues. We also considered but excluded those federal programs that are not directly related to broadband expansion and deployment, such as the Department of Housing and Urban Development's Choice Neighborhoods Program, whose primary purpose is housing related. We compiled total funding data for these four federal programs and the amount of funding provided to tribes and tribal entities for broadband deployment projects for 2010 to 2017. We took steps to assess the reliability of the data—such as cross-checking the data, following up with agency officials, and reviewing documentation—and found the data were sufficiently reliable for the purposes of summarizing total funding and the amount provided to tribes and tribal entities. Because we relied upon titles or names of grant recipients to identify those grants awarded to tribes and tribally owned providers, our analysis may not include some grants awarded to broadband providers that deploy infrastructure to larger service areas that may also include tribal lands.

To determine stakeholder-identified barriers that tribal entities face in obtaining federal funding for broadband deployment and federal government efforts to address those barriers, we interviewed FCC and RUS officials and the tribal government officials, tribally owned broadband providers, and tribal associations listed in Table 1. We interviewed representatives from 17 tribes in different locations with varying population sizes and levels of broadband deployment. Additionally, we interviewed officials from 9 tribally owned and 7 private broadband providers operating on tribal lands. We selected these broadband providers to interview because they received federal support to serve on tribal lands or because they were a designated eligible telecommunications carrier (ETC) serving tribal interests. Furthermore, we identified and interviewed industry stakeholders such as research groups and telecommunications associations on their views regarding barriers to obtaining federal program assistance for broadband deployment on tribal lands. These stakeholders were selected based on their exposure to broadband issues on tribal lands such as representing tribally owned broadband providers. The views obtained from these interviews are not generalizable to all tribes, all broadband providers, or all industry stakeholders. We also reviewed a report from the Broadband Opportunity Council directing agencies to identify and address regulatory barriers that may unduly impede broadband deployment and assessed RUS's efforts to address the regulatory barriers tribes may face in attempting to obtain RUS funding for broadband deployment. For a complete list of entities we interviewed see Table 1.

We conducted this performance audit from September 2017 to September 2018 in accordance with generally accepted government auditing standards. Those standards require that we plan and perform the audit to obtain sufficient, appropriate evidence to provide a reasonable basis for our findings and conclusions based on our audit objectives. We believe that the evidence obtained provides a reasonable basis for our findings and conclusions based on our audit objectives.

Table 1. List of Entities Interviewed

Representatives from Tribal Governments
Choctaw Nation of Oklahoma (OK)
Confederated Tribes of the Colville Reservation (WA)
Coeur d'Alene Tribe (ID)
Fond du Lac Band of Lake Superior Chippewa (MN)
Fort Belknap Indian Community (MT)
Jamestown S'Klallam Tribe (WA)
Karuk Tribe (CA)
Leech Lake Band of Ojibwe (MN)
Makah Tribe (WA)
Nez Perce Tribe (ID)
Pueblo of Acoma (NM)
Pueblo of Pojoaque (NM)
Pueblo of San Ildefonso (NM)
Taos Pueblo (NM)
Saint Regis Mohawk Tribe (NY)
Spokane Tribe of Indians (WA)
Yurok Tribe (CA)
Representatives from Tribally Owned Broadband Providers
Gila River Telecommunications, Inc. (AZ)
Hopi Telecommunications, Inc. (AZ)
Mohawk Networks, LLC (NY)
Mescalero Apache Telecommunications (NM)
Navajo Tribal Utility Authority (AZ, NM, UT)
Red Spectrum Communications (Coeur d'Alene Tribe (ID))
San Carlos Apache Telecommunications Utility, Inc. (AZ)
Standing Rock Telecommunications, Inc. (SD)
Warm Springs Telecommunications Co. (OR)
Representatives from Tribal Associations
Affiliated Tribes of Northwest Indians
National Congress of American Indians
National Tribal Telecommunications Association
Native American Finance Officers Association (NAFOA)
Southern California Tribal Chairmen's Association - Tribal Digital Village Network (CA)
Representatives from Companies Groups that Work with Tribes
AMERIND Risk
Mobius Legal Group PLLC

Cooperatives	
Kit Carson Electric Cooperative	
Regional Consortia	
Middle Rio Grande Pueblo Consortium	
North Central New Mexico Economic Development District	
Community Access Providers	
Northwest Open Access Network	
REDI Net	
Representatives from Providers and Trade Associations	
Carlson Wireless	
Century Link	
Commnet	
Connected Nation	
Frontier	
Inland Cellular	
NTCA – The Rural Broadband Association	
Pine Telephone Company	
Rural Wireless Association	
Verizon	

Source: GAO. | GAO-18-682

In: Broadband
Editor: Orlando Harrison

ISBN: 978-1-53616-409-1

Chapter 9

TRIBAL BROADBAND: FCC'S DATA OVERSTATE ACCESS, AND TRIBES FACE BARRIERS ACCESSING FUNDING*

United States Government Accountability Office

Chairman Hoeven, Vice Chairman Udall, and Members of the Committee:

I am pleased to be here today to discuss our September 2018 reports on the Federal Communications Commission's (FCC) data regarding broadband access on tribal lands[1] and barriers tribes face in obtaining federal funding for broadband deployment.[2] Broadband infrastructure is critical for economic development, educational and job opportunities, and public health and safety.[3] However, residents of tribal lands continue to have lower levels of broadband access than other Americans.[4]

---

* This is an edited, reformatted and augmented version of United States Government Accountability Office; Report to Congressional Requesters, Publication No. GAO-19-134T, dated October 2018.

[1] GAO, *Broadband Internet: FCC's Data Overstate Access on Tribal Lands*, GAO-18-630 (Washington D.C.: Sept. 7, 2018).

[2] GAO, *Tribal Broadband: Few Partnerships Exist and the Rural Utilities Service Needs to Identify and Address Any Funding Barriers Tribes Face*, GAO-18-682 (Washington D.C.: Sept. 28, 2018). GAO also has ongoing work related to spectrum use on tribal lands for this Committee, which will be issued later in 2018.

[3] Broadband service may be "fixed"—that is, providing service to a single location, such as a customer's home—or "mobile," that is, providing service wherever a customer has access to a mobile wireless network, including while on the move, through a mobile device, such as a smartphone.

[4] For the purposes of this testimony, we use the definition of "tribal lands" from FCC's 2018 Broadband Deployment Report. That report uses the following definition of tribal lands: (1) Joint Use Areas; (2) legal federally recognized American Indian area consisting of reservation and associated off-reservation trust land; (3) legal federally recognized American Indian area consisting of reservation only; (4) legal federally recognized American Indian area consisting of off-reservation trust land only; (5) Statistical American Indian area defined for a federally recognized tribe that does not have reservation or off- reservation trust land, specifically a Tribal Designated Statistical Area (TDSA) or Oklahoma Tribal Statistical Area (OTSA); (6) Alaskan Native village statistical area; and (7) Hawaiian Home Lands established by the Hawaiian Homes Commission Act of 1921. *See* 33 FCC Rcd 1660 (2018).

Policy-makers have noted the need for accurate information in order to target funding to areas lacking broadband access, and FCC has identified the need to work with tribes to ensure such information is accurate for tribal lands. Currently, the primary source of information regarding where broadband is and is not available is the FCC, which collects this information from broadband providers. FCC collects this data by requiring that fixed and mobile broadband providers report on their broadband deployment by filing a form twice a year (Form 477). FCC uses data from this form to determine which areas qualify for broadband funding.

One barrier to increasing access to broadband on tribal lands is the cost to providers of deploying infrastructure to tribal lands located in rugged, sparsely populated areas. In an attempt to address this and other issues, the federal government administers a number of programs to subsidize broadband deployment in areas in which the return on investment has not attracted private investment. For example, FCC administers the Connect America Fund—a Universal Service Fund program—which provides subsidies to fixed and mobile providers of telecommunications and broadband services in rural, insular, and other remote areas where the cost of providing service is high. To be eligible to receive subsidies under the Connect America Fund, a provider must be designated an eligible telecommunications carrier. In addition, the Rural Utilities Service (RUS) has a current program and had a prior program and the National Telecommunications and Information Administration (NTIA) had a prior program that provided funding to improve broadband service in unserved or underserved areas.[5] The RUS and NTIA prior programs were authorized by the American Recovery and Reinvestment Act of 2009 (Recovery Act) to expand high-speed Internet service in unserved areas, and there is no current funding for these programs.[6]

My statement today discusses: (1) the extent to which FCC's approach to collecting broadband availability data accurately captures the ability of Americans living on tribal lands to access broadband Internet services; (2) the extent to which FCC obtains tribal input on the data; (3) examples of partnership arrangements that tribal entities have used to increase broadband deployment on tribal lands; and (4) barriers that tribal entities face in obtaining federal funding for broadband deployment. This statement is based on two reports that we issued in September 2018.[7] To perform the work for our report on FCC's broadband data, we analyzed FCC's broadband availability data for tribal lands as well as FCC's processes for collecting and using those data. We interviewed FCC officials as well as a non-generalizable sample of tribal and industry stakeholders and reviewed relevant FCC rulemaking proceedings.[8] To perform the work for our report on tribal

[5] Other federal programs can also be used to fund broadband deployment, including additional RUS programs. A list of funding resources is available at: https:// broadbandusa.ntia.doc.gov/funding-list.

[6] American Recovery and Reinvestment Act of 2009, Pub. L. No. 111-5, 123 Stat. 115, 118-119 (2009).

[7] GAO-18-630 and GAO-18-682.

[8] These interviews included representatives from 25 tribal governments or tribally owned providers, including visits to 9 tribal lands, and 10 organizations that include tribal entities or work with tribes on broadband issues.

partnerships and barriers to federal funding, we reviewed program documentation from FCC, RUS, and NTIA. We also interviewed FCC, RUS, and NTIA officials and a non-generalizable sample of representatives from tribal governments, tribally owned broadband providers, and tribal associations. More detailed information about our scope and methodology can be found in our reports.

The work upon which this testimony is based was conducted in accordance with generally accepted government auditing standards.

FCC's Data Overstate Broadband Access on Tribal Lands

In our September 2018 report on broadband access on tribal lands, we found that FCC collects broadband availability data from broadband providers, but its method for collecting the data does not accurately or completely capture broadband access—the ability to obtain service—on tribal lands.[9] Specifically, FCC directs fixed broadband providers to submit a list of census blocks where service is available on their Form 477 filings. In the Form 477 instructions, FCC defines "available"[10] as whether the provider does—or could, *within a typical service interval or without an extraordinary commitment of resources*—provide service to at least one end-user premises in a census block.[11] Thus, in its annual reports and maps of fixed broadband service, FCC considers an entire block to be served if a provider reports that it does, or could offer, service to at least one household in the census block. As shown in Figure 1, FCC's definition of availability leads to overstatements of fixed broadband availability on tribal lands by: (1) counting an entire census block as served if only one location has broadband, and (2) allowing providers to report availability in blocks where they do not have any infrastructure connecting homes to their networks if the providers determine they could offer service to at least one household. FCC has noted that overstatements of availability can be particularly problematic in rural areas, where census blocks cover larger areas.

According to FCC officials, FCC requires providers to report fixed broadband availability where they could provide service to: (1) ensure that it captures instances in which a provider has a network nearby but has not installed the last connection to the homes, and (2) identify where service is connected to homes, but homes have not subscribed. FCC officials also told us that FCC measures availability at the census block level because sub-census block data may be costly to collect. However, FCC

---

[9] GAO-18-630.

[10] We use the term broadband availability to refer to broadband deployment. FCC officials noted that the data collected by the Form 477 reflect broadband deployment. We use the term broadband availability because FCC's Form 477 instructs fixed broadband providers to report fixed broadband deployment by submitting a list of census blocks in which the filer makes broadband connections available.

[11] A "typical service interval" refers to the amount of time between when a customer requests service, and when a provider is able to begin providing service.

acknowledged that by requiring a provider to report where it could provide service, it is not possible to tell whether the provider would be unable or unwilling to take on additional subscribers in a census block it lists as served.[12] In addition, when reporting on broadband access in tribal lands,[13] FCC uses the broadband availability data described above, and does not collect information on factors that FCC and tribal stakeholders have stated can affect broadband access.[14] These factors include affordability, service quality, and service denials.

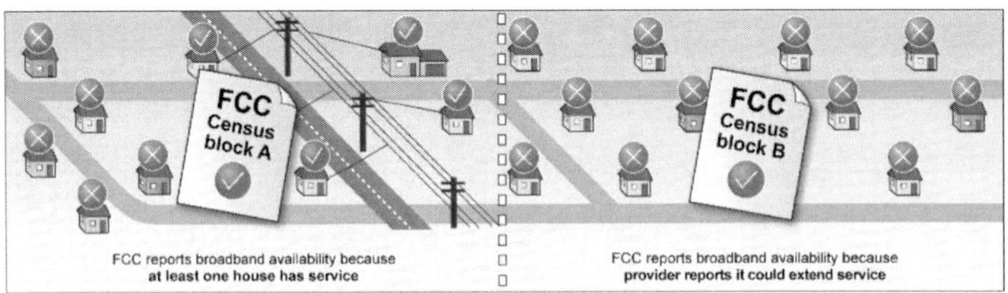

Source: GAO analysis of Federal Communication Commission (FCC) documents. | GAO-19-134T

Figure 1. Overstatement of Broadband Availability in FCC's Form 477 Data.

By developing and implementing methods for collecting and reporting accurate and complete data on broadband access specific to tribal lands, FCC would be better able to target federal broadband funding to tribal areas that need it the most. We recommended FCC develop and implement methods for collecting and reporting accurate and complete data on broadband access specific to tribal lands. FCC agreed with this recommendation and stated that it is exploring methods to collect more granular broadband deployment data.

FCC Does Not Have a Formal Process to Obtain Tribal Input on Its Broadband Data

As we reported in September 2018, FCC does not have a formal process to obtain input from tribes on the accuracy of the data and tribal stakeholders can face difficulties

---

[12] *Modernizing the FCC Form 477 Data Program*, Further Notice of Proposed Rulemaking, 32 FCC Rcd 6329 (2017).

[13] *In the Matter of Inquiry Concerning the Deployment of Advanced Telecommunications Capability to All Americans in a Reasonable and Timely Fashion,* 2018 Broadband Deployment Report, 33 FCC Rcd 1660 (2018).

[14] FCC officials we interviewed stated that FCC has not defined the term "broadband access," and noted that the use of the term may vary across FCC documents. However, FCC and tribal stakeholders have noted that broadband access can be affected by factors such as the affordability and quality of the broadband services being offered and the extent to which providers deny service to those who request it. For example, see 2016 Broadband Progress Report 31 FCC Rcd 699 ¶ 62 (2016); FCC, *National Broadband Plan*; FCC, Strategic Plan 2018-2022. FCC officials also identified the cost of deployment and regulatory barriers as important factors when determining whether an area has access to broadband.

obtaining information from providers.[15] FCC's 2010 *National Broadband Plan* noted the need for the federal government to improve the quality of data regarding broadband on tribal lands and recommended that FCC work with tribes to ensure that any information collected is accurate and useful.[16] Although the *Plan* also noted that tribal representatives should have the opportunity to review mapping data and offer supplemental data or corrections, FCC lacks a formal process to obtain tribal input on its broadband data. FCC officials told us that they address questions and concerns regarding providers' coverage claims submitted to FCC's Office of Native Affairs and Policy. However, about half of the tribal representatives we spoke to stated that they were not aware of the Form 477 data or corresponding maps, or raised concerns about a lack of outreach from FCC to inform tribes about the data. Most of the tribal stakeholders we interviewed told us that FCC should work more directly with tribes to obtain information from them to improve the accuracy of FCC's broadband deployment data for tribal lands. These stakeholders identified several ways in which FCC could work with tribes on this issue, including onsite visits, increased outreach and technical training, and opportunities for tribes to collect their own data or submit feedback regarding the accuracy of FCC's data.

FCC's *National Broadband Plan* also noted the importance of supporting tribal efforts to build technical expertise with respect to broadband issues. A few of the stakeholders we interviewed noted that tribes have faced difficulties when they attempt to challenge FCC's broadband availability data. For example, in 2013, all of the tribal entities that challenged FCC's data on mobile service availability were unsuccessful in increasing the number of eligible areas. A few tribal stakeholders provided varying reasons for this, one of which was the need for more technical expertise to help the tribes meet FCC's requirements regarding the information needed to support a challenge. Because FCC lacks a formal process to obtain tribal input on its broadband data, FCC is missing an important source of information regarding areas in which the data may overstate broadband service on tribal lands.

By establishing a process to obtain input from tribal governments on the accuracy of provider-submitted broadband data as recommended in the *National Broadband Plan*, FCC could help tribes develop and share locally-specific information on broadband access and improve FCC's data for tribal lands. However, the success of such an effort may rely on the tribes' knowledge of, and technical ability to participate in, the process. Thus, we recommended FCC develop a formal process to obtain tribal input on the accuracy of provider-submitted broadband data that includes outreach and technical assistance to help tribes participate in the process. FCC agreed with this recommendation and stated that it will work with stakeholders to explore options for implementing such a process.

---

[15] GAO-18-630.

[16] FCC, *Connecting America: The National Broadband Plan* (Mar. 16, 2010).

Finally, some tribes face challenges accessing data from providers. In 2011, FCC required that providers receiving funds to serve tribal lands meaningfully engage with the tribes and discuss broadband deployment planning.[17] In 2012, FCC issued guidance on meeting this requirement and stated that the guidance would evolve over time based on the feedback of both tribal governments and broadband providers.[18] However, FCC has taken limited steps to obtain such feedback and has not updated the guidance. About half of the tribal stakeholders we interviewed raised concerns about difficulties accessing information from providers regarding broadband deployment on their tribe's lands (which providers may consider proprietary), and some providers told us that they attempt to engage with tribes, but the level of responsiveness they receive from tribes varies. Thus, we recommended, and FCC agreed, that FCC obtain feedback from tribal stakeholders and providers to determine whether it needs to clarify its tribal engagement guidance.

FEW TRIBAL BROADBAND PARTNERSHIPS EXIST

In our September 2018 report on tribal partnerships, we found that partnership arrangements between tribes and other entities to increase broadband deployment on tribal lands are not widespread.[19] Because of the greater costs associated with deploying broadband on unserved tribal lands that are generally rural, with possibly rugged terrain, there may be little to no private sector incentive to deploy broadband or enter into a partnership arrangement to do so. The partnership examples we identified were ones that obtained federal funding under past programs funded by the Recovery Act. Among these examples, tribes partnered with several different types of entities, including private providers, a community access network provider, an electric cooperative, a regional consortium, and tribally owned providers.

TRIBES FACE BARRIERS TO OBTAIN FEDERAL FUNDING FOR BROADBAND DEPLOYMENT

We also reported in September 2018 that FCC and RUS are the primary sources of federal funding to deploy broadband infrastructure in rural and remote areas where the cost of providing service is high, including tribal lands.[20] Based on our review of the

---

[17] *In the Matter of Connect America Fund*, Report and Order and Further Notice of Proposed Rulemaking, 26 FCC Rcd 17663 (2011).

[18] *Office of Native Affairs and Policy, Wireless Telecommunications Bureau, and Wireline Competition Bureau Issue Further Guidance on Tribal Government Engagement Obligation Provisions of the Connect America Fund*, Public Notice, 27 FCC Rcd 8176 (2012).

[19] GAO-18-682.

[20] GAO-18-682.

funding provided by four federal programs targeted to increase deployment in unserved areas, very little has gone directly to tribes or to tribally owned broadband providers. Specifically, we found that from 2010 to 2017, less than 1 percent of FCC funding and about 14 percent of RUS funding went directly to tribes and tribally owned providers. Combined, FCC and RUS funding totaled $34.6 billion during that time period and tribes and tribally owned providers received $235 million, or about 0.7 percent.

FCC's 2010 *National Broadband Plan* stated that tribes needed substantially greater financial support than was available to them at the time and that accelerating tribal broadband deployment would require increased funding. Furthermore, the National Congress of American Indians expressed concerns that the needs for federally funded broadband projects are greater on tribal lands but tribes do not receive the appropriate share of federal funding aimed at increasing broadband deployment.[21] Several of the tribes we visited told us they were trying to deploy broadband infrastructure or offer service because the private providers were not building out on their lands.

Through our analysis, we found that from 2010 to 2017, 14 tribal entities received federal funding from FCC and RUS to increase broadband deployment (see Figure 2).

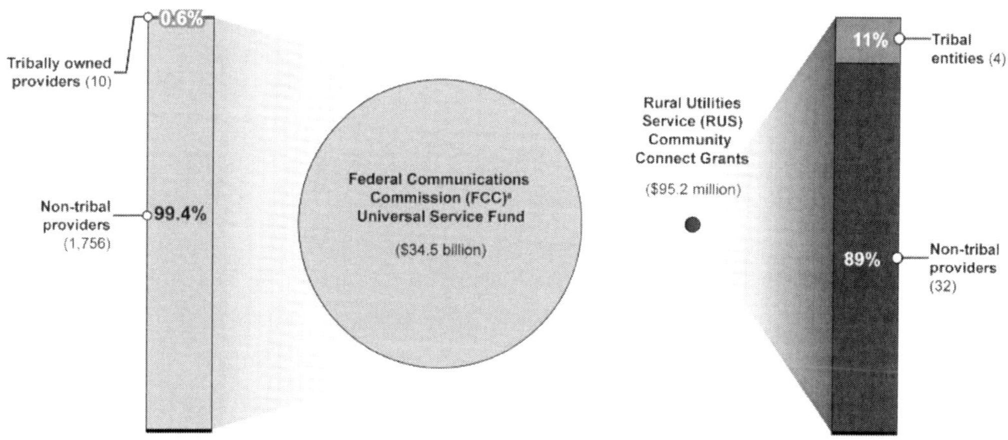

Sources: GAO analysis of FCC and RUS data. | GAO-19-134T

[a] FCC funding includes the Mobility Fund Phase I, Tribal Mobility Fund Phase I, and Connect America Fund.

Figure 2. Percentage of Tribal Entities and Non-Tribal Broadband Providers Receiving Funds from Ongoing FCC and RUS Programs to Increase Broadband Deployment, 2010-2017.

The tribal officials, tribal associations, and tribally owned broadband providers we interviewed cited several barriers that tribes may face when seeking federal funding for broadband deployment. The two primary barriers these interviewees cited were (1) the statutory requirement for the eligible telecommunications carrier (ETC) designation and

[21] According to its website, the National Congress of American Indians is the oldest, largest, and most representative American Indian and Alaska Native organization serving the broad interests of tribal governments and communities.

(2) grant application requirements. Regarding the statutory requirement for ETC designation, FCC officials told us there were 11 tribes that have providers designated as ETCs and therefore would be eligible to receive support from FCC's Connect America Fund (CAF)—the largest source of federal funding for broadband deployment in unserved and underserved areas. Although FCC adopted rules in 2011 to create CAF and modernize the program so that it could support broadband capable networks, FCC officials told us that most ETCs are the telephone companies that were in existence when the Telecommunications Act of 1996 was enacted into law.[22] According to FCC officials, FCC has explored whether it has authority to allow non-ETC providers to receive CAF support payments but determined that the statute is clear that only ETCs can receive program support. Between 2012 and 2017, FCC officials said FCC received nine ETC applications, four of which were from tribally owned providers. Of those four, only one tribally owned provider was designated as an ETC.

According to representatives from a tribal association we contacted, FCC has provided ETCs with billions of dollars to deploy service to unserved areas, but FCC's efforts have not always been successful in the hardest to reach areas, particularly tribal lands. The representatives stated that FCC's competitive market approach does not work where competition cannot be supported and that there needs to be a different approach. Similarly, tribal officials from Idaho told us that although the provider in their area has received millions of dollars in CAF subsidies, it has not deployed broadband on the tribal lands. Other tribal officials from Idaho told us that although private providers received CAF subsidies to deploy broadband service to their reservation, the private providers told the tribe it would be years before they offer service on tribal lands.

Additionally, the tribal officials, tribal associations, and tribally owned broadband providers we interviewed said tribes may face barriers completing federal grant applications to obtain funding for broadband deployment. For example, they said tribes face regulatory barriers in applying for RUS's grant funding, including preparing existing and proposed network design, demonstrating financial sustainability of the broadband project within 5 years, and obtaining matching funds.

The *National Broadband Plan* recommended that federal agencies facilitate tribal access to broadband funding opportunities. Furthermore, recognizing the need to reduce barriers to expand broadband deployment, the Broadband Opportunity Council, established in March 2015, issued a memorandum stating that federal agencies should use all available and appropriate authorities to identify and address regulatory barriers that may unduly impede either broadband deployment or the infrastructure to augment broadband deployment.[23] However, according to RUS officials, RUS has not taken steps

---

[22] Telecommunications Act of 1996, Pub. L. No. 104-104, § 706, 110 Stat. 56, 153 (1996).

[23] The Broadband Opportunity Council was tasked with producing specific recommendations to increase broadband deployment, competition, and adoption through executive actions within the scope of agency programs, mission, and budgets.

to identify or address the barriers tribes face when applying for RUS grant funding due to limited resources and multiple competing priorities for those resources. We recommended that RUS identify any regulatory barriers that may unduly impede efforts by tribes to obtain RUS grant funds for broadband deployment on tribal lands and implement any steps necessary to address the identified barriers. By doing so, RUS could help tribes obtain funding to expand broadband deployment on tribal lands. RUS neither agreed nor disagreed with this recommendation.

Chairman Hoeven, Vice Chairman Udall, and Members of the Committee, this completes my prepared statement. I would be pleased to respond to any questions that you may have.

In: Broadband
Editor: Orlando Harrison

ISBN: 978-1-53616-409-1

Chapter 10

TRIBAL BROADBAND: FCC SHOULD UNDERTAKE EFFORTS TO BETTER PROMOTE TRIBAL ACCESS TO SPECTRUM*

United States Government Accountability Office

ABBREVIATIONS

AWS	Advanced Wireless Services
CBRS	Citizens Broadband Radio Service
FCC	Federal Communications Commission
GHz	gigahertz
Mbps	megabits per second
MHz	megahertz
NPRM	Notice of Proposed Rulemaking
ONAP	Office of Native Affairs and Policy
PCS	Personal Communications Service
WCS	Wireless Communication Services.

WHY GAO DID THIS STUDY

In 2018, FCC estimated that 35 percent of Americans living on tribal lands lack broadband service compared to 8 percent of Americans overall. Broadband service can be

* This is an edited, reformatted and augmented version of United States Government Accountability Office, Report to Congressional Requesters, Publication No. GAO-19-75, dated November 2018.

delivered through wireless technologies using radio frequency spectrum. According to FCC, increasing tribal access to spectrum would help expand broadband service on tribal lands. GAO was asked to review spectrum use by tribal entities— tribal governments and tribally owned telecommunications providers.

This chapter examines (1) tribal entities' ability to obtain and access spectrum to provide broadband services and the reported barriers that may exist, and (2) the extent to which FCC promotes and supports tribal efforts to obtain and access spectrum. GAO interviewed 16 tribal entities that were using wireless technologies. Selected entities varied geographically, among other characteristics. GAO analyzed FCC's license and auction data as of September 2018, reviewed FCC's rulemakings on spectrum for broadband services, and interviewed other tribal and industry stakeholders and FCC officials. The information presented is not generalizable to all tribes or industry participants.

WHAT GAO RECOMMENDS

FCC should (1) collect data on tribal access to spectrum; (2) analyze unused licensed spectrum over tribal lands; and (3) make information available in a more accessible manner that would promote tribes' ability to purchase or lease spectrum licenses over their lands from other providers. FCC agreed with the recommendations.

WHAT GAO FOUND

The tribal entities GAO contacted cited various barriers to obtaining spectrum licenses in bands that can be used to provide broadband services. According to its analysis of data from the Federal Communications Commission (FCC), GAO identified 18 tribal entities that held active spectrum licenses in such bands. For example, of these 18 tribal entities, 4 obtained licenses through secondary market transactions—that is, they bought or leased the license from another provider, and 2 obtained a license through an FCC spectrum auction. Licensed spectrum is generally preferred because it offers better quality of service compared to unlicensed spectrum; however, almost all of the tribal entities GAO contacted said that they are accessing unlicensed spectrum to provide Internet service. They identified barriers to obtaining licensed spectrum through auctions and secondary market transactions, barriers such as high costs and, in the case of secondary market transactions, a lack of information on who holds licenses over tribal lands. Because most spectrum allocated for commercial use has already been assigned, the secondary market is one of very few avenues available to tribal entities that would like to access licensed spectrum.

FCC has taken steps to promote and support tribal access to spectrum. For example, FCC issued proposed rulemakings in 2011 and 2018 that sought comment on tribal-specific proposals, such as establishing tribal-licensing priorities and initiating processes to transfer unused spectrum licenses to tribal entities. However, FCC has not finalized these rules and is in the process of responding to comments to the 2018 rulemaking. Also, while FCC has made additional spectrum available for broadband use in recent years, tribal stakeholders cited limitations with the spectrum FCC has made available. For example, FCC allows broadband providers to operate in unused television broadcast bands on an unlicensed basis. While stakeholders GAO interviewed cited some advantages of these bands, such as being useful to reach remote customers, they also noted technical and cost limitations that reduced the potential to improve tribal access to spectrum. FCC stated that it is implementing spectrum initiatives and recognizes the importance of promoting a robust secondary market to improve communications throughout the United States, including tribal lands. However, GAO found that FCC has not collected data related to tribal access to spectrum, analyzed unused licensed spectrum that exists over tribal lands, or made data available to tribal entities in an accessible and easy manner that could be beneficial in their efforts to obtain spectrum licenses from other providers. By collecting data on the extent that tribal entities are obtaining and accessing spectrum, FCC could better understand tribal spectrum issues and use this information as it implements ongoing spectrum initiatives. Further, given that the secondary market is one of few ways for tribal entities to access licensed spectrum to be able to provide Internet service, FCC could promote a more robust secondary market by analyzing unused licensed spectrum over tribal lands and using that information to inform FCC's oversight of the secondary market. Additionally, by making information available on who holds spectrum licenses over tribal lands, FCC could remove a barrier tribes may face in attempting to obtain spectrum through the secondary market.

November 14, 2018

The Honorable John Hoeven
Chairman

The Honorable Tom Udall
Vice Chairman
Committee on Indian Affairs
United States Senate

The Honorable John Barrasso The Honorable Maria Cantwell The Honorable Steve Daines The Honorable Martin Heinrich The Honorable Heidi Heitkamp The Honorable Brian Schatz The Honorable Jon Tester United States Senate Broadband service is viewed as vital to economic growth and improved quality of life across the country. In 2018, the Federal Communications Commission (FCC) reported that an estimated 35 percent of Americans living on tribal lands lack access to broadband services, compared to 8 percent of all Americans.[1] Furthermore, the gap in broadband access between rural areas and rural tribal lands is even larger. In particular, FCC reported in 2018 that nearly 60 percent of Americans living on rural tribal lands nationwide lack broadband access, compared to about 31 percent of rural Americans overall.[2] We have previously reported that tribal lands can have challenging terrain and low population densities that increase the cost and reduce business incentives for high-speed Internet deployment.[3] According to FCC, this lack of service on tribal lands could impede efforts by Indian tribes to achieve self-governance and promote economic opportunity, education, public safety, and cultural preservation. Broadband service can be delivered through wireless technologies using radio frequency spectrum.[4] According to FCC, wireless technologies are cost-effective for some remote and sparsely populated areas compared to wireline broadband technologies, such as buried fiber optic or copper cables, which can be costly to install where there is challenging terrain.[5]

According to FCC's 2010 *National Broadband Plan*, some tribes have successfully used wireless technologies to deliver Internet access, and increasing tribal access to and use of spectrum would create additional opportunities to expand broadband service on tribal lands.[6] Congress has delegated responsibility for regulating commercial and other nonfederal spectrum use to FCC, and as part of its responsibilities, FCC assigns spectrum licenses through auctions and other mechanisms; oversees secondary market transactions,

---

[1] *In the Matter of Inquiry Concerning the Deployment of Advanced Telecommunications Capability to All Americans in a Reasonable and Timely Fashion*, 2018 Broadband Deployment Report, 33 FCC Rcd 1660 (2018). As of September 24, 2018, there were 573 federally recognized Indian tribes. Federally recognized tribes have a government-togovernment relationship with the United States and are eligible to receive certain protections, services, and benefits by virtue of their status as Indian tribes. In this chapter, the term "tribal lands" refers to any federally recognized Indian tribe's reservation, off-reservation trust lands, pueblo, or colony; land held in trust by the federal government for Indian(s); and Alaska Native regions established pursuant to the Alaska Native Claims Settlement Act, but do not include Oklahoma Tribal Statistical Areas.

[2] Levels of broadband access may vary between specific tribal lands, which may not be reflected in these nationwide figures. We previously reported that FCC's broadband data overstate broadband access on tribal lands. As such, it is possible that the percentage of Americans living on tribal lands that lack broadband access is higher than reported. See GAO, *Broadband Internet: FCC's Data Overstate Access Tribal Lands,* GAO-18-630 (Washington, D.C.: Sept. 7, 2018).

[3] GAO, *Telecommunications: Additional Coordination and Performance Measurement Needed for High-Speed Internet Access Programs on Tribal Lands*, GAO-16-222 (Washington, D.C.: Jan. 29, 2016).

[4] Spectrum is the part of the natural spectrum of electromagnetic radiation lying between the frequencies of 3 kilohertz and 300 gigahertz.

[5] For additional information on the types of broadband technologies, see, https://www.fcc.gov/general/types-broadband-connections.

[6] In March 2010, an FCC task force issued the *National Broadband Plan* that included a centralized vision for achieving affordability and maximizing use of high-speed broadband. See FCC, *Connecting America: The National Broadband Plan* (Washington, D.C.: 2010).

such as leasing a spectrum license;[7] and promulgates regulations for the use of licensed and unlicensed spectrum.[8] FCC has asserted that its authority to regulate nonfederal spectrum use applies to the spectrum over tribal lands. In FCC's 2000 policy statement on establishing its relationship with tribes, FCC stated that it recognizes that the federal government has a fiduciary responsibility in its dealings with tribes and has a longstanding policy of promoting tribal self-sufficiency and economic development.[9] As a result, FCC has recognized its own general responsibility to tribes.

You asked us to review issues related to spectrum use by tribal entities— tribal governments and telecommunications providers owned by tribes. This chapter examines (1) tribal entities' ability to obtain and access spectrum to provide broadband services on tribal lands and the reported barriers that may exist, and (2) the extent to which FCC promotes and supports tribal efforts to obtain and access spectrum for broadband services.

To address these objectives, we reviewed relevant statutes and regulations,[10] FCC documents, including FCC's *Statement of Policy on Establishing a Government-to-Government Relationship with Indian Tribes*,[11] the *National Broadband Plan*,[12] FCC's current strategic plan,[13] and academic and government publications identified through a literature search of spectrum-related issues on tribal lands.

To obtain information on tribal entities' ability to obtain and access spectrum, we identified tribal entities that have applied to participate in FCC's spectrum auctions or that have obtained spectrum licenses for frequency bands that can be used for broadband services by analyzing FCC data on (1) the participants in all relevant auctions, and (2) spectrum license holders that were active as of September 6, 2018—the date that FCC downloaded this data for our review. To identify tribal entities in these data, we reviewed the list of federally recognized tribes and identified search terms related to these tribes. We then identified partial matches between the identified search terms and the FCC data on auction applicants and license holders, and manually reviewed these matches based on information from interviews, Internet research, and professional judgment. Because tribal entities may have applied to participate in a spectrum auction or may hold spectrum licenses under names that did not include the search terms we identified based on the names of federally recognized tribes, there may be additional tribal entities that we did

---

[7] FCC promulgates regulations for the secondary market, and its rules permit licensees to lease portions of the licensed spectrum rights to others.

[8] FCC implements its policy initiatives through a process known as rulemaking, which is the government-wide process for creating rules or regulations that implement, interpret, or prescribe law or policy. 5 U.S.C. § 551(4), (5).

[9] *Statement of Policy on Establishing a Government-to-Government Relationship with Indian Tribes*, Policy Statement, 16 FCC Rcd 4078 (2000).

[10] Communications Act of 1934, Pub. L. No. 73-416, 48 Stat. 1064 (1934), *as amended by Telecommunications Act of* 1996, Pub. L. No. 104-104, § 706, 110 Stat. 53, 153 (1996) (codified as amended at 47 U.S.C. § 151).

[11] 16 FCC Rcd 4078 (2000).

[12] FCC, *Connecting America: The National Broadband Plan* (Washington, D.C.: 2010).

[13] FCC, *Strategic Plan 2018—2022* (Washington, D.C.).

not include in our analysis. We also analyzed FCC's license data to determine how tribal entities obtained these licenses, including through an FCC auction, administrative assignment, or a secondary market transaction. To assess the reliability of FCC's data, we manually reviewed the data and interviewed FCC officials. Based on the results of our analysis, we determined the data to be reliable for our purpose to describe the extent that tribal entities have participated in auctions and obtained spectrum licenses; however, our analysis does not capture the extent that tribal entities may have obtained a license that is no longer active. In addition, we obtained stakeholder views on the barriers that tribal entities may face in obtaining spectrum licenses by interviewing 24 tribal entities, 3 tribal associations, 7 private providers that deliver Internet services over tribal lands, 3 industry associations that represent rural and urban telecommunications providers, 3 regional consortia, 3 companies that work with tribal entities, and 1 academic group. Of the 24 tribal entities we selected to interview, 16 were using wireless technologies to provide Internet service.[14] We selected tribal entities to have variation in geographic location, level of broadband deployment, population size and density, and urban or rural distinction. We selected other stakeholders to represent a range of views and those with experience working with Indian tribes and broadband service. The views presented in our report are not generalizable to those of all stakeholders.

To determine the extent that FCC promotes and supports tribal entities' efforts to obtain and access spectrum, we reviewed relevant FCC rulemaking proceedings. We summarized stakeholder perspectives on these rulemakings by interviewing the tribal entities, tribal associations, regional consortia, industry associations, and private providers noted above, and by reviewing public comments submitted by private providers, industry and tribal associations, and tribal governments and providers.

We identified FCC's efforts to provide tribal entities with spectrum-related assistance and communications by interviewing FCC officials and reviewing documentation, such as presentations provided at FCC-led tribal training workshops, e-mail communications with tribal entities, and public notices related to spectrum use over tribal lands. We also interviewed FCC officials on the information that they collect, analyze, and report related to tribal use of spectrum and reviewed related documentation, including the FCC Office of Native Affairs and Policy's 2012 Annual Report and FCC's license and auction data.[15] We compared FCC's efforts to increase tribal entities' abilities to obtain and access spectrum against recommendations made in FCC's *National Broadband Plan*,[16] FCC's

---

[14] While FCC defines broadband as Internet connection speeds of a certain threshold, we were not able to identify many tribal entities that were able to provide services at that speed. FCC, in its 2018 *Broadband Deployment Report,* sets a benchmark speed for wireline data rates of 25 megabits per second (Mbps) download and 3 Mbps upload, but does not have a similar benchmark for mobile broadband services. Mbps is a measure of the network's data transfer rate (speed) and refers to the number of bits per second that travel to a user's device (the download speed) and from a user's device (the upload speed). 33 FCC Rcd 1660 (2018).

[15] FCC's Office of Native Affairs and Policy, *2012 Annual Report*, (Washington, D.C.). This is the only report that the Office of Native Affairs and Policy has issued on tribal issues.

[16] FCC, *Connecting America: The National Broadband Plan* (Washington, D.C.: 2010).

current strategic plan,[17] and *Standards for Internal Control in the Federal Government* related to using quality information.[18] Appendix I describes our scope and methodology in greater detail.

We conducted this performance audit from November 2017 to November 2018 in accordance with generally accepted government auditing standards. Those standards require that we plan and perform the audit to obtain sufficient, appropriate evidence to provide a reasonable basis for our findings and conclusions based on our audit objectives. We believe that the evidence obtained provides a reasonable basis for our findings and conclusions based on our audit objectives.

BACKGROUND

The federal government has recognized 573 Indian tribes as distinct, independent political communities with certain powers of sovereignty and self-government, including some power to manage the use of their territory and resources and control economic activity within their jurisdiction.[19] Some tribal lands include reservations—land set aside by treaty or other agreement with the United States, executive order, or federal statute or administrative action for the residence or use of an Indian tribe.[20] Some tribal lands include parcels with different ownership; for example, parcels may be held in trust by the federal government for the benefit of a tribe or an individual tribal citizen. Trust and restricted lands can affect a tribe's ability to use their land as collateral to obtain a loan. Tribal lands vary in size, demographics, and location. For example, the smallest in size are less than one square mile, and the largest, the Navajo Nation, is more than 24,000 square miles (the size of West Virginia). Tribal land locations can range from extremely remote, rural locations to urban areas. Indian tribes may form governments and subsidiaries to help manage tribal affairs including schools, housing, health, and economic enterprises.

Internet access in the United States is generally privately financed. Broadband providers build infrastructure and sell broadband services to individual consumers. We previously reported that tribal lands can have conditions that increase the cost of broadband deployment, such as remote areas with challenging terrain, which increases construction costs, as well as relatively low population densities and incomes that make it

---

[17] FCC, *Strategic Plan 2018-2022* (Washington, D.C.).

[18] GAO, *Standards for Internal Control in the Federal Government*, GAO-14-704G (Washington, D.C.: September 2014).

[19] *See e.g., New Mexico v. Mescalero Apache Tribe*, 462 U.S. 324, 335 (1983) (*citing Merrion v. Jucarilla Apache Tribe*, 455 U.S. 130, 137 (1982); *see Montana v. United States*, 450 U.S. 544 (1981).

[20] The land within the reservation may include a mixture (or checkerboard) of tribal trust land, individual Indian trust land, and non-Indian land.

difficult to recoup deployment costs.[21] These conditions may make it less likely that a service provider will build or maintain a network. Some tribal governments provide Internet access to their members, through an information-technology or utility department, and others have created their own telecommunications companies to provide services. FCC has reported that in many instances, tribal governments must build and pay for their own communications infrastructure to ensure Internet access will be "delivered across Indian Country."[22]

The term "broadband" commonly refers to Internet access that is high speed and provides an "always-on" connection, so users do not have to reestablish a connection each time they access the Internet. Telecommunications providers use a range of technologies to provide broadband service, including cable, fiber, satellite, and wireless. Wireless broadband connects users to the Internet using spectrum to transmit data between the customer's location and the service provider's facility, and can be transmitted using fixed wireless and mobile technologies, as shown in Figure 1.

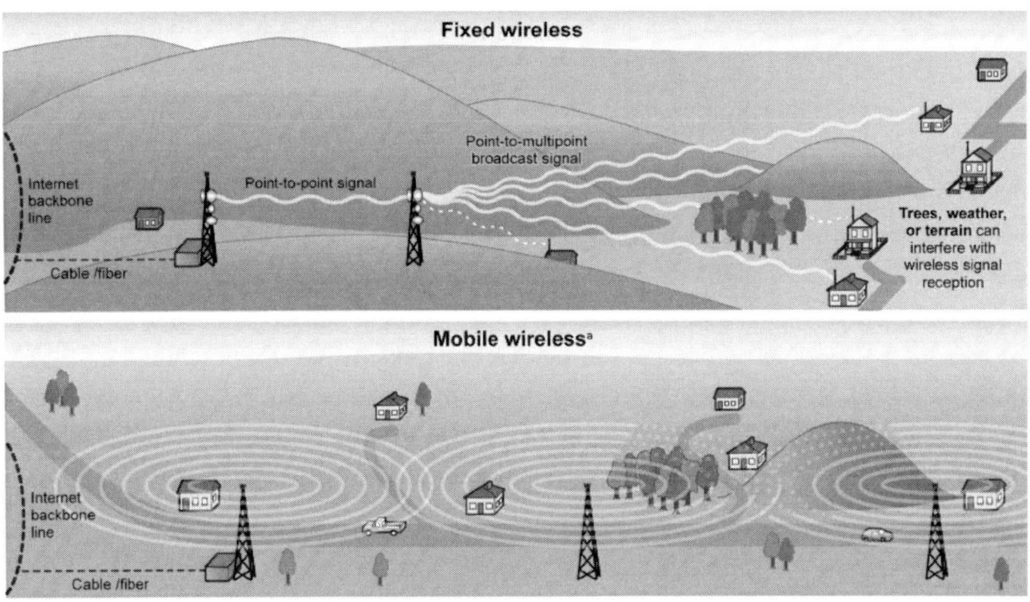

Source: GAO. | GAO-19-75.

[a]Satellite communication is also a wireless technology that exchanges data using a satellite in the sky and a dish on earth.

Figure 1. Fixed and Mobile Wireless Broadband.

---

[21] GAO, *Telecommunications: Challenges to Assessing and Improving Telecommunications for Native Americans on Tribal Lands*, GAO-06-189 (Washington, D.C.: Jan. 11, 2006) and GAO-16-222.

[22] FCC's Office of Native Affairs and Policy, *2012 Annual Report* (Washington, D.C.). The term "Indian country" refers to all land within the limits of any Indian reservation under the jurisdiction of the U.S. government, all dependent Indian communities within U.S. borders, and all Indian allotments, the Indian titles to which have not been extinguished including any rights-of-way running through an allotment. *See* 18 U.S.C. § 1151.

Fixed wireless broadband technologies establish an Internet connection between fixed points, such as from a radio or antenna that may be mounted on a tower, to a stationary wireless device located at a home. This technology generally requires a direct line of sight, and can be delivered two ways: (1) as a point-to-point transmission—between two fixed points—or (2) as a point-to-multipoint transmission—from one point to multiple users. Mobile wireless broadband technologies also establish a connection to the Internet that requires the installation of antennas, but this technology provides connectivity to customers wherever they are covered by service, including while on the move, such as with a cell phone.

Spectrum is the resource that makes wireless broadband connections possible. Spectrum frequency bands each have different characteristics that result in different levels of ability to cover distances, penetrate physical objects, and carry large amounts of information. For example, lower frequency bands are able to transmit signals that travel greater distances, thus requiring the use of fewer antennas, and are able to penetrate solid objects. Higher frequency bands are able to transmit more data, but are more easily obstructed.

FCC administers spectrum for nonfederal users—such as state, local government, and commercial entities—through a system of frequency allocation and assignment.[23] Allocation involves segmenting the radio spectrum into bands of frequencies designated for use by particular types of radio services or classes of users, such as commercial and nonfederal broadband services. Examples of some of the frequency bands that can be used by commercial and nonfederal entities for broadband services are shown in Figure 2. Appendix II presents a full list of the auctioned licensed frequency bands that FCC told us could be used to provide broadband services.

The frequency bands that can be used for broadband services are either licensed or unlicensed. For licensed spectrum, FCC can assign licenses through auctions, in which prospective users bid for the exclusive rights to transmit on a specific frequency band within geographic areas, ensuring that interference does not occur. License holders may sell or lease their license, in whole or in part, to another provider, a process that is known as a secondary market transaction, with FCC's approval. FCC requires license holders to meet specified buildout requirements within a specified amount of time or face penalties, typically termination of all or part of the license.[24] These buildout requirements are designed to ensure that licensees put spectrum to use within a specific period rather than let it sit idle and vary based on the type of license.[25]

---

[23] The Department of Commerce's National Telecommunications and Information Administration (NTIA) is responsible for managing the federal government's use of spectrum. FCC and NTIA jointly determine the amount of spectrum allocated for federal, nonfederal, and shared use. FCC and NTIA also specify service rules, which outline the technical and operating requirements for stations using specific frequency bands.

[24] 47 C.F.R § 1.946(c).

[25] For example, as of January 2014, FCC's buildout requirements for Broadband PCS licenses require licensees to provide service to at least one-third of the population in the licensed area within 5 years of initial license grant, and two-thirds of the population within 10 years of initial license grant. Alternatively, licensees may provide

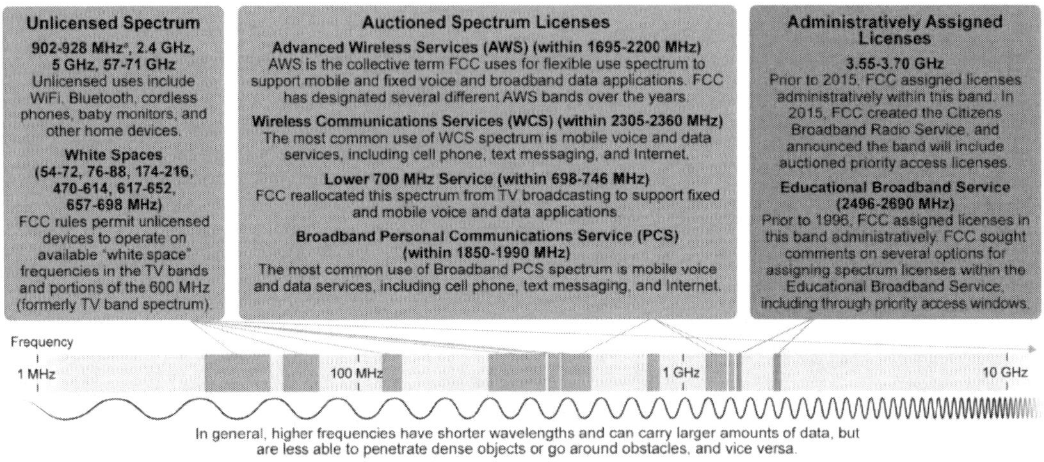

Source: GAO analysis of Federal Communications Commission (FCC) data. | GAO-19-75.

[a]Radio frequencies are measured in units of Hertz, or cycles per second. The term megahertz (MHz) refers to millions of Hertz and gigahertz (GHz) to billions of Hertz.

Figure 2. Examples of Spectrum Frequency Bands Available for Commercial Broadband Services.

FCC has also assigned licenses administratively in two frequency bands that can be used for broadband services. Specifically, prior to 1996 FCC assigned geographic licenses for exclusive use in the Educational Broadband Service (2496-2690 megahertz (MHz)),[26] and from 2005 to 2015, FCC assigned non-exclusive nationwide licenses in the 3650-3700 MHz band, where use of the band may be shared by other license holders.[27] FCC also authorizes the use of some spectrum for broadband services without a license on a non-exclusive basis. With unlicensed spectrum, an unlimited number of users can share frequencies using wireless equipment certified by FCC, such as wireless microphones, baby monitors, and garage door openers. In contrast to users of licensed spectrum, unlicensed users have no regulatory protection from interference by other licensed or unlicensed users in the bands.[28] If multiple users are operating simultaneously on the same frequency band, the transmissions may be susceptible to interference, which reduces the quality of service.

FCC's rulemaking process includes multiple steps as outlined by law with opportunities for the public to participate during each step.[29] In general, FCC initiates a

---

substantial service to their licensed area within the appropriate 5- and 10-year benchmarks. For entities that hold 39 GHz licenses, FCC requires that substantial service must be provided within the license area within 10 years of the initial license grant.

[26] FCC suspended the processing of Educational Broadband Service applications in 1993 and only twice since then opened filing windows—the last one in 1996. Since then, FCC has granted a number of requests for waiver and special temporary authority to permit use of these frequencies.

[27] In June 2018, FCC made the 37-38.6 GHz band, also known as the Lower 37 GHz band, available to nonfederal users for broadband services on a non-exclusive basis, but is licensed by rule. *In the Matter of Use of Spectrum Bands Above 24 GHz For Mobile Radio Services*, Third Report and Order 32 FCC Rcd 10988 (2018).

[28] While there are no regulatory protections against interference for users of unlicensed spectrum, FCC has certification rules and standardized protocols that help to mitigate interference and users must accept any interference caused by all compliant devices in these bands.

[29] *See* 5 U.S.C. § 553.

rulemaking in response to statutes, petitions for rulemaking, or its own initiative, and releases a Notice of Proposed Rulemaking (NPRM) to propose new rules or to change existing rules.[30] Any interested person may submit comments as part of the public record through electronic filings and meetings with FCC officials. Following internal analysis of the public record, FCC staff may propose actions for consideration for a vote, such as adopting final rules, amending existing rules, or stating that there will be no changes. All of FCC's sitting commissioners vote on these items.

The American Recovery and Reinvestment Act of 2009 directed FCC to develop a plan to ensure every American had access to high-speed Internet.[31] In March 2010, an FCC task force issued the *National Broadband Plan* that included a centralized vision for achieving affordability and maximizing use of high-speed Internet.[32] The plan made many recommendations to FCC, including that FCC should take into account the unique spectrum needs of tribal communities when implementing spectrum policies and evaluate its policies and rules to address obstacles to spectrum access by tribal communities. With regard to tribal lands, the plan recommended that FCC increase its commitment to government-to-government consultation with tribal leaders and consider increasing tribal representation in telecommunications planning. FCC established the Office of Native Affairs and Policy (ONAP) in July 2010 to promote the deployment and adoption of communication services and technologies to all native communities, by, among other things, ensuring consultation with tribal governments pursuant to FCC policy.

SELECTED TRIBAL ENTITIES REPORTED BARRIERS TO OBTAINING LICENSED SPECTRUM

Few Tribal Entities Have Obtained Spectrum Licenses, Although Representatives from Selected Tribal Entities Emphasized the Importance of Licensed Spectrum

Through our analysis of FCC license data as of September 2018, we identified 18 tribal entities that held active spectrum licenses in bands that can be used to provide broadband services.[33] Because tribal entities may hold licenses using entity names that do not include the search terms we identified in our review of the list of tribes in the *Federal Register*, there may be additional tribal entities that we have not identified. We found that most of the tribal entities obtained the licenses through FCC administrative assignment rather than through an FCC spectrum auction or secondary market transaction.

---

[30] 47 C.F.R. § 1.412.

[31] American Recovery and Reinvestment Act of 2009, Pub. L. No. 111-5, 123 Stat 115 (2009).

[32] FCC, *Connecting America: The National Broadband Plan* (Washington, D.C.: 2010).

[33] As of September 2018, there were over 27,000 active spectrum licenses held by over 4,400 licensees.

Thirteen of the tribal entities we identified in FCC's license data held administratively assigned licenses, and these licenses are subject to certain limitations and were only available to applicants for limited time periods.[34] Eleven of these administratively assigned licenses are nonexclusive nationwide licenses in the 3.65 GHz frequency band (3550- 3700 MHz) and were available between 2005 and 2015, when FCC issued a new rule for this band and stopped accepting new applications for these licenses. Two of the tribal entities we identified held administratively assigned Educational Broadband Service licenses in the 2.5 GHz frequency band (2496-2690 MHz). These licenses allow for the transmission of educational materials by accredited educational institutions and government organizations, including tribes, engaged in formal education and require that licensees use the spectrum for educational purposes for a certain amount of time each week. Both of these tribal entities obtained these licenses after the last filing window closed in 1996 through a waiver and special temporary authority permit.[35]

Four tribal entities we identified in FCC's license data held a total of 13 active licenses obtained through secondary market transactions, such as leases and sales of portions of partitioned licenses.[36] Of these 13 secondary market transactions, 2 involved nationwide providers.

Two of the tribal entities we identified held active licenses in bands available for broadband deployment that they obtained through an FCC spectrum auction. One of these tribal entities won with a winning bid of over $800,000 in a 2015 auction, and the other won two licenses with winning bids of under $50,000 in a 2002 auction. This second tribal entity also qualified for but did not win a 2003 auction. In addition to these two tribal entities, we identified the following four tribal entities that had applied to participate in auctions with varying results but did not hold active licenses in frequency bands available for broadband deployment as of September 2018:[37]

- Two tribal entities each won a single spectrum license. The first won its license, which has since expired, in 2000, and the second won its license, which it has since been transferred to a nationwide provider through a secondary market transaction, in 2003. The first tribal entity also applied but did not qualify to participate in a 2001 auction.[38]

---

[34] One of these thirteen tribal entities also held a license obtained through a secondary market transaction.

[35] FCC has granted a number of requests for waiver and special temporary authority to permit use of Educational Broadband Service frequencies.

[36] One of these four tribal entities also held an administratively assigned license.

[37] As with our analysis of FCC license data, there may be additional tribal entities that applied to participate in FCC spectrum auctions not reflected in our report if the applicant name used did not include a search term we identified in our review of the list of federally recognized tribes in the *Federal Register*.

[38] Requirements to qualify for auctions may include limits on combined total assets and combined gross revenues in a certain number of years preceding the auction.

- One tribal entity qualified to participate but did not win in a 2003 auction, and another tribal entity applied but did not qualify to participate in a 2008 auction.

In addition, representatives from 2 of the 16 tribal entities we interviewed that were using wireless technologies told us that they use licensed spectrum that is owned by a private provider through a partnership relationship. We have previously reported that some tribes have formed partnership arrangements with other entities to increase broadband access on tribal lands.[39]

Most (14 of 16) of the tribal entities we contacted that were using wireless technologies told us that they are accessing various unlicensed bands, such as the 2.4 GHz and 5 GHz bands, to provide service. Representatives from eight of these tribal entities reported using only unlicensed spectrum for their fixed wireless networks. Representatives from 13 tribal entities told us that unlicensed spectrum had the advantage of being free, and representatives from one tribal entity told us that the equipment needed to access these spectrum bands is less expensive than equipment for accessing other spectrum bands. Representatives from some tribal entities reported success in using unlicensed spectrum in certain circumstances. For example, one tribal entity reported using unlicensed spectrum for homes in remote areas where the only potential signal degradation is from trees as well as to set up local hot spots that can serve 5 to 10 users at a time. Another tribal entity reported using primarily unlicensed spectrum to carry signals to end users together with non-exclusive licensed spectrum (3.65 GHz band) for locations where there is congestion in the unlicensed bands.

Table 1. Limitations of Unlicensed Spectrum Identified by Tribal Entities

Limitation of unlicensed spectrum	Number of tribal entities that identified this limitationa	Examples of tribal entities' experiences with this limitation
Interference	11	Representatives reported that users experience interference from a variety of devices that use unlicensed spectrum such as Bluetooth devices, baby monitors, remote controls, cordless phones, smart televisions, and highway signage.
Speed or capacity	12	Representatives from one tribal entity reported that there is insufficient unlicensed bandwidth available to provide high-speed Internet for a whole town. The entity is currently able to offer 25 megabits download speed, but as it gets more subscribers, it will be unable to maintain this speed. Representatives from another tribal entity indicated that unlicensed spectrum lacks the capacity and speed to provide residential broadband.

Source: GAO analysis of information provided by tribal entities. | GAO-19-75.

[a]Tribal entities may have identified more than one limitation of licensed spectrum.

[39] GAO, *Tribal Broadband: Few Partnerships Exist and the Rural Utilities Service Needs to Identify and Address Any Funding Barriers Tribes Face*, GAO-18-682 (Washington, D.C.: Sept. 28, 2018).

However, representatives from the tribal entities we contacted that were using wireless technologies emphasized the advantages of licensed spectrum and discussed their experiences with the limitations of unlicensed spectrum. As described earlier, exclusive-use spectrum licenses protect license holders from interference from other users, whereas unlicensed spectrum provides no protection against interference. Representatives from 13 of 16 tribal entities identified the fact that unlicensed spectrum is available at no cost as an advantage of this type of spectrum. However, representatives from 15 of the 16 tribal entities identified limitations associated with unlicensed spectrum, such as interference, as described in Table 1.

Tribal associations, an academic group, a tribal consortium, and FCC have all highlighted the importance of exclusive-use licensed spectrum for tribal entities. Specifically, both a tribal association and an academic group we contacted discussed interference and other challenges of unlicensed spectrum. Representatives from one tribal association pointed out that unlicensed spectrum might not be available in the future if it is allocated for other purposes. Representatives from a tribal consortium we contacted told us that they are already using all of the available unlicensed spectrum for providing Internet access and that they cannot expand service without encountering interference and capacity limitations. Lastly, ONAP reported in 2012 that unlicensed spectrum is not an option across all tribal lands and that tribal access to robust licensed spectrum is a critical need.[40]

Representatives from the stakeholders we interviewed told us that there are also non-technological benefits for tribal entities to obtain greater access to licensed spectrum. For example:

- Enhanced ability to deliver additional Internet service. Representatives from one of the tribal associations, an academic group, and six of the tribal entities said that increased access to licensed spectrum would enable them to deliver their own Internet services and bridge service gaps, thus improving Internet access to their members. For example, representatives from three of these tribal entities said that such access would enable them to deploy in areas where providers that currently hold licenses were not willing to deliver services. In addition, representatives from another tribal entity said that having access to licensed spectrum is one factor that would enable the tribe to establish its own telecommunications company.
- Ability to sell or lease spectrum for profit. Representatives from one tribal association, an academic group, and two tribal entities told us that holding spectrum licenses would enable tribal governments to sell or lease their licenses. For example, we heard from one of these tribal entities that it was able to sell

[40] FCC's Office of Native Affairs and Policy, *2012 Annual* Report, (Washington, D.C.).

portions of its license that did not cover tribal lands and to use the profits from the sale to invest in its own network infrastructure.

- Opportunities for federal funding. Access to licensed spectrum may also provide tribal entities with more opportunity to obtain federal funding, specifically through two Universal Service Fund programs— the Mobility Fund and the Tribal Mobility Fund. These programs provide funding to broadband service providers to expand service in areas where it is not available, including tribal lands. However, service providers must hold, lease, or show they have access to licensed spectrum to participate in these programs, among other requirements.[41] For example, the National Congress of American Indians stated that two tribal entities submitted applications to participate in the Mobility Fund program but were not eligible to participate in part because they did not hold a spectrum license. Moreover, representatives from two of the tribal entities we interviewed told us that they considered applying for one of these programs but realized they were ineligible because they did not have access to licensed spectrum.

Furthermore, representatives from one of the tribal associations, an academic group, and seven of the tribal entities told us that having access to licensed spectrum would enable tribes to exercise their rights to sovereignty and self-determination. Representatives from three of the tribal entities we contacted said that they view spectrum as a natural resource that should be managed by the tribe.[42] FCC officials, however, told us that spectrum is not considered a reserved right under treaties with Indian tribes, as it is not explicitly stated. In addition, representatives from four of the tribal entities told us that having access to licensed spectrum would ensure that spectrum is being used in a way that aligns with tribal goals and community needs, further supporting their rights to self-determination.

Representatives from Selected Tribal Entities Reported Cost and Other Barriers to Accessing Licensed Spectrum

Representatives from the tribal entities we contacted identified several barriers to accessing licensed spectrum through spectrum auctions and secondary market transactions.[43] Regarding spectrum auctions, representatives from tribal entities that

---

[41] In order to apply to participate in the Mobility Fund program, an applicant must be designated as an eligible telecommunications carrier, with one narrow exception for tribally owned or controlled entities. To qualify, a tribally owned or controlled entity may have an application pending for eligible telecommunications carrier designation. FCC allowed this exception to afford tribes an increased opportunity to participate at auction.

[42] We did not analyze this issue as it was outside the scope of the report.

[43] In addition, representatives from six tribal entities identified barriers to accessing licenses in the 3.65 GHz band and in the Educational Broadband Service band, both of which are administratively assigned licenses. For

provide wireless Internet service most frequently (13 of 16) indicated that spectrum licenses are too expensive for tribal entities. For example, over 60 percent (983 of 1,611) of the winning bids from a 2015 spectrum auction, including bids for spectrum over non-tribal lands, were over $1 million. Representatives from one tribal entity explained that auction licenses are often too expensive for tribal entities because these licenses cover large geographic areas that may include non-tribal urban areas as well as rural tribal areas. Moreover, representatives from eight tribal entities stated that they are unable to obtain financing to participate in auctions because tribal governments cannot use tribal lands as collateral to obtain loans. In addition, representatives from eight tribal entities mentioned that participating in spectrum auctions requires auction-specific expertise that tribal entities may not have.

Tribal entities also face barriers obtaining spectrum through secondary market transactions. Most of the spectrum allocated for commercial use has already been assigned through spectrum auctions and other mechanisms to private providers, including licensees that may not be providing service on tribal lands. In a single geographic area, several frequency bands could be used to deploy broadband services, as shown in Figure 2, and licenses for these various frequency bands may be held by different providers. There may be tribal areas where providers hold licenses for bands but are not using the spectrum to provide Internet access. In other tribal areas, services may be offered using one or two of the spectrum licenses with the other licenses in the area remaining fallow and inaccessible to tribal entities. All three of the tribal associations we contacted confirmed that there are unused spectrum licenses over tribal lands, and representatives from a nationwide provider indicated that they only deploy services if there is a business case to support doing so.[44] Accordingly, the secondary market is one of few avenues available to tribal entities that would like to access licensed spectrum. However, representatives from tribal entities we contacted identified the following challenges related to participating in the secondary market:

- Lack of willing sellers. Representatives from eight of the tribal entities, one of the tribal representative groups, and an industry association we contacted indicated that spectrum license holders are often unwilling to participate in secondary market transactions, citing a variety of reasons. For example, representatives from one tribal entity stated that large carriers have no business

---

example, representatives from two tribal entities told us that accessing spectrum in the Educational Broadband Service band was challenging because most licenses are held by other providers. In addition, representatives from two tribal entities stated that the equipment required to deploy wireless services in the 3.65 GHz band is more expensive than equipment for unlicensed spectrum and was not widely available. Representatives from two of the tribal entities we contacted were unaware that these licenses were available.

[44] As described above, FCC uses buildout requirements to ensure that spectrum is put to use. However, representatives from one of the tribal associations we interviewed told us that because the geographic size of licenses are often very large, providers are able to meet their buildout requirements without delivering services across tribal lands by targeting their deployment in larger, urban areas.

incentive to negotiate secondary market agreements with tribal entities and that tribal entities do not have the resources to make such transactions sufficiently lucrative for license holders. Representatives from another tribal entity stated that license holders may lack knowledge about the areas covered by their licenses, including tribal areas, and therefore may be unwilling to consider secondary market transactions. Representatives from a tribal representative group told us that license holders may be unwilling to consider secondary market transactions with tribal entities because spectrum is a valuable resource that may become even more valuable over time, and a representative from an industry association indicated that transaction costs such as legal fees outweigh any potential income from such transactions. None of the private providers we contacted reported entering into a secondary market transaction with tribal entities, but one of these providers stated that it had never been approached by a tribal entity interested in a secondary market transaction and was unaware of challenges that are unique to tribal entities.

- License holders unknown. Representatives from eight of the tribal entities we contacted stated that it is difficult to determine who holds spectrum licenses. For example, two tribal entities had to hire consultants to identify who held licenses for spectrum over the tribes' lands, and another tribal entity relied on the expertise of its non-tribal partner to identify the license holders.
- Unaware of secondary market transactions. Representatives from six of the tribal entities we contacted were unaware of the possibility of accessing licensed spectrum through a secondary market transaction prior to our contacting them.

Accordingly, secondary market transactions involving tribal entities are rare. As discussed above, our analysis of FCC license data identified four tribal entities that have successfully accessed licensed spectrum in this manner. Regarding one of these tribal entities' experiences with the secondary market, the tribal representative we contacted stated that an Indian-owned telecommunications consulting company was pivotal in identifying the license holder and facilitating the transaction and that the transaction would not have happened without the consulting company. Representatives from this company told us that they conducted an analysis to identify unused spectrum licenses over the tribe's land. The company identified three providers holding such licenses, but only one of those providers was willing to participate in a secondary market transaction. Representatives from another of the tribal entities that accessed licensed spectrum through the secondary market told us that they relied on the expertise of their non-tribal partner to facilitate these transactions.

FCC HAS SOME EFFORTS TO ENHANCE TRIBAL ACCESS TO SPECTRUM, BUT FCC DOES NOT COLLECT OR COMMUNICATE KEY INFORMATION TO TRIBAL ENTITIES

FCC Has Taken Steps to Promote and Support Tribal Entities' Ability to Obtain Spectrum, However These Efforts Are Not Likely to Address Tribal Spectrum Needs

We found that FCC has taken the following actions to increase tribal access to and use of spectrum: (1) initiated proposed rulemakings on promoting tribal access to spectrum, (2) adopted rules to increase spectrum available for broadband use, and (3) conducted outreach and training for tribal entities on spectrum-related issues.

Initiated Proposed Rulemakings on Promoting Tribal Access to Spectrum

FCC issued two NPRMs—one in March 2011 and one in May 2018—that included policy options intended to enhance tribal access to spectrum. At the time of our report, FCC had not adopted new rules or taken further action on the 2011 rulemaking,[45] and FCC had not taken further actions since the comments period ended on September 7, 2018, on the May 2018 rulemaking.[46] According to FCC officials, the 2011 NPRM addressed several recommendations made in the *National Broadband Plan* to promote the greater use of spectrum over tribal lands.[47] Among other things, the 2011 NPRM sought comments on three proposals to create new spectrum access opportunities for tribal entities (see Figure 3).[48] FCC officials told us that they have reviewed public comments to the proposed rulemaking, but have no current plans to take further actions.

We reviewed the public comments FCC received that pertained to the three proposals, which included comments from tribal associations, tribal governments, rural and nationwide industry associations, and tribal and private providers.[49] Based on our analysis of the comments that included positions on the proposal for a tribal licensing priority, eight stakeholders—including industry associations, private providers, and a

---

[45] *In the Matter of Improving Communications Services for Native Nations by Promoting Greater Utilization of Spectrum over Tribal Lands*, Notice of Proposed Rulemaking 26 FCC Rcd 2623 (2011).

[46] *In the Matter of Transforming the 2.5 GHz Band*, Notice of Proposed Rulemaking, WT Docket No. 18-120 (2018).

[47] FCC, *Connecting America: The National Broadband Plan* (Washington, D.C.: 2010).

[48] The 2011 NPRM included other proposals not related to enhancing tribal access to spectrum, but rather to incentivize build-out in tribal areas by license holders, such as making modifications to the Tribal Lands Bidding Credit Program and creating Construction Safe Harbor provisions.

[49] We identified 16 stakeholders that provided public comments on these proposals, including 3 tribal associations, 2 rural industry associations, 4 nationwide associations, 4 private providers, 1 rural non-tribal provider, 1 tribal provider, and 1 tribal government. Not all stakeholders provided comments on each proposal.

tribal government—were supportive of this proposal.[50] However, we found that stakeholder views differed on implementing good faith negotiations and on the build-or-divest processes.

Proposal	Description
Tribal licensing priority 	Establish a licensing priority for tribal entities to obtain not-yet-assigned licenses, such as at future auctions for fixed and mobile wireless services, for unserved or underserved tribal lands. The NPRM proposed two ways to provide tribal entities with priority access, including (1) provide a tribal priority application window after FCC has announced a spectrum auction but before the window for filling auction applications opens or (2) provide a tribal priority application window before FCC has announced a spectrum auction.
Good faith negotiations 	Establish a process to help tribal entities secure access to spectrum over tribal lands through secondary market opportunities. The NPRM proposed creating a formal negotiation process that would enable a tribal entity to require a licensee holder to enter into good faith negotiations regarding a secondary market transaction for any geographic portion of the license that is covered by unserved or underserved tribal lands.
Build-or-divest process 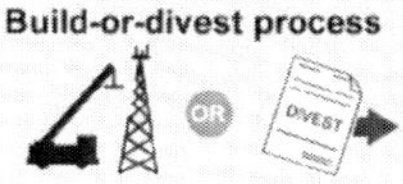	Establish a process where a tribal entity could require a licensee to build or divest a geographic area covering unserved or underserved tribal lands within its license area. The NPRM proposed initiating such a process where an existing licensee has satisfied the applicable construction requirements for the license yet tribal land areas remain unserved or underserved.

Source: GAO analysis of FCC information. | GAO-19-75.

Figure 3. Proposals Made in Federal Communications Commission's (FCC) 2011 Notice of Proposed Rulemaking (NPRM) to Promote Tribal Access to Spectrum.

In general, the tribal stakeholders indicated that they were supportive of these proposals, while the industry associations and private providers were not. In addition to reviewing the public comments, we asked representatives from the tribal and industry associations and private providers that we interviewed about their views of these proposals.[51] Representatives from the three tribal associations and two rural industry associations were generally supportive of all three of the proposals, while representatives from one of the private providers that we interviewed told us they did not support any of the three proposals, because, for example, they said that there are more effective ways to increase broadband service over tribal lands.[52] Representatives from another private provider said that they supported the tribal priority process but did not indicate their views on the other two proposals.[53] Representatives from six tribal entities and a representative from a tribal consortium told us that these types of proposals would help them obtain spectrum.

In May 2018, FCC issued an NPRM that sought comments on establishing a tribal priority window for tribal nations located in rural areas as part of a process to re-license the Educational Broadband Service spectrum band.[54] As described above, FCC originally

[50] Stakeholders' views differed on how the tribal licensing priority provision should be structured.

[51] Representatives from the nationwide industry association that we contacted referred us to the comments they submitted to the 2011 NPRM, which indicated their support for the implementation of a tribal priority process, but not for the implementation of a good faith negotiation process or the build-or-divest proposal.

[52] Representatives from the rural industry associations both supported the tribal priority process and the build-or-divest process but differed in their support on the good faith negotiations process; however both agreed that some of the proposals should be expanded to include non-tribal rural providers.

[53] Representatives from the other private provider did not indicate their position on any of the proposals.

[54] *In the Matter of Transforming the 2.5 GHz Band*, Notice of Proposed Rulemaking, DA 18-394 (2018).

allocated this band to qualifying educational institutions and government organizations for the transmission of educational materials. While FCC permitted licensees to lease their excess capacity to commercial providers, FCC reported that significant portions of this band were not being used, primarily in rural areas. In an effort to make additional spectrum available for broadband use, FCC issued this NPRM seeking comments on options to promote the use of this spectrum over tribal lands. One of the options included implementing a local priority filing window so that tribal entities could get access to unassigned spectrum prior to an FCC auction. In a June 2018 order, FCC extended the comment deadline for the NPRM to August 8, 2018, partly in response to a request for a deadline extension. As a result, FCC also extended the deadline to respond to those comments to September 7, 2018. Because FCC was in the process of responding to these comments at the time of our review, we did not analyze these comments.

Adopted Rules to Allocate Additional Spectrum for Broadband Use

FCC has made additional unlicensed and licensed spectrum available for broadband use and has implemented rules that according to FCC, may make it easier for rural providers to obtain licenses.[55] However, these efforts were not targeted to tribal entities, and according to ONAP's 2012 report, allocating additional unlicensed spectrum may not be a technically feasible solution for all tribal entities, and such spectrum may not have the necessary capacity to handle an increase in users.[56] In addition, representatives from the tribal associations and entities we contacted told us that there are limitations to the extent that these efforts can address the spectrum needs of tribal entities. In particular, they discussed the effect of FCC's changes to the rules on the use of TV white space spectrum and the Citizens Broadband Radio Service spectrum:

- *TV white space spectrum:* In 2010, FCC made additional unlicensed spectrum available for broadband use by allowing providers to operate in the TV bands at locations where those frequencies were not in use, known as TV white space,[57] but none of the tribal entities we interviewed was using this spectrum. A representative from a tribal consortium said that it used TV white space spectrum, and representatives from three of the tribal entities said that they were considering using it in the future because TV white space spectrum can better

[55] FCC has issued two NPRMs on making spectrum in certain bands above 95 GHz and between 3.7 GHz and 4.2 GHz available for broadband services on an unlicensed basis. See, *In the Matter of Spectrum Horizons*, Notice of Proposed Rulemaking and Order, 33 FCC Rcd 2438 (2018) and *In the Matter of Expanding Flexible Use of the 3.7 to 4.2 GHz Band*, Notice of Proposed Rulemaking, 2018 LEXIS 1827 (2018).

[56] FCC's Office of Native Affairs and Policy, *2012 Annual* Report, (Washington, D.C.).

[57] *In the Matter of Unlicensed Operation in the TV Broadcast Bands; Additional Spectrum for Unlicensed Devices Below 900 MHz and in the 3GHz Band*, Second Memorandum Opinion and Order, 25 FCC Rcd 18661 (2010). FCC adopted additional rules on the use of TV white space in 2012, rules that included increasing antenna height limitations to make it easier for rural providers to deploy broadband services. See, *In the Matter of Unlicensed Operation in the TV Broadcast Bands; Additional Spectrum for Unlicensed Devices Below 900 MHz and in the 3GHz Band*, Third Memorandum Opinion and Order, 28 FCC Rcd 5330 (2012).

pass through some environmental barriers, such as trees, reaching more remote customers. However, representatives from five tribal entities, one tribal consortium, one academic group, and three companies that we interviewed told us about several limitations to the use of TV white space spectrum. For example:

- limited bandwidth capacity, which causes lower speeds, high latency,[58] and limits the number of households that can be served;
- equipment needed to access TV whitespace spectrum is expensive and less available;
- the spectrum may not always be available; and
- similar to other unlicensed frequency bands, as described above, there is potential for interference and difficulty to pass through extreme terrain.

- *Citizens Broadband Radio Service (CBRS) Spectrum:* In 2015, FCC made additional licensed spectrum available for broadband use when it issued a new rule for the 3.65 GHz frequency band (3550-3700 MHz).[59] However the tribal entities who held licenses in this band indicated there are limitations to their ability to use this band and their future use of this spectrum remains unknown. As described earlier, FCC had allocated non-exclusive nationwide licenses in this band. In the 2015 rule, FCC created the CBRS, increased the amount of spectrum allocated for commercial broadband use, and implemented a new licensing scheme. This three-tier priority licensing scheme for spectrum sharing included auctioning exclusive-use geographic licenses and allowing non-exclusive use of the band where a license holder is not operating, an approach that is intended to provide a low-cost entry point for users, but will have no protections from interference. Representatives from four of the five tribal entities that we contacted that held licenses in this band said that there were technical advantages to using it, such as the ability for a signal to pass through dense forests. However, representatives from two tribal entities said that the high cost of the equipment needed to access this spectrum prevented them from either using the frequency band extensively or at all. In addition, representatives from two tribal entities said that they were not sure about their ability to access this band in the future given the changes made in FCC's 2015 rulemaking.[60] FCC's 2015 rule also created small-sized and shorter-termed licenses, which FCC stated would decrease the costs of obtaining a license and help rural providers access it. However, FCC issued an NPRM in 2017 that sought comments on suggested changes to CBRS,

---

[58] Latency refers to the amount of time it takes for data to travel from a computer to a server and back again.

[59] *In the Matter of Amendment of the Commission's Rules with Regard to Commercial Operations in the 3550-3650 MHz Band*, Report and Order and Second Further Notice of Proposed Rulemaking 30 FCC Rcd 3959 (2015).

[60] In the 2015 rules, FCC, with some exceptions allowed entities with existing licenses to continue operations until April 17, 2020 or until the expiration date of their authorization. Existing licensees whose licenses expire before April 17, 2020 are allowed to request a one-time renewal not to exceed April 17, 2020. Entities that registered their site, constructed, began operating, and were in compliance with FCC's rules prior to April 17, 2015, were granted exclusive use the band until the end of their license term.

including *increasing* the geographic area covered by licenses and *lengthening* the license term.[61] In October 2018, FCC adopted rules that, among other changes, increased the license area from census tracks to counties and extended the license term from 3 to 10 years, which FCC officials told us were modest changes made to accomplish FCC's goals of creating incentives for investment, including in urban and rural areas, encouraging efficient spectrum use, and promoting robust network deployments.[62]

Conducted Outreach and Training for Tribal Entities on Spectrum-Related Issues

FCC's ONAP conducts training, consultation, and outreach to tribal entities on spectrum-related issues. For example, ONAP officials told us that they have conducted 21 training and consultation workshops for tribal entities on broadband and telecom since 2012, where spectrum has been discussed in general in the introduction and has been addressed specifically in separate sessions in some of the workshops. These officials also told us that they communicate with tribal entities prior to when FCC holds auctions or when implementing regulatory actions or policies that will affect tribal governments and spectrum over their lands. While representatives from 9 of the 16 tribal entities using wireless technologies told us that they had received some outreach on spectrum-related issues from FCC, representatives from 2 of these entities said that they had not.[63] In addition, ONAP issued a report in 2012 to provide FCC with a review of its work with tribal governments and organizations, including information on its tribal broadband efforts, priorities, and tribal consultations.[64] Among other things, the report included case-study information on tribal entities' efforts to access spectrum. Although the report stated that this would be the first of such annual reporting, this is the only report that ONAP has issued on tribal issues. According to ONAP officials, ONAP has not published subsequent reports because it provides FCC with information on its work with tribal governments and organizations, including spectrum-related matters, through more frequent informal briefings and regular updates.

---

[61] *In the Matter of Promoting Investment in the 3550-3700 MHz Band; Petitions for Rulemaking Regarding the Citizens Broadband Radio Service*, Notice of Proposed Rulemaking and Order Terminating Petitions, 2017 FCC LEXIS 3329 (2017).

[62] *In the Matter of Promoting Investment in the 3550-3700 MHz Band*, Report and Order, FCC 18-149 (2018). FCC officials told us that the new rules also provided licensees with the ability to renew, new performance requirements, expanded secondary market rights, bidding credits for small and rural entities, including a tribal land bidding credit, and other changes to the technical rules and auction methodology. In addition, FCC officials said that because the new rules are intended to increase investment, the cost of equipment should decrease.

[63] Five of the tribal entities we contacted did not provide us with responses on whether they had received outreach on spectrum-related issues from FCC.

[64] FCC's Office of Native Affairs and Policy, *2012 Annual* Report, (Washington, D.C.).

FCC Does Not Collect Key Information Related to Spectrum over Tribal Lands or Communicate It to Tribal Entities

FCC has not consistently collected information related to tribal access to spectrum. For example, FCC does not collect data on whether holders of spectrum licenses or auction applicants are tribal entities even though it collects self-reported data on licensee type, such as corporation and government entity. To obtain this information, FCC could include an option for the licensee type, along with the other options, in applications for future licenses and auctions that allows an applicant to identify as a tribal government or tribally owned entity. FCC officials told us that they use information on licensee type to determine eligibility for a license. Because eligibility is not based on whether the applicant is a tribal entity, FCC officials said this information is not needed. However, without this information, FCC does not have a comprehensive understanding of the extent that tribal entities are attempting to obtain or access licensed spectrum or have been successful at obtaining and accessing it.

Additionally, FCC does not analyze information on unused licensed spectrum that exists over tribal lands, even though FCC has information—broadband availability data from providers and information on geographic areas covered by spectrum licenses—that could be used for such analysis. As we described earlier, representatives from all three of the tribal associations we contacted reported that there are unused spectrum licenses over tribal lands that could present opportunities through the secondary market for tribal entities to obtain spectrum. When we asked FCC officials why they do not analyze the extent that unused spectrum licenses exists over tribal lands, they told us that the spectrum data noted above is not specific enough to allow for a license by license analysis of unused spectrum. For example, they said that broadband availability data from providers is aggregated across wide spectrum bands to minimize reporting burdens on the wireless industry, and the data are not sufficiently detailed to identify which spectrum blocks and licenses are being used in particular areas. However, FCC could use this data to conduct, at a minimum, high-level analysis that would result in useful information on the extent to which unused spectrum exists over tribal lands. In addition, FCC officials told us that they evaluate the effectiveness of FCC's secondary markets policies, which FCC views as a mechanism to promote the increased use of unused spectrum licenses, but this approach does not include an analysis of unused spectrum licenses as part of these efforts. As a result, FCC's evaluations of the secondary market may not accurately reflect how these policies affect tribal access to spectrum. Because the secondary market is one of few ways for tribal entities to access licensed spectrum, an analysis of unused licensed spectrum that exists over tribal lands would enable FCC to better promote a robust secondary market that provides additional opportunities for tribes to access spectrum.

FCC's 2010 *National Broadband Plan* stated that ongoing measurement of spectrum utilization should be developed to better understand how spectrum resources are being used because some studies indicated that spectrum goes unused in many places much of the time. The plan also stated that any spectrum utilization studies that FCC conducts should identify tribal lands as distinct entities. In FCC's February 2018 strategic plan, FCC stated that it will implement ongoing initiatives that will assist in spectrum policy planning and decision making, promote a robust secondary market in spectrum, and improve communications services in all areas of the United States, including tribal areas. Additionally, *Standards for Internal Control in the Federal Government* state that agencies should use quality information, including information that is complete, to inform the decision-making processes.[65]

FCC also does not make information on spectrum-license holders available in an easy or accessible manner; such information could be beneficial to the tribes in their efforts to obtain spectrum in the secondary market. As described earlier, the secondary market is a significant mechanism for tribal entities to obtain spectrum licenses, but representatives from the tribal entities we interviewed reported challenges related to participating in the secondary market, such as not knowing whom to contact should they wish to engage in a secondary market transaction to obtain a spectrum license. In July 2014, FCC stopped updating its spectrum dashboard, which provided the public with a way to identify who holds licenses in what areas, including features that allowed users to identify spectrum allocated and assigned in tribal lands. ONAP stated in its 2012 report that this feature represented the first step for individual tribal entities to reach out to licensees and seek leasing, partnership, or other arrangements that could ultimately result in the provision of service over tribal lands.[66] FCC officials told us that the public may view electronic records of all wireless spectrum licenses in FCC's Universal Licensing System, using a wide range of license and geographic parameters, such as licensee names, radio services, spectrum bands, and geographic locations. However, we attempted to navigate the Universal Licensing System to determine spectrum-license holders for specific tribal lands using geographic parameters, but we were unable to successfully do so because the system is so difficult to use. Furthermore, as described above, representatives from eight of the tribal entities that we contacted stated that it is difficult to determine who holds spectrum licenses. When we asked FCC officials why they do not communicate information to tribes about spectrum-related transactions over tribal lands, FCC officials also told us that they issue public notices on applications for all proposed spectrum transactions and on the winning bidders of all auctions, but they have not made it a practice to reach out directly to tribes to make them aware of when providers have obtained spectrum licenses that cover tribal lands.

---

[65] GAO-14-704G.

[66] FCC's Office of Native Affairs and Policy, *2012 Annual Report*, (Washington, D.C.).

The *National Broadband Plan* stated that FCC should make data available that would promote a robust secondary market for spectrum licenses, such as information on how and to whom spectrum is allocated on tribal lands. Additionally, *Standards for Internal Control in the Federal Government* state the need for federal agencies to communicate with external entities and to enable these entities to provide quality information to the agency that will help it achieve its objectives.[67] Tribal governments are an example of such external entities. The ability of tribal governments to make informed spectrum planning decisions and to participate in secondary market transactions is diminished without information from FCC on the spectrum transactions that occur over tribal lands. Providing this information in a manner that is accessible and easy for tribal entities to obtain could enable them to enter into leasing, partnership, or other arrangements to obtain spectrum.

CONCLUSION

Broadband service on tribal lands continues to lag behind the rest of the country, especially on rural tribal lands, which could hinder tribal efforts to promote self-governance, economic opportunity, education, public safety, and cultural preservation. FCC has reported that wireless technologies that access spectrum to deliver broadband services are cost-effective for remote and sparsely populated areas, such as tribal lands. However, FCC's efforts to promote and support tribal entities' access to spectrum have done little to increase tribal use of spectrum, as only very few tribes are accessing spectrum to be able to provide Internet service. Additionally, FCC lacks information that could help inform its decision-making processes related to spectrum policy planning, which is intended to improve communications services in all areas of the United States, including tribal lands. By collecting data on the extent that tribal entities are obtaining and accessing spectrum, FCC could better understand tribal spectrum issues and use this information as it implements ongoing spectrum initiatives. Furthermore, the secondary market is one of few ways for tribal entities to access licensed spectrum to be able to provide Internet service, and FCC has recognized the importance of promoting a robust secondary market. FCC could promote a more robust secondary market by analyzing data to better understand how much unused licensed spectrum exists over tribal lands and using that information to promulgate regulations and to evaluate how FCC policies affect tribal participation in the secondary market. Additionally, by making information on who holds spectrum licenses over tribal lands more accessible and easy to understand, FCC could remove a barrier tribes may face in attempting to obtain spectrum through the secondary market.

---

[67] GAO-14-704G.

Recommendations for Executive Action

We are making the following three recommendations to the Chairman of FCC.

- The Chairman of FCC should collect data on the extent that tribal entities are obtaining and accessing spectrum and use this information as FCC implements ongoing spectrum initiatives (Recommendation 1).
- The Chairman of FCC should analyze data to better understand the extent that unused spectrum licenses exist over tribal lands, such as by analyzing the data for a sample of tribal lands, and as appropriate use this information to inform its oversight of the secondary market (Recommendation 2).
- The Chairman of FCC should make information on spectrum-license holders more accessible and easy to understand for interested parties, including tribal entities, to promote their ability to purchase or lease spectrum licenses from other providers (Recommendation 3).

Agency Comments

We provided a draft of this chapter to FCC for comment. In its comments, reproduced in appendix III, FCC agreed with the recommendations. FCC also provided technical comments, which we incorporated as appropriate.

We are sending copies of this chapter to the appropriate congressional committees, the Chairman of FCC, and other interested parties.

Mark L. Goldstein
Director, Physical Infrastructure.

Appendix I: Objectives, Scope, and Methodology

This chapter examines (1) what is known about the ability of tribal entities to obtain and access spectrum to provide broadband services on tribal lands and the reported barriers that may exist; and (2) the extent to which the Federal Communications Commission (FCC) promotes and supports tribal entities' ability to obtain and access spectrum for broadband services.

To address both objectives, we reviewed relevant statutes and regulations[68] and FCC documents, including FCC's Statement of Policy on Establishing a Government-to-Government Relationship with Indian Tribes,[69] the *National Broadband Plan*,[70] and FCC's current strategic plan.[71] We interviewed FCC officials and representatives from 3 tribal associations, 7 private providers that deliver Internet services over tribal lands, 3 industry associations that represent rural and urban telecommunications providers, 3 regional consortia, 3 companies that work with tribal entities, and 1 academic group. In addition, we selected 24 tribal entities—13 Indian tribes and nations and 11 tribally owned providers—to interview.

To identify tribal entities that were using wireless technologies, we obtained recommendations from stakeholders, reviewed data on relevant federal grants, such as the Broadband Technology Opportunities Program, and conducted Internet research. We then selected 16 tribal entities considering (1) stakeholder suggestions, (2) population, (3) population density, and (4) urban or rural designation. We visited 7 of these tribes in Idaho, New Mexico, and Washington State. The views represented in our report are not generalizable to those of all stakeholders. See Table 2 for a complete listing of the entities we interviewed. We also conducted a literature review to identify relevant academic, government, and media publications that were published between January 1, 2013, and January 11, 2018, that discuss the importance of and options to enhance tribal access to spectrum.

To identify tribal entities that applied to participate in spectrum auctions or that held active spectrum licenses in bands that can be used to provide broadband service, we analyzed (1) FCC data on entities that applied to participate in auctions for spectrum in these bands and (2) FCC data on spectrum licenses in these bands that were active as of September 6, 2018. We also analyzed FCC license data, together with license information publicly available through FCC's Universal Licensing System, to determine whether the tribal entities that held active licenses obtained those licenses through an FCC spectrum auction, administrative assignment, or the secondary market. We then reviewed the list of federally recognized tribes in the *Federal Register* and identified search terms related to these tribes. For example, we identified the following search terms based on the federally recognized tribe, Yurok Tribe of the Yurok Reservation, California, "Reservation, Tribe, and Yurok." We then used the identified search terms to search for tribal entities in FCC's data on auction participants and spectrum license holders.

---

[68] Communications Act of 1934, Pub. L. No. 73-416, 48 Stat. 1064 (1934), *as amended by Telecommunications Act of* 1996, Pub. L. No. 104-104, § 706, 110 Stat. 53, 153 (1996) (codified as amended at 47 U.S.C. § 151).

[69] *Statement of Policy on Establishing a Government-to-Government Relationship with Indian Tribes*, Policy Statement, 16 FCC Rcd 4078 (2000).

[70] FCC, *Connecting America: The National Broadband Plan* (Washington, D.C.: 2010).

[71] FCC, *Strategic Plan 2018-2022* (Washington, D.C.).

We manually reviewed these matches to identify tribal entities based on information from interviews, Internet research, and professional judgment. Because tribal entities may have applied to participate in spectrum auctions or may hold spectrum licenses under names not associated with their tribes, there may be additional tribal entities that we were unable to identify. Through interviews with FCC officials and review of related documentation, we determined that the license and auction participant data are sufficiently reliable for our purpose of identifying some tribal entities that have applied to participate in a spectrum auction or held active spectrum licenses as of September 2018. However, our analysis does not capture the extent that tribal entities may have obtained a license that is no longer active. To identify tribal entities that used unlicensed spectrum to deliver unlicensed service, we interviewed the tribal entities identified above. In addition, we obtained stakeholder views on the advantages and disadvantages of using unlicensed and licensed frequency bands and any barriers that tribal entities face in obtaining spectrum licenses by interviewing the selected stakeholders noted above.

To determine the extent to which FCC promotes and supports tribal entities' efforts to obtain and access spectrum, first, we reviewed FCC's proposals in its ongoing 2011 Notice of Proposed Rulemaking In the Matter of Improving Communications Services for Native Nations by Promoting Greater Utilization of Spectrum over Tribal Lands and its ongoing 2018 Notice of Proposed Rulemaking In the Matter of Transforming the 2.5 GHz Band. We summarized public comments submitted, as of August 2018, by private and tribal providers, rural and nationwide industry associations, tribal associations, and tribal governments on FCC's 2011 proposed rulemaking. We did not analyze comments on FCC's 2018 Notice of Proposed Rulemaking because FCC was in the process of responding to these comments at the time of our review. Second, we reviewed rules that FCC officials identified that increased the availability of unlicensed and licensed frequency bands for broadband use and that may be particularly useful for tribal entities. These rules included FCC's 2010 and 2012 rules related to TV white space spectrum and its 2015 rule and 2017 Notice of Proposed Rulemaking related to the Citizens Broadband Radio Services (CBRS) spectrum.[72]

[72] *In the Matter of Unlicensed Operation in the TV Broadcast Bands; Additional Spectrum for Unlicensed Devices Below 900 MHz and in the 3GHz Band*, Second Memorandum Opinion and Order, 25 FCC Rcd 18661 (2010); *In the Matter of Unlicensed Operation in the TV Broadcast Bands; Additional Spectrum for Unlicensed Devices Below 900 MHz and in the 3GHz Band*, Third Memorandum Opinion and Order, 28 FCC Rcd 5330 (2012); *In the Matter of Amendment of the Commission's Rules with Regard to Commercial Operations in the 3550-3650 MHz Band*, GN Docket No. 12-354, Report and Order and Second Further Notice of Proposed Rulemaking (2015); *In the Matter of Promoting Investment in the 3550-3700 MHz Band; Petitions for Rulemaking Regarding the Citizens Broadband Radio Service*, Notice of Proposed Rulemaking and Order Terminating Petitions, 2017 FCC LEXIS 3329 (2017).

Table 2. List of Entities Interviewed

Tribal entities (tribal governments and telecommunications providers owned by tribes)
Choctaw Nation of Oklahoma (OK)
Confederated Tribes of the Colville Reservation (WA)
Fond du Lac Band of Lake Superior Chippewa (MN)
Fort Belknap Indian Community (MT)
Gila River Telecommunications, Inc. (AZ)
Hopi Telecommunications, Inc. (AZ)
Jamestown S'Klallam Tribe (WA)
Karuk Tribe (CA)
Makah Indian Tribe (WA)
Navajo Tribal Utility Authority (AZ, NM, UT)
Nez Perce Tribe (ID)
Osage Nation (OK)
Pueblo of Acoma (NM)
Pueblo of Pojoaque (NM)
Pueblo of San Ildefonso (NM)
Pueblo of Santa Clara (NM)
Red Spectrum Communications (ID)
Saint Regis Mohawk Tribe (NY)/Mohawk Networks, LLC
San Carlos Apache Telecommunications Utility, Inc. (AZ)
Standing Rock Telecommunications, Inc. (SD)
Spokane Tribe of Indians (WA)/Spokane Tribe Telecom Exchange
Taos Pueblo (NM)
Warm Springs Telecommunications Co. (OR)
Yurok Tribe (CA)/Yurok Connect
Tribal associations
National Congress of American Indians
National Tribal Telecommunications Association
Native American Finance Officers Association (NAFOA)
Regional consortia
Middle Rio Grande Pueblo Consortium
REDINet
Southern California Tribal Chairmen's Association - Tribal Digital Village Network
Groups that work with tribes
Carlson Wireless Technologies
Microsoft
Turtle Island Communications
Industry associations
CTIA
NTCA
Rural Wireless Association
Private providers
AT&T
CenturyLink
Commnet
Frontier
Inland Cellular
Verizon

Source: GAO | GAO-19-75.

We identified tribal entities that had been using CBRS frequency bands by reviewing FCC licensed data and TV white space frequency bands through interviews with tribal entities and regional consortia.

Third, we identified FCC's outreach activities to provide tribal entities with assistance on spectrum-related issues by interviewing FCC officials and reviewing documentation on the content of FCC-led trainings and workshops, e-mail correspondences, and related publications, such as public notices. Lastly, we interviewed FCC officials on the information that they collect, analyze, and report related to tribal use of spectrum and reviewed related documentation, including FCC's Office of Native Affairs and Policy 2012 Annual Report and FCC's license and auction data.[73] We interviewed stakeholders, as noted above, and summarized their views of FCC efforts. We also compared FCC's efforts against FCC's 2018-2022 strategic plan, recommendations made in FCC's *National Broadband Plan*, and *Standards for Internal Control in the Federal Government* related to using quality information.[74]

We conducted this performance audit from November 2017 to November 2018 in accordance with generally accepted government auditing standards. Those standards require that we plan and perform the audit to obtain sufficient, appropriate evidence to provide a reasonable basis for our findings and conclusions based on our audit objectives. We believe that the evidence obtained provides a reasonable basis for our findings and conclusions based on our audit objectives.

APPENDIX II: AUCTIONED LICENSED SPECTRUM AVAILABLE FOR COMMERCIAL BROADBAND SERVICES

We identified the spectrum frequency bands that the Federal Communications Commission (FCC) has made available for commercial broadband services and that FCC assigns licenses through auctions. Table 3 describes these licenses, including the number and date of related auctions.

---

[73] FCC's Office of Native Affairs and Policy, *2012 Annual Report*, (Washington, D.C.).

[74] FCC, *Strategic Plan 2018-2022* (Washington, D.C.), FCC, *Connecting America: The National Broadband Plan* (Washington, D.C.: 2010); GAO, *Standards for Internal Control in the Federal Government*, GAO-14-704G (Washington, D.C.: September 2014).

Table 3. Auctioned Licensed Spectrum Frequency Bands the Federal Communications Commission (FCC) Has Made Available for Commercial Broadband Services

FCC radio service code	Frequency band, megahertz (MHz)/gigahertz (GHz)a	Radio service description	Auction(s)	Auction date closed
AH	1915-1920 MHz and 1995-2000 MHz	Advanced Wireless Service	96	2/27/2014
AT	1695-1710 MHz, 1755-1780 MHz, and 2155-2180 MHz	Advanced Wireless Service-3 (AWS-3)	97	1/29/2015
AW	1710-1755 MHz and 2110-2155 MHz	Advanced Wireless Service (AWS-1)	66 78	9/18/2006 8/20/2008
BB	1390-1392 MHz Block A: 1392-1393.5 and 1432-1433.5 MHz Block B: 1393.5-1395 and 1433.5-1435 MHz (1.4GHz)	Wireless Communications Services	69	3/8/2007
BC	1670-1675 MHz (1.6 GHz)	Wireless Communications Services	46	4/30/2003
BR	2496-2502 MHz, 2602-2615 MHz, 2616-2673.5 MHz (2.5 GHz)	Broadband Radio Service	06 86	3/28/1996 10/30/2009
CJ	800 MHz Block C: 849-850.5/894-895.5 MHz Block D: 850.5-851/895.5-896 MHz	Commercial Aviation Air Ground Radiotelephone	65	6/2/2006
CL	800 MHz Block A: Mobile: 824-835 MHz, 845-846.5 MHz Base: 869-880 MHz, 890-891.5 MHz Mobile: 835-845 MHz, 846.5-849 MHz Base: 880-890 MHz, 891.5-894 MHz	Cellular Services	77	6/17/2008
CW	Block A: 1850-1865, 1930-1945 MHz Block B: 1870-1885, 1950-1965 MHz Block C: 1895-1910, 1975-1990 MHz Block D: 1865-1870, 1945-1950 MHz Block E: 1885-1890, 1965-1970 MHz Block F: 1890-1895, 1970-1975 MHz	Broadband Personal Communications Service (PCS)	45 10 11 22 35 58 71 78	3/13/1995 5/06/1996 7/16/1996 1/14/1997 4/15/1999 1/26/2001 2/15/2005 5/21/2007 8/20/2008
CY	1915-1920 MHz and 1995-2000 MHz	Narrowband PCS	96	2/27/2014
WS	Block A: 2305-2310 MHz paired with 2350-2355 MHz Block B: 2310-2315 MHz paired with 2355-2360 MHz Block C: 2315-2320 MHz Block D: 2345-2350 MHz (2.3 GHz)	Wireless Communications Services	14	4/25/1997
WT	600 MHz	600 MHz Band	1002	3/30/2017

Table 3. (Continued)

FCC radio service code	Frequency band, megahertz (MHz)/gigahertz (GHz)a	Radio service description	Auction(s)	Auction date closed
WU	700 MHz Upper Band (Block C: 710-716 MHz and 740-746 MHz)	Wireless Communications Services	60	7/26/2005
WY	700 MHz Lower Band Block A: 698-704 and 728-734 MHz Block B: 704-710 and 734-740 MHz Block C: 746-757 and 776-787 MHz Block D: 758-763 and 788-793 MHz Block E: 722-728 MHz	Wireless Communications Services	73 92	3/18/2008 7/25/2011
WZ	700 MHz Lower Band, Guard Bands Block C: 710-716 and 740-746 MHz	Wireless Communications Services	60	7/26/2005
YC	800 MHz Specialized Mobile Radio 851.0125-854.7375 MHz / 806.0125-809.7375 MHz Block A: 401-420 and 861.0-861.5 MHz paired with 816.0-816.5 MHz Block B: 421-480 and 861.5-863.0 MHz paired with 816.5-818.0 MHz Block C: 481-600 and 863.0-866.0 MHz paired with 818.0-821.0 MHz	Specialized Mobile Radio, 806-821/851-866 MHz, Auctioned	16 34 43	12/8/1997 9/1/2000 1/17/2002
YD	900 MHz Specialized Mobile Radio Blocks A thru T: half in 935-940 MHz and half in the 896-901 MHz	Specialized Mobile Radio, 896-901/935-940 MHz, Auctioned	7 55	4/15/1996 2/25/2004
UU	24.25-24.45 GHz and 24.75-25.25 GHz (24 GHz)	Upper Microwave Flexible Use Service	54 102	7/28/2004 Future
DV	12.2-12.7 GHz	Multichannel Video and Distribution and Date Service	53 63	1/14/2004 12/7/2005
UU	27.5-28.35 GHz	Upper Microwave Flexible Use Service	17 23 101	3/25/1998 5/12/1999 Future
LD	29.1-29.25 GHz and 31-31.3 GHz	Local Multipoint Distribution Service	17 23	3/25/1998 5/12/1999
UU	37.6-38.6 GHz (Upper 37 GHz)	Upper Microwave Flexible Use Service	Future	Future
UU	38.6-40 GHz (39 GHz)	Upper Microwave Flexible Use Service	30	5/8/2000
UU	47.2-48.2 GHz (47 GHz)	Upper Microwave Flexible Use Service	Future	Future

Source: GAO analysis of FCC information.|GAO-19-75.

[a]Radio frequencies are grouped into bands and are measured in units of Hertz, or cycles per second. The term megahertz (MHz) refers to millions of Hertz and gigahertz (GHz) to billions of Hertz.

APPENDIX III: COMMENTS FROM THE FEDERAL COMMUNICATIONS COMMISSION

Federal Communications Commission
Washington, D.C. 20554

November 2, 2018

Dr. Mark L. Goldstein
Director, Physical Infrastructure Issues
Government Accountability Office
441 G St., NW
Washington, DC 20548

Dear Mr. Goldstein:

Thank you for the opportunity to review GAO's draft report, "Tribal Broadband, FCC Should Undertake Efforts to Better Promote Tribal Access to Spectrum." The Commission has long recognized the particular challenges of providing communications services on Tribal lands, and the role of Tribal access to spectrum in meeting those challenges. As you have noted in the draft report, the Commission has taken a variety of actions to increase Tribal access to and use of spectrum, including rulemaking, outreach, and training efforts. We appreciate your recommendations, and we plan to address them as described below.

Collection of data on tribal acquisition of spectrum. GAO's first recommendation is that the FCC Chairman should "collect data on the extent that tribal entities are obtaining and accessing spectrum and use this information as FCC implements ongoing spectrum initiatives." As noted in the draft report at page 22, the FCC has been collecting information on licensee type in order to determine eligibility to hold a particular license, but since eligibility is not based on Tribal entity status, we have not required reporting of Tribal entity status. Recognizing GAO's concern about whether the FCC has a full understanding of Tribal entity interest and success in obtaining licensed spectrum, the Commission will consider ways to collect data on the extent that tribal entities are obtaining and accessing spectrum, including updating application forms to provide for voluntary submission of this information.

Existence of unused spectrum. GAO's second recommendation is that the FCC Chairman "should analyze data to better understand the extent that unused spectrum licenses exist over tribal lands, such as by analyzing the data for a sample of tribal lands, and as appropriate use this information to inform its oversight of the secondary market." Following this recommendation, we plan to take a sample of spectrum licenses on Tribal lands and then analyze these data to inform our spectrum policies.

Data to assist Tribes. GAO's third recommendation is that the FCC Chairman should "make information on spectrum license holders more accessible and easy to understand for interested parties, including tribal entities, to promote their ability to purchase or lease spectrum licenses from other providers." Our official licensing data is available to Tribal entities and others through our Universal Licensing System (ULS). The Commission is currently engaged in a multi-year project to modernize ULS, transitioning to a new platform that will provide more consistent performance, easier access to information, and enhanced functionality. In the meantime, our Office of Native Affairs and Policy (ONAP) will, of course, continue its outreach and educational efforts toward Tribal entities. Both ONAP and the staff of our Wireless

Telecommunications Bureau are always available to assist any individual Tribe that has questions or requests assistance accessing information.

Thank you for the opportunity to review GAO's recommendations. We look forward to working with GAO in the future.

Sincerely,

Donald Stockdale
Chief, Wireless Telecommunications Bureau

Patrick Webre
Chief, Consumer and Governmental Affairs Bureau

In: Broadband
Editor: Orlando Harrison
ISBN: 978-1-53616-409-1

Chapter 11

TRIBAL BROADBAND: STATUS OF DEPLOYMENT AND FEDERAL FUNDING PROGRAMS*

Lennard G. Kruger

ABSTRACT

Tribal areas and communities continue to lag behind other areas and segments of American society with respect to broadband and telecommunications services. High poverty rates and low income levels in tribal lands—along with the fact that many tribal communities are located in remote rural areas (often with rugged terrain)—are major factors that may explain why tribal areas have comparatively poor levels of broadband access, and why providers may lack an economic incentive to serve those areas.

Until recently, data on tribal broadband deployment had been scarce. However, the Federal Communications Commission (FCC) and the Department of Commerce have begun to collect and compile data on tribal broadband deployment. The most recent data show that, as of December 31, 2017, approximately 32% of Americans living on tribal lands lacked access to broadband at speeds of at least 25 Mbps download/3 Mbps upload. This compares unfavorably to about 6% of all Americans lacking access to broadband at those speeds. Tribal areas that are the most lacking in broadband service are rural Alaskan villages and rural tribal lands in the lower 48 states.

Because the presence of robust broadband and improved digital connectivity in tribal areas could play a significant role in revitalizing many tribal communities, the federal government continues to provide some financial assistance to tribal lands for broadband deployment. The Government Accountability Office, in its 2016 report, *Challenges to Assessing and Improving Telecommunications for Native Americans on Tribal Lands*, identified programs in two federal agencies that serve as the primary source of funding for deploying broadband infrastructure in tribal lands and communities. These federal

* This is an edited, reformatted and augmented version of Congressional Research Service, Publication No. R44416, dated January 9, 2019.

agencies are the FCC and the Rural Utilities Service (RUS) in the U.S. Department of Agriculture.

Tribal entities and projects in tribal areas are eligible for virtually all federal broadband programs. With a few exceptions, however, there are no carve-outs or dedicated funding streams specifically for tribal applicants or nontribal entities proposing to serve tribal lands. Thus, annual amounts of federal financial assistance vary, depending on the number and quality of tribal-related applications received, and the number of tribal-related broadband awards subsequently made by the funding agencies.

Debate has centered on whether federal funding for tribal broadband is sufficient, and the extent to which portions of federal funds available for broadband should be specifically targeted for tribal broadband. In the 115th Congress, bills were introduced to direct federal funding specifically for tribal broadband. Notwithstanding whether federal broadband funding programs target tribal lands, whether or not tribal lands will receive additional funding for broadband in the 116th Congress will likely be determined by the ongoing trajectory of overall federal funding for broadband.

Background

Broadband—whether delivered via fiber, cable modem, copper wire, satellite, or wirelessly—is increasingly the technology underlying telecommunications services such as voice, video, and data.[1] Since the initial deployment of high-speed internet in the late 1990s, broadband technologies have been deployed primarily by the private sector throughout the United States. While the number of new broadband subscribers continues to grow, studies and data suggest that the rate of broadband deployment in urban/suburban and high-income areas is outpacing deployment in rural and low-income areas.[2] In particular, tribal communities stand out as being among the most unserved or underserved populations with respect to broadband deployment.

According to the Federal Communications Commission (FCC), "[b]y virtually any measure, communities on tribal lands have historically had less access to telecommunications services than any other segment of the population."[3] According to Census data, about 28.3% of Native Americans live in households below the poverty

[1] The term "broadband" is typically used interchangeably with "high speed internet" or "advanced telecommunications." Section 706 of the Telecommunications Act of 1996 (P.L. 104-104) defined advanced telecommunications capability as "high-speed, switched, broadband telecommunications capability that enables users to originate and receive high-quality voice, data, graphics, and video telecommunications using any technology."

[2] See for example Federal Communications Commission, *Communications Marketplace Report,* GN Docket No. 18- 231, FCC 18-181, adopted December 12, 2018, released December 26, 2018, available at https://docs.fcc.gov/public/attachments/FCC-18-181A1.pdf. Also see John B. Horrigan and Maeve Duggan, Pew Research Center, *Home Broadband 2015*, December 21, 2015, available at http://www.pewinternet.org/files/2015/12/Broadband-adoption-full.pdf.

[3] Federal Communications Commission, "In the Matter of Extending Wireless Telecommunications Services to Tribal Lands," *Report and Order and Further Notice of Proposed Rulemaking*, WT Docket No. 99-266, FCC 00-209, Adopted June 8, 2000, p. 5, available at http://wireless.fcc.gov/auctions/general/releases/fc000209.pdf.

level (compared to 15.5% nationally), and tribal communities often lack basic infrastructure such as water and sewer systems, and telecommunications.[4]

High poverty rates and low income levels in tribal lands—along with the fact that many tribal communities are located in remote rural areas (often with rugged terrain)—are major factors that explain why tribal areas have comparatively poor levels of broadband access, and why providers may lack an economic incentive to serve those areas. According to the FCC's Office of Native Affairs and Policy (ONAP):

> Understanding the complexity of the digital divide in Indian Country requires an appreciation of the unique challenges facing Tribal Nations, which include deployment, adoption, affordability, and access to spectrum, as well as lack of investment dollars and access to credit and start-up or gap financing. Barriers to the deployment of communications services include rural, remote, rugged terrain, areas that are not connected to a road system, and difficulty in obtaining rights-of-way to deploy infrastructure across some Tribal lands—all of which increase the cost of installing, maintaining, and upgrading infrastructure. Affordability of communications services is affected by often endemic levels of poverty. Because Tribal Nations cannot easily collateralize assets that are held in trust by the federal government, and cannot easily access investment dollars, the ability to obtain credit and financing is limited.[5]

The presence of robust broadband and improved digital connectivity in tribal areas could play a significant role in revitalizing many tribal communities. The FCC's 2010 National Broadband Plan[6] identified broadband as a basic infrastructure necessary for improving economic growth, job creation, global competitiveness, and a better way of life. According to ONAP, "[t]he lack of robust communications services presents serious impediments to Tribal Nations' efforts to preserve their cultures and build their internal structures for self-governance, economic opportunity, health, education, public safety, and welfare."[7]

STATUS OF TRIBAL BROADBAND

Until recently, data on tribal broadband had been scarce. The Government Accountability Office (GAO) noted in 2006 that "[t]he rate of Internet subscribership for

---

[4] Government Accountability Office, *Additional Coordination and Performance Measurement Needed for High-Speed Internet Access Programs on Tribal Lands*, GAO-16-222, January 2016, p. 5, available at http://www.gao.gov/assets/ 680/674906.pdf.

[5] Federal Communications Commission, Office of Native Affairs and Policy, *2012 Annual Report*, released March 19, 2013, p. 7, available at http://transition.fcc.gov/cgb/onap/ONAP-AnnualReport03-19-2013.pdf.

[6] Federal Communications Commission, *Connecting America: The National Broadband Plan*, March 2010, 360 pages, available at https://transition.fcc.gov/national-broadband-plan/national-broadband-plan.pdf.

[7] FCC, Office of Native Affairs and Policy, *2012 Annual Report*, p. 6.

Native American households on tribal lands is unknown because neither the Census Bureau nor FCC collects this data at the tribal level."[8]

The FCC and the Department of Commerce (Census and the National Telecommunications and Information Administration) have begun to collect and compile data on tribal broadband deployment.[9] According to December 2017 FCC deployment data, 32% of Americans living on tribal lands lacked access to terrestrial fixed broadband at speeds of 25 Mbps download/3 Mbps upload. This is an improvement over 2014 data (42.8% without broadband) and 2013 data (62.9%).[10] Table 1 shows the percentages of Americans with access to fixed terrestrial broadband service with respect to tribal lands, rural areas, urban areas, and the United States as a whole.

Table 1. Percentage of Americans with access to fixed terrestrial broadband at minimum speed of 25 Mbps/3 Mbps

	2013	2014	2015	2016	2017
All U.S.	83.6%	89.4%	89.9%	91.9%	94.0%
Urban	92.3%	96.4%	96.7%	97.7%	98.5%
Rural	47.6%	60.4%	61.5%	67.8%	75.7%
Tribal	37.1%	57.2%	57.8%	63.1%	68.0%

Source: FCC, Communications Marketplace Report, p. 132.

Table 2 shows broadband availability within the various categories of tribal lands. Areas that are the most lacking in broadband service are rural Alaskan villages and rural tribal lands in the lower 48 states. Table 3 shows tribal lands with access to fixed terrestrial broadband by state.

Table 4 shows 2013-2017 fixed broadband adoption rates for tribal lands and the United States as a whole. Broadband adoption in this table reflects the percentage of households that actually subscribe to broadband service offering speeds of at least 25 Mbps/3 Mbps.

Finally, Table 5 shows that Native Americans have a lower rate of internet usage than other races and ethnicities.

---

[8] Government Accountability Office, *Challenges to Assessing and Improving Telecommunications for Native Americans on Tribal Lands*, GAO-06-189, January 2006, p.4, available at http://www.gao.gov/assets/250/248920.pdf.

[9] According to GAO, the Census Bureau began collecting internet adoption data beginning in 2013. Five years of these data are required to accurately profile areas with small populations. Data will be released in late 2018, and will contain an estimate for internet adoption in Native American populations. See GAO, *Additional Coordination and Performance Measurement Needed for High-Speed Internet Access Programs on Tribal Lands*, p. 25.

[10] FCC, *Consolidated Market Report*, p. 132.

Table 2. Percentage of population on tribal lands with access to fixed terrestrial 25 Mbps/3Mbps services and mobile LTE with a minimum advertised speed of 5 Mbps/1Mbps

	2013	2014	2015	2016	2017
All Tribal Lands	35.5%	56.2%	57.0%	62.4%	67.7%
- Rural Areas	14.1%	29.5%	30.1%	37.8%	45.6%
- Urban Areas	57.9%	84.5%	85.6%	88.8%	91.6%
Alaskan Villages	28.2%	44.4%	42.7%	51.5%	57.0%
- Rural Areas	13.1%	25.8%	23.7%	36.2%	42.4%
- Urban Areas	54.9%	77.4%	76.7%	79.0%	83.3%
Hawaiian Home Lands	90.6%	96.9%	88.9%	88.6%	89.4%
- Rural Areas	45.0%	83.0%	43.9%	43.5%	47.7%
- Urban Areas	99.4%	99.8%	98.0%	98.0%	98.2%
Lower 48 States	30.0%	38.8%	41.5%	46.1%	53.5%
- Rural Areas	18.9%	25.8%	28.4%	32.3%	41.7%
- Urban Areas	51.9%	64.8%	67.8%	74.1%	78.1%
Tribal Statistical Areas	37.8%	64.2%	64.5%	70.2%	74.6%
- Rural Areas	11.2%	32.1%	32.0%	41.5%	48.6%
- Urban Areas	58.8%	89.7%	90.3%	93.0%	95.4%

Source: FCC, Consolidated Marketplace Report, pp. 142-143.

Table 3. Americans living on tribal lands with access to fixed terrestrial 25 Mbps/3Mbps services by state (as of December 31, 2017)

	Total Population	Population with Access	Percentage of Population with Access
Tribal Lands	4,017,350	2,731,250	68.0%
Alaskan Villages	265,340	153,960	58.0%
Hawaiian Homelands	33,560	30,000	89.4%
Lower 48 States	1,117,110	606,710	54.3%
Alabama	280	150	51.0%
Alaska	1,540	0	0%
Arizona	198,260	20,490	10.3%
California	72,240	49,610	68.7%
Colorado	16,260	9,760	60.0%
Connecticut	350	350	100%
Florida	5,030	4,140	82.2%
Idaho	33,740	8,320	24.7%
Iowa	960	680	70.6%
Kansas	5,690	3,520	61.9%
Louisiana	770	220	28.3%
Maine	2,370	1,870	79.0%
Massachusetts	80	70	97.3%
Michigan	34,660	33,820	97.6%
Minnesota	39,370	32,040	81.4%
Mississippi	7,070	4,790	67.8%
Montana	69,720	38,130	54.7%
Nebraska	8,660	3,870	44.7%
Nevada	14,140	5,450	38.6%

Table 3. (Continued)

	Total Population	Population with Access	Percentage of Population with Access
New Mexico	148,070	53,610	36.2%
New York	13,610	13,610	100%
North Carolina	9,230	210	2.2%
North Dakota	25,090	20,090	80.1%
Oklahoma	90,380	62,610	69.3%
Oregon	9,490	3,340	35.2%
Rhode Island	0	0	100%
South Carolina	1,010	1,010	100%
South Dakota	65,050	41,590	63.9%
Texas	1,900	1,410	74.0%
Utah	36,800	17,460	47.4%
Washington	139,170	128,250	92.2%
Wisconsin	39,970	32,410	81.1%
Wyoming	26,160	13,850	52.9%
Tribal Statistical Areas	2,601,350	1,940,590	74.6%
California	3,190	3,180	99.7%
New York	2,710	2,710	100%
Oklahoma	2,555,790	1,895,050	74.1%
Washington	39,650	39,650	100%

Source: FCC, *Communications Marketplace Report*, *Appendix D*, pp. 203-204, available at https://docs.fcc.gov/public/attachments/FCC-18-181A9.pdf.

Table 4. Broadband adoption rates for fixed terrestrial services, 2013-2017 (25 Mbps/3 Mbps)

	2013	2014	2015	2016	2017
United States	29.7%	38.5%	48.1%	53.5%	59.8%
- Non-Urban Core Areas	28.5%	34.4%	43.2%	48.9%	54.5%
- Urban Core Areas	30.4%	41.3%	51.5%	56.9%	63.9%
Tribal Lands	31.9%	27.2%	31.7%	33.4%	37.9%
- Non-Urban Core Areas	27.8%	23.3%	28.5%	30.2%	34.5%
- Urban Core Areas	36.6%	33.9%	37.1%	39.4%	45.1%

Source: FCC, Consolidated Marketplace Report, p. 144.

Table 5. Internet use by race or ethnicity (percentage of age 3+ civilian persons)

	Total U.S.	White	African American	Hispanic	American Indian or Alaska Native
Internet Use (any location)	77.7%	80.2%	73.4%	72.1%	62.7%
Internet Use at Home	71.9%	75.2%	65.3%	64.5%	51.5%

Source: Digital Nation Data Explorer, National Telecommunications and Information Administration, November 2017 data.

The GAO September 2018 report, *FCC's Broadband Internet: Data Overstate Access on Tribal Lands*, found that the FCC's tribal broadband data relies exclusively on provider-reported data, and overstates broadband access because the FCC considers broadband to be available for an entire census block if the provider could serve at least one location in that census block.[11] GAO also found that the FCC does not collect data on factors significant to tribal broadband, such as affordability, quality, and denials of service.[12] GAO recommended that the FCC develop and implement methods, such as targeted data collection, for collecting and reporting more accurate broadband data on tribal lands; develop a formal process, including outreach and technical assistance, to obtain tribal input on the accuracy of provided-submitted broadband data; and obtain feedback from tribal stakeholders and providers on the effectiveness of the FCC's 2012 statement to providers on how to fulfill their tribal engagement requirements.[13]

FEDERAL FUNDING FOR TRIBAL BROADBAND

A precise accounting of federal funding for tribal broadband is problematic. A comprehensive listing of all federal funding programs for broadband is found in the publication, *Guide to Federal Funding of Broadband Projects*, compiled by the National Telecommunications and Information Administration (NTIA).[14] Tribal entities or projects are eligible for virtually all of these programs, but with a few exceptions,[15] there are no carve-outs or dedicated funding streams *specifically* for tribal applicants or nontribal entities proposing to serve tribal lands. Thus, annual amounts of federal financial assistance vary depending on the number and quality of tribal-related applications received, and the number of tribal-related broadband awards made by the funding agencies. Compounding the challenge in assessing federal funding for tribal broadband, some programs may not formally track funding to tribal areas, making it difficult to come up with an accurate overall number from year to year.

GAO, in its 2016 report, *Challenges to Assessing and Improving Telecommunications for Native Americans on Tribal Lands*, identified programs in two federal agencies that serve as the primary source of funding for deploying broadband infrastructure in tribal lands and communities: the FCC and the Rural Utilities Service (RUS) in the U.S. Department of Agriculture. Subsequently, the September 2018 GAO

[11] Government Accountability Office, *Broadband Internet: Data Overstate Access on Tribal Lands*, GAO-18-630, September 2018, pp. 14-20, available at https://www.gao.gov/assets/700/694386.pdf.

[12] Ibid., pp. 20-24.

[13] Ibid., p. 35.

[14] U.S. Department of Commerce, National Telecommunications and Information Administration, *BroadbandUSA: Guide to Federal Funding of Broadband Projects*, June 2017, 44 p., available at https://broadbandusa.ntia.doc.gov/sites/default/files/resource-files/ntia_guidetofedfunding_062317.pdf. Additionally, an online search tool is available at https://broadbandusa.ntia.doc.gov/funding-list.

[15] Most notably, the Tribal Mobility Fund, which is part of the FCC's Universal Service/Connect America Fund.

report, *Tribal Broadband: Few Partnerships Exist and the Rural Utilities Service Needs to Identify and Address Any Funding Barriers Tribes Face*, found that few federal funds were provided to tribal entities to increase broadband deployment from 2010 to 2017. GAO found that "less than 1 percent of FCC funding and about 14 percent of RUS funding[16] went directly to tribes and tribally owned providers," and that combined, "FCC and RUS funding totaled $34.6 billion during that time period and tribes and tribally owned providers received $235 million, or about 0.7 percent."[17] The GAO analysis excluded nontribally owned broadband providers that serve tribal areas.

FCC

The FCC has established a Universal Service Fund (USF) which provides financial support to ensure that telecommunications services are available to all Americans.[18] The USF currently administers four programs: the High Cost/Connect America Fund (CAF) Program; the Schools and Libraries Program (E-Rate); the Rural Health Care Program/Health Connect Fund; and the Low Income Program (Lifeline and Link-up).[19] GAO has identified three of those programs (CAF, E-Rate, and Rural Health) as subsidizing telecommunications carriers providing broadband to areas that include tribal lands. Additionally, on March 31, 2016, the FCC adopted an Order that modernized the Lifeline Program and reoriented its focus on broadband services.[20]

On February 8, 2018, the FCC, through its Office of Native Affairs and Policy (ONAP), announced it was seeking applications for membership on a renewed FCC Native Nations Communications Task Force. The renewed Task Force's mission will be to make recommendations to the FCC on communications-related issues that affect tribal interests. The issues to be considered by the Task Force "may include, but are not limited to: (i) executing the Commission's Tribal Consultation policy; (ii) identifying barriers to broadband deployment that are unique to Tribal lands; (iii) ensuring Tribal concerns are considered in all Commission proceedings related to broadband and other Commission

---

[16] GAO only considered Community Connect grants in its report. RUS broadband loan programs were excluded from consideration because GAO determined that loan programs are often not a feasible option for tribes.

[17] Government Accountability Office, *Tribal Broadband: Few Partnerships Exist and the Rural Utilities Service Needs to Identify and Address Any Funding Barriers Tribes Face*, GAO-18-682, September 2018, p. 16, available at https://www.gao.gov/assets/700/694810.pdf. The GAO data does not include nontribal recipients that provide service in tribal areas.

[18] For more information on the USF, see CRS Report RL30719, *Broadband Internet Access and the Digital Divide: Federal Assistance Programs*, by Lennard G. Kruger and Angele A. Gilroy.

[19] The Low Income Program (which includes the Lifeline and Link-Up programs) has traditionally subsidized telephone service for low-income residents, including those in tribal lands. For more information, see CRS Report R44487, *Federal Lifeline Program: Frequently Asked Questions*, by Angele A. Gilroy.

[20] FCC, "In the Matter of Lifeline and Link Up Reform and Modernization," *Third Report and Order, Further Report and Order, and Order on Reconsideration*, WC Docket No. 11-42, FCC 16-38, adopted March 31, 2016, released April 27, 2016, 224 pages., available at https://docs.fcc.gov/public/attachments/FCC-16-38A1.pdf.

undertakings that affect Tribal interests regarding communications services and facilities."[21] On October 24, 2018, the FCC announced appointments to the Native Nations Communications Task Force.[22]

High Cost/Connect America Fund Program

The High Cost Fund Program, which is transitioning into the Connect America Fund (CAF), provides subsidies to telecommunications providers offering broadband in rural areas. According to the 2016 GAO report, "the High Cost and Connect America Fund distributed about $20 billion in subsidies to providers between 2010 and 2014, portions of which went to providers that serve tribal lands."[23] Of the total, GAO was unable to determine the amount of funding that went to tribal lands. However, in its 2018 report, GAO reported that between 2010 and 2017, nine tribally owned providers received high cost support funding totaling $218.1 million.[24]

In order to be eligible to receive USF support, a telecommunications provider must be designated as an eligible telecommunications carrier (ETC) by the FCC.[25] According to GAO, the statutory requirement for ETC designation is a major barrier for tribes to obtain USF support, with only "11 tribes that have providers that are designated as ETCs and therefore would be eligible to receive CAF funding."[26]

As part of the CAF, the FCC established a Mobility Fund which consists of two phases. Phase I of the Mobility Fund ($300 million) included $50 million for a Tribal Mobility Fund to extend wireless voice and broadband infrastructure into tribal lands. On February 28, 2014, the FCC announced completion of the Tribal Mobility Fund Phase I auction, with five wireless providers becoming eligible to receive a total of up to approximately $50 million in one-time support. Since July 2014, $16.6 million in initial disbursements have been made.[27] According to GAO, one tribally owned provider received support totaling $3.3 million under the Mobility Fund Phase 1, and no tribal providers have received funding under the Tribal Mobility Fund Phase 1.[28]

---

[21] FCC, *Public Notice*, "FCC Seeks Nominations for Tribal Government Representatives to Serve on Renewed FCC Native Nations Communications Task Force," February 8, 2018, available at https://apps.fcc.gov/edocs_public/ attachmatch/DA-18-127A1.pdf.

[22] FCC, Public Notice, "Chairman Pai Announced New Appointments to the Native Nations Communications Task Force," October 24, 2018, available at https://docs.fcc.gov/public/attachments/DA-18-1083A1.pdf.

[23] GAO, *Additional Coordination and Performance Measurement Needed for High-Speed Internet Access Programs on Tribal Lands*, p. 17.

[24] GAO, *Tribal Broadband: Few Partnerships Exist and the Rural Utilities Service Needs to Identify and Address Any Funding Barriers Tribes Face*, p. 17.

[25] An ETC must meet various criteria with respect to its service and capabilities. See GAO, *Tribal Broadband: Few Partnerships Exist and the Rural Utilities Service Needs to Identify and Address Any Funding Barriers Tribes Face*, p. 8.

[26] Ibid., pp. 18-19.

[27] Federal Communications Commission, *2016 Broadband Progress Report*, FCC 16-6, released January 29, 2016, p. 55, available at https://www.fcc.gov/document/fcc-releases-2016-broadband-progress-report.

[28] GAO, *Tribal Broadband: Few Partnerships Exist and the Rural Utilities Service Needs to Identify and Address Any Funding Barriers Tribes Face*, p. 17.

Phase II of the Mobility Fund ($453 million per year for 10 years) will designate up to an estimated $34 million of annual support for deploying wireless mobile broadband service on eligible tribal lands.[29]

On March 27, 2018, the FCC adopted a *Report and Order* that will "increase the amount of operating costs that carriers that predominantly serve Tribal lands can recover from the universal service fund (USF) in recognition that they are likely to have higher costs than carriers not serving Tribal lands. This action will provide additional funding to these carriers to provide both voice and broadband services to their customers."[30]

Also, on March 14, 2018, the FCC adopted a *Notice of Proposed Rulemaking* asking for comment on its proposal to incorporate a Tribal Broadband Factor into the CAF model.[31]

Schools and Libraries (E-Rate) Program

The E-rate Program subsidizes discounts to providers offering telecommunications services, internet access, and internal connections to schools and libraries. According to the 2016 GAO report, "the E-rate program provided about $13 billion in discounts to schools and libraries between 2010 and 2014, portions of which went to schools and libraries on tribal lands."[32] Of that total, "at least $1 billion of that amount supports tribal institutions."[33]

Lifeline Program

The Lifeline Program provides a subsidy to providers serving low-income households, thereby eliminating or significantly reducing the monthly cost to low-income households for telecommunications service. While traditionally geared toward subsidizing telephone service, a March 31, 2016, FCC Order transitions Lifeline toward subsidizing broadband service. While low-income nontribal households are eligible for a $9.25 per month subsidy, low-income households in tribal areas are eligible for a subsidy of $34.25 per month plus a one-time initiation of service discount of up to $100 for Link-Up support.

---

[29] FCC, "In the Matter of Connect America Fund Universal Service Reform—Mobility Fund," *Report and Order and Further Notice of Proposed Rulemaking*, WC Docket No. 10-90, FCC 17-11, adopted February 23, 2017, released March 7, 2017, pp. 13-17, available at https://apps.fcc.gov/edocs_public/attachmatch/FCC-17-11A1.pdf.

[30] FCC, "In the Matter of Connect America Fund," *Report and Order*, WC Docket No. 10-90, FCC 18-37, adopted March 27, 2018, released April 5, 2018, 19 pages, available at https://transition.fcc.gov/Daily_Releases/Daily_Business/2018/db0405/FCC-18-37A1.pdf.

[31] FCC, "In the Matter of Connect America Fund," *Report and Order, Third Order on Reconsideration, and Notice of Proposed Rulemaking*, WC Docket No. 10-90, FCC 18-29, adopted March 14, 2018, released March 23, 2018, paragraph 120, p. 47, available at https://ecfsapi.fcc.gov/file/0323601741 01/FCC-18-29A1.pdf.

[32] GAO, *Additional Coordination and Performance Measurement Needed for High-Speed Internet Access Programs on Tribal Lands*, p. 17.

[33] Ibid., p. 27.

On November 16, 2017, the FCC adopted an Order[34] "limiting enhanced Tribal Lifeline support—$25 monthly in addition to the standard $9.25 per household—to facilities-based providers," and "limiting enhanced Tribal support to rural areas and eliminating enhanced support in urban areas, where the additional $25 a month is not required to make service affordable or to promote deployment."[35]

Rural Health Care Program/Healthcare Connect Fund

The Rural Health Care Support Mechanism provides discounts to rural care providers for broadband connectivity. According to GAO, "[a]lthough the Healthcare Connect Fund does not specifically target tribal institutions, assistance may be provided to a service provider (or group of providers) that serve tribal lands."[36] The Healthcare Connect Fund provided $52 million in 2014, "a portion of which went to tribal lands."[37]

RUS Broadband Funding Programs

The Rural Utilities Service of the U.S. Department of Agriculture maintains a portfolio of telecommunications programs to finance broadband deployment and infrastructure in rural areas.[38] Since 2010, RUS has provided a total of $6.1 billion in loans and grants to build out broadband in rural areas.[39] Between 2010 and 2017, RUS invested approximately $512 million in telecommunications projects serving Tribal Lands, Tribal Organizations, American Indians, and Alaska Natives.[40]

RUS broadband programs include the Community Connect Grant Program, the Distance Learning and Telemedicine Grant Program, the Rural Broadband Access Loan and Loan Guarantee Program (Farm Bill Broadband Loans), and the Telecommunications Infrastructure Loan and Loan Guarantee Program. Additionally, a new broadband loan and grant pilot program—called the ReConnect Program—was established by the Consolidated Appropriations Act, 2018 (P.L. 115-141). Table 6 shows the total amounts awarded for each program and the amounts awarded to Tribal Lands, Tribal Organizations, American Indians, and Alaska Natives.

---

[34] See https://www.fcc.gov/document/fcc-action-transform-lifeline-program-low-income-americans.

[35] FCC, *FCC News Release*, "FCC Takes Major Steps to Transform Lifeline Program for Low-Income Americans," November 16, 2017, available at https://apps.fcc.gov/edocs_public/attachmatch/DOC-347792A1.pdf.

[36] GAO, *Additional Coordination and Performance Measurement Needed for High-Speed Internet Access Programs on Tribal Lands*, p. 17.

[37] Ibid.

[38] See CRS Report RL33816, *Broadband Loan and Grant Programs in the USDA's Rural Utilities Service*, by Lennard G. Kruger.

[39] Andrew Hayes, General Field Representative, USDA Rural Development, *Introduction to the USDA and Overview of Rural Utilities Service Programs*, power point presentation to the National Association of Regional Councils' 52 Annual Conference, Orlando, FL, June 4, 2018, p. 30, available at http://narc.org/wp-content/uploads/Mon_5-7_1015- 1130_Broadband_USDA_Hayes.pdf.

[40] Ibid., p. 31.

Table 6. RUS telecommunications awards since 2010 ($ millions)

Program	Funds Awarded, Tribal Areas[a]	Funds Awarded, Total
Telecommunications Infrastructure	246.5	2800
Farm Bill Broadband	25.3	205.7
Distance Learning and Telemedicine	42.4	208.7
Community Connect Grants	18.6	111.9
Broadband Initiatives Program (ARRA)	179.2	2900
Grand Total	512.0	6200

Source: USDA Rural Development, *Introduction to the USDA and Overview of Rural Utilities Service Programs*, June 4, 2018, pp. 30-31, available at http://narc.org/wp-content/uploads/Mon_5-7_1015-1130_Broadband _ USDA_Hayes.pdf.

[a]Includes both tribally-owned telecommunications providers and nontribal providers serving tribal areas.

Community Connect Grant Program

The Community Connect Program[41] provides grant money to applicants proposing to provide broadband on a "community-oriented connectivity" basis to currently unserved rural areas.

Federally recognized tribes are eligible to apply for Community Connect grants. According to GAO, between 2010 and 2017, four tribal entities received $13.5 million in Community Connect grants, comprising 11% of the total grant recipients over that period.[42]

In its 2018 report, GAO found that meeting application requirements was a barrier to tribes receiving Community Connect grants. Specific regulatory requirements that made it difficult for tribes to receive grant funding included preparing existing and proposed network design; demonstrating financial sustainability within five years; and obtaining matching funds.[43] GAO recommended that "the Secretary of Agriculture should direct the Administrator of RUS to undertake an assessment to identify any regulatory barriers that may unduly impede efforts by tribes to obtain RUS federal grant funds for broadband deployment on tribal lands and implement any steps necessary to address the identified barriers."[44]

Distance Learning and Telemedicine Program

Distance Learning and Telemedicine (DLT) grants[45] serve as initial capital assets for equipment (e.g., video conferencing equipment, computers) that operate via telecommunications to rural end-users of telemedicine and distance learning. Broadband facilities (if owned by the applicant) are also eligible. Federally recognized tribes are eligible to apply for DLT grants.

---

[41] For more information, see http://www.rd.usda.gov/programs-services/community-connect-grants.

[42] GAO, *Tribal Broadband: Few Partnerships Exist and the Rural Utilities Service Needs to Identify and Address Any Funding Barriers Tribes Face*, pp. 17-18.

[43] Ibid., pp. 20-21.

[44] Ibid., p. 23.

[45] For more information, see http://www.rd.usda.gov/programs-services/distance-learning-telemedicine-grants.

Rural Broadband Access Loan and Loan Guarantee Program

The Rural Broadband Access Loan and Loan Guarantee Program (also known as the Farm Bill Broadband Loan and Loan Guarantee Program)[46] provides loans for the costs of construction, improvement, or acquisition of facilities and equipment needed to provide broadband service in eligible rural areas. Indian tribes or tribal organizations are eligible to apply.

Telecommunications Infrastructure Loans and Loan Guarantee Program

The Telecommunications Infrastructure Loan and Loan Guarantee Program[47] provides loans and loan guarantees for the construction, maintenance, improvement, and expansion of telephone service and broadband in rural areas. The program was first authorized in 1949 to finance rural telephone service. Since 1995, RUS has required that networks funded by this program offer broadband service as well. Federally recognized tribes are eligible for these loans and loan guarantees.

Substantially Underserved Trust Areas (SUTA)

The 2008 Farm Bill directed USDA to establish an initiative to identify and improve the availability of loan programs for communities in substantially underserved trust areas.[48] Section 6105 of the Food, Conservation, and Energy Act of 2008 (P.L. 110-234) authorized RUS to make loans and guarantee loans with interest rates as low as 2% and with extended repayment terms; waive nonduplication restrictions,[49] matching fund requirements, or credit support requirements[50] to facilitate construction, acquisition, or improvements of infrastructure; and give highest priority to designated projects in substantially underserved trust areas. The Final Rule, developed in consultation with tribal communities and governments, was released on June 13, 2012 (7 C.F.R. 1700 Subpart D). The SUTA rules apply to the Rural Broadband Access Loan and Loan Guarantee Program and the Telecommunications Infrastructure Loan and Loan Guarantee Program; the rule does not apply to the Community Connect Grant Program or the Distance Learning and Telemedicine Grant Program. Recent examples of Telecommunications Infrastructure loans financed using SUTA are Mescalero Apache Telecommunications (NM): $5.4 million (FY2015) and Sacred Wind Communications (NM): $13.8 million (FY2016).

---

[46] For more information, see http://www.rd.usda.gov/programs-services/farm-bill-broadband-loans-loan-guarantees.

[47] For more information, see http://www.rd.usda.gov/programs-services/telecommunications-infrastructure-loans-loan- guarantees.

[48] For more information, see http://www.rd.usda.gov/about-rd/initiatives/substantially-underserved-trust-area-suta.

[49] Nonduplication generally means a restriction on financing projects for services in a geographic area where reasonably adequate service already exists as defined by the applicable program.

[50] Credit support means equity, cash requirements, letters of credit, and other financial commitments provided in support of a loan or loan guarantee.

ReConnect Program

The Consolidated Appropriations Act, 2018 (P.L. 115-141) appropriated $600 million to RUS to conduct a new broadband loan and grant pilot program. Projects in rural areas without sufficient access to broadband are eligible, specifically those rural areas where at least 90% of the households to be served by a project do not have sufficient access to broadband, defined as 10 Mbps downstream, and 1 Mbps upstream.

On December 14, 2018, RUS released the *Funding Opportunity Announcement* (FOA) and solicitation of applications for the new pilot broadband loan and grant program, which is named the "ReConnect Program."[51] As set forth in the statute, at least 90% of the households to be served by a project receiving a loan or grant under the pilot program shall be in a rural area without sufficient access to broadband at a minimum speed of 10 Mbps/1 Mbps. RUS defines "sufficient access to broadband" as any rural area that has fixed, terrestrial broadband service delivering at least 10 Mbps downstream and 1 Mbps upstream. Mobile and satellite service will not be considered in making the determination that households in the proposed funded service area do not have sufficient access to broadband.

Approximately $560 million has been set aside for funding opportunities under the FOA, with additional budget authority available for a reserve which may be used for additional loans or grants. While there are no specific carve-outs for projects serving tribal areas, the FOA addresses tribal broadband in the following ways:

- Entities eligible for awards include "an Indian tribe" (as defined in section 4 of the Indian Self-Determination and Education Assistance Act (25 U.S.C. 450b)). (Other eligible entities are: states or local governments, U.S. territories, nonprofit entities, for-profit corporations, limited liability companies, and cooperative or mutual organizations.)
- If service is being proposed on tribal land, a certification is required from the proper tribal official that they are in support of the project and will allow construction to take place on tribal land. If no certification is provided, then this area will be ineligible for funding.
- With respect to ranking applications by priority, for applications where at least 50% of the geographical area of the proposed funded service area is to provide service on tribal lands, 5 points shall be awarded. (Up to 150 priority points can be awarded, taking account of such factors as rurality, state broadband activity, and number of critical community facilities served.)

The ReConnect Program consists of three funding categories:

---

[51] U.S. Department of Agriculture, Rural Utilities Service, "Broadband Pilot Program," *Federal Register*, vol. 83, no. 240, December 14, 2018, pp. 64315-64325, available at https://www.govinfo.gov/content/pkg/FR-2018-12- 14/pdf/2018-27038.pdf.

100% Loan

- Up to $200 million is available.
- The maximum amount that can be requested is $50 million.
- Interest rate is set at a fixed 2%.
- Eligible areas are where 90% of households do not have sufficient access to broadband at 10 Mbps/1 Mbps.
- Applicants must propose to build a network capable of providing service to every premise in the proposed funded service area at a minimum speed of 25 Mbps/3 Mbps.
- Applications accepted on a rolling basis through June 28, 2019.

50% Loan/50% Grant Combination

- Up to $200 million is available.
- The maximum amount that can be requested is $25 million for the loan and $25 million for the grant. Loan and grant amounts will always be equal.
- Interest rate for the loan will be set at the Treasury rate.
- Eligible areas are where 90% of households do not have sufficient access to broadband at 10 Mbps/1 Mbps.
- Applicants must propose to build a network capable of providing service to every premise in the proposed funded service area at a minimum speed of 25 Mbps/3 Mbps
- Applications accepted on a rolling basis through May 29, 2019.

100% Grant

- Up to $200 million is available.
- The maximum amount that can be requested is $25 million.
- Applicants must provide a matching contribution equal to 25% of the cost of the overall project.
- Eligible areas are where 100% of households do not have sufficient access to broadband at 10 Mbps/1 Mbps.
- Applicants must propose to build a network capable of providing service to every premise in the proposed funded service area at a minimum speed of 25 Mbps/3 Mbps.
- Applications accepted on a rolling basis through April 29, 2019.

According to RUS, the online application portal will be open in late February.

Stimulus Broadband Grants and Loans

Broadband provisions of the American Recovery and Reinvestment Act (ARRA, P.L. 111-5) provided a total of $6.9 billion for broadband grants, loans, and loan/grant combinations. The total consisted of $4.4 billion to NTIA/DOC for a newly established Broadband Technology Opportunities Program (BTOP grants) and $2.5 billion to the RUS/USDA Broadband Initiatives Program (BIP grants, loans, and grant/loan combinations).[52] In 2009 and 2010, NTIA awarded funding for 233 projects and RUS awarded funding for 297 broadband infrastructure projects.[53] Virtually all projects are now completed and closed; no new funding is available.

While there was no set-aside for tribal broadband, a number of ARRA broadband awards were made to tribal entities or providers serving tribal lands. According to RUS, awarded BIP projects overlapped with 31 tribal lands, and nine awards were made to Indian Tribes.[54] Out of a total of $2.9 billion awarded for BIP, $179.2 million was awarded to projects serving Tribal Lands, Tribal Organizations, American Indians, and Alaska Natives. According to NTIA, six tribal authorities received BTOP grants and at least 65 BTOP projects will directly benefit tribal communities.[55]

The September 2018 GAO report, *Tribal Broadband: Few Partnerships Exist and the Rural Utilities Service Needs to Identify and Address Any Funding Barriers Tribes Face*, looked at partnership arrangements between tribes and other entities to increase broadband deployment on tribal lands. GAO found that while it could find no partnerships leveraged from funding by current federal grant programs (CAF or Community Connect), there were seven partnerships identified that had been funded by BIP or BTOP.[56]

Other Federal Funding Programs

Aside from the programs listed above, the NTIA report, *Guide to Federal Funding of Broadband Projects*, cites several other federal funding programs as specific to tribal broadband.

---

[52] For information on broadband stimulus programs, see CRS Report R41775, *Background and Issues for Congressional Oversight of ARRA Broadband Awards*, by Lennard G. Kruger.

[53] A small portion of these project awards were ultimately rescinded; see ibid., p. 5.

[54] U.S. Department of Agriculture, Broadband Initiatives Program, Awards Report, *Advancing Broadband: A Foundation for Strong Rural Communities*, January 2011, p. 3, available at http://www.rd.usda.gov/files/reports/RBBreportV5ForWeb.pdf.

[55] Department of Commerce, National Telecommunications and Information Administration, *The Broadband Technology Opportunities Program: Expanding Broadband Access and Adoption in Communities Across America, Overview of Grant Awards*, December 14, 2010, p. 16, available at http://www.ntia.doc.gov/reports/2010/NTIA_Report_on_BTOP_12142010.pdf.

[56] GAO, *Tribal Broadband: Few Partnerships Exist and the Rural Utilities Service Needs to Identify and Address Any Funding Barriers Tribes Face*, p. 9.

The Department of Housing and Urban Development (HUD) contains an Office of Native American Programs (ONAP). According to NTIA, ONAP has three programs that could potentially be used to fund broadband projects

- Indian Community Development Block Grant (ICDBG)—"The ICDBG program provides funds to eligible grantees for housing rehabilitation, land acquisition, community facilities, infrastructure construction and economic development activities that benefit primarily low- and moderate-income persons."[57]
- Indian Housing Block Grant (IHBG)—"Eligible activities include housing development, assistance to housing developed under the Indian Housing Program, housing services to eligible families and individuals, crime prevention, safety and model activities that provide creative approaches to solving affordable housing problems."[58] There is also a Native Hawaiian Housing Block Grant program.
- Tribal Housing Activities Loan Guarantee Program (Title VI)—the program "assists IHBG recipients (borrowers) who want to finance eligible affordable housing activities but are unable to secure financing without the assistance of a federal guarantee."[59]

Another broadband-related source of funding specifically targeted to Native Americans is the Native American and Native Hawaiian Library Services Grant programs at the Office of Library Services, Institute of Museum and Library Services. Programs include Native American Library Services Basic Grants, Native American Library Services Enhancement Grants, and Native Hawaiian Library Services Grants.[60]

An additional source of federal funding potentially available for tribal broadband is the Economic Development Assistance programs at the Economic Development Administration (EDA) in the Department of Commerce. For example, in September 2017, EDA announced an Economic Adjustment Assistance project award of $144,000 to the Confederated Tribes of the Umatilla Indian Reservation in Pendleton, OR, to support the development of a broadband fiber optics network near Pendleton to be located on the Umatilla Reservation.

---

[57] *BroadbandUSA: Guide to Federal Funding of Broadband Projects,* p. 24. As an example, in 2005 the Coquille Tribe of Oregon received an ICDBG grant of $421,354 for broadband infrastructure deployment (see https://www.ntia.doc.gov/files/ntia/publications/broadband_fed_funding_guide.pdf, p. 18).

[58] Ibid., p. 25.

[59] Ibid., p. 26.

[60] Ibid., pp. 35-36.

BroadbandUSA

While not providing funding, BroadbandUSA at the NTIA offers one-to-one technical assistance to communities (including tribal communities) seeking to plan and implement broadband initiatives.[61] BroadbandUSA also organizes regional events and workshops bringing together broadband stakeholders and publishes guides and tools (including toolkits for local and tribal governments) that can serve as resources for communities seeking to launch broadband initiatives.[62]

Infrastructure Initiative and Broadband

On February 12, 2018, the Trump Administration released its *Legislative Outline for Rebuilding Infrastructure in America*.[63] The plan does not dedicate any funding exclusively for broadband, but does include rural broadband among the types of infrastructure projects that would be eligible for funding. Proposed funding streams include $50 billion for a Rural Infrastructure Program, $20 billion for a Transformative Projects Program, $14 billion for expanding existing federal credit programs that address infrastructure, and $6 billion for expanding the scope of Public Activity Bonds (PABs). It will be up to Congress to determine the extent to which the Administration infrastructure proposal will be implemented, and how an infrastructure initiative will be legislated.

ACTIVITIES IN THE 114TH CONGRESS

On April 27, 2016, an amendment to S. 2644 (the FCC Authorization Act of 2016) offered by Senator Cantwell and adopted by the Senate Committee on Commerce, Science, and Transportation, would have directed the FCC to develop metrics to measure the impact of universal service support on tribal lands and would have required the FCC to prepare a biennial report to Congress on the impact of universal service support on tribes and tribal lands. S. 2644 was reported by the Senate Committee on Commerce, Science, and Transportation on September 20, 2016 (S.Rept. 114-355) but was not enacted by the 114th Congress.

Meanwhile, on April 27, 2016, the Senate Committee on Indian Affairs held an oversight hearing on the GAO report, *Telecommunications: Additional Coordination and*

---

[61] For more information on the types of technical assistance BroadbandUSA offers, see http://www2.ntia.doc.gov/technical_assistance.

[62] See http://www2.ntia.doc.gov/publications.

[63] The White House, *Legislative Outline for Rebuilding Infrastructure in America*, released February 12, 2018, 53 pages, available at https://www.whitehouse.gov/wp-content/uploads/2018/02/INFRASTRUCTURE-211.pdf.

Performance Measurement Needed for High-Speed Internet Access Programs on Tribal Lands. Testimony was heard from the RUS, FCC, GAO, and private witnesses.[64]

LEGISLATION IN THE 115TH CONGRESS

The Consolidated Appropriations Act, 2018 (P.L. 115-141) appropriated $600 million to RUS to conduct a new broadband loan and grant pilot program. While no funding was specifically set aside for tribal broadband, projects in rural tribal areas would be eligible for funding. Also in P.L. 115-141, Section 508 of Division P required the FCC to submit a report to Congress evaluating broadband coverage in Indian country and on land held by a Native Corporation pursuant to the Alaska Native Claims Settlement Act, with the FCC required to complete a proceeding to address the unserved areas identified in the report.

Other introduced legislation specifically related to tribal broadband includes the following:

- H.R. 800 (Huffman), introduced on February 1, 2017, as the New Deal Rural Broadband Act of 2017, includes a provision (§4) that would have established a Tribal Broadband Assistance Program. The Department of Agriculture would have been authorized to make grants, loans, or loan guarantees to entities to (1) plan, construct, acquire, or improve facilities or equipment for the purpose of providing broadband service on tribal lands; (2) provide broadband service on tribal lands; (3) develop among tribal members technical expertise related to broadband service; and (4) improve the adoption of broadband service by individuals on tribal lands. The bill would have authorized $25 million for each of fiscal years 2017 through 2022. Referred to the Committee on Agriculture, and in addition to the Committees on Natural Resources and Energy and Commerce.
- H.R. 1581 (Ruiz), introduced on March 16, 2017, as the Tribal Digital Access Act of 2017, would have amended the Communications Act of 1934 to add access to telecommunications and information services in Indian country and areas with high populations of Indian people to the universal service principle relating to access to such services in rural, insular, and high cost areas. Referred to the Committee on Energy and Commerce.
- H.R. 3268 (Aderholt), the FY2018 Agriculture Appropriations bill, funds Community Connect grants and broadband loans, and is accompanied by the

---

64 Testimony is available at http://www.indian.senate.gov/hearing/oversight-hearing-gao-report-telecommunications- additional-coordination-and-performance.

House Appropriations Committee report (H.Rept. 115-232), which includes the following:

> Tribal Communities.—The Committee notes that tribal communities continue to struggle with gaining access to broadband service. The Committee encourages the Secretary to provide a report that identifies the specific challenges Indian Tribal Organizations (ITOs) have in gaining access to broadband service and provide a plan for addressing these challenges, including how the Community Connect program can assist ITOs.

- H.R. 4506 (Torres), introduced on November 30, 2017, as the Jobs for Tribes Act, would have directed GAO to conduct a study assessing a range of federal programs (including broadband and telecommunications programs) available to assist Indian communities with business and economic development. Referred to the Committees on Natural Resources; Foreign Affairs; and Education and the Workforce.
- H.R. 5007 (Ruiz), introduced on February 13, 2018, would have directed the FCC to submit to Congress a report evaluating broadband coverage in Indian country and on land held by a Native Corporation and to complete a proceeding to address the unserved areas identified in the report. Referred to the Committee on Energy and Commerce.
- H.R. 5172 (O'Halleran), introduced on March 6, 2018, would have assisted Indian tribes in maintaining, expanding, and deploying broadband systems. Referred to the Committee on Agriculture, and in addition to the Committee on Energy and Commerce.
- H.R. 5961 (Aderholt), introduced on May 24, 2018, as the FY2019 Agriculture Appropriations bill, funds Community Connect grants and broadband loans, and is accompanied by the House Appropriations Committee report (H.Rept. 115-706), which includes the following:

> Tribal Communities.—The Committee notes that tribal communities continue to struggle with gaining access to broadband service. The Committee encourages the Secretary to provide a report that identifies the specific challenges Indian Tribal Organizations (ITOs) have in gaining access to broadband service and provide a plan for addressing these challenges, including how the Community Connect program can assist ITOs.

- S. 1116 (Hoeven), introduced on May 11, 2017, as the Indian Community Economic Enhancement Act of 2017, would have directed GAO to conduct a study assessing a range of federal programs (including broadband and telecommunications programs) available to assist Indian communities with business and economic development. Referred to the Committee on Senate Indian Affairs; reported by committee on October 17, 2017 (S.Rept. 115-174).

- S. 2205 (Heinrich), introduced on December 7, 2017, as the Tribal Connect Act of 2017, would have improved access by Indian tribes to support from the Schools and Libraries Universal Service Support program (E-rate) of the Federal Communications Commission. Referred to the Committee on Indian Affairs.

CONCLUSION

With respect to broadband and telecommunications access and adoption, tribal areas and communities continue to lag behind other areas and segments of American society. Many contend that without federal assistance, tribal lands will continue to be on the wrong side of the digital divide. At issue is what role the federal government can play to most effectively and efficiently support broadband deployment on tribal lands.

Aside from providing funding for broadband deployment, the federal government has pursued other policies relevant to tribal broadband. These include mechanisms for effective coordination and consultation with tribes on broadband issues,[65] spectrum policies to promote wireless broadband deployment on tribal lands,[66] addressing permitting and environmental review issues for deploying broadband infrastructure on tribal lands,[67] and rights-of-way policies to enable broadband infrastructure deployment on public lands.[68] On January 31, 2017, FCC Chairman Ajit Pai announced the formation of a Broadband Deployment Advisory Committee, which is developing advice and recommendations to the FCC on how to accelerate the deployment of broadband by reducing and/or removing regulatory barriers to infrastructure investment.[69]

---

[65] The FCC's Office of Native Affairs and Policy (ONAP) was established in 2010 and was charged with "ensuring robust government-to-government consultation with Federally recognized tribal governments and other native organizations; working with Commissioners, Bureaus, and Offices, as well as with other government agencies and private organizations, to develop and implement policies for assisting native communities; and ensuring that Native concerns and voices are considered in all relevant Commission proceedings and initiatives." FCC, *In the Matter of Establishment of the Office of Native Affairs and Policy in the Consumer and Governmental Affairs Bureau*, Order, FCC 10-141, released August 12, 2010, p. 1, available at https://apps.fcc.gov/edocs_public/attachmatch/FCC-10- 141A1.pdf. Subsequently in 2011, the FCC-Native Nations Broadband Task Force was established, see https://apps.fcc.gov/edocs_public/attachmatch/DA-14-1558A1.pdf.

[66] See, for example, FCC Tribal Lands Bidding Credit Program, http://wireless.fcc.gov/auctions/default.htm?job=tribal_bidding&page=1.

[67] See, for example, https://www.fcc.gov/document/fcc-acts-speed-deployment-next-gen-wireless-infrastructure-0. Also see https://www.fcc.gov/document/fcc-streamlines-requirements-utility-pole-replacements-0.

[68] See FCC, "In the Matter of Acceleration of Broadband Deployment: Expanding the Reach and Reducing the Cost of Broadband Deployment by Improving Policies Regarding Public Rights of Way and Wireless Facilities Siting," Notice of Inquiry, WC Docket No. 11-59, FCC 11-51, April 7, 2011. Also see FCC, "In the Matter of Acceleration of Broadband Deployment by Improving Wireless Facilities Siting Policies," *Report and Order*, FCC 14-153, October 21, 2014.

[69] See https://www.fcc.gov/broadband-deployment-advisory-committee.

Regarding funding, debate has centered on whether federal funding for tribal broadband is sufficient, and the extent to which portions of federal funds available for broadband generally should be specifically targeted for tribal broadband. The 2010 National Broadband Plan (NBP) found that "[t]ribes need substantially greater financial support than is presently available to them, and accelerating tribal broadband deployment will require increased funding."[70] The NBP recommended that Congress establish a Tribal Broadband Fund, which would be administered by NTIA in consultation with the FCC and the Bureau of Indian Affairs. To date, no legislation has been enacted by Congress that would specifically establish a Tribal Broadband Fund.

Currently, the largest overall source of federal funding for telecommunications services is the FCC's Universal Service Fund programs. As these programs transition toward a broadband- centric orientation (e.g., the Connect America Fund), the issue for tribal broadband is how this transition will affect broadband funding to tribal lands, and to what extent these programs might be configured toward addressing the relatively low levels of broadband deployment and adoption in tribal lands.[71] In the 116th Congress, notwithstanding whether federal broadband funding programs target tribal lands, whether or not tribal lands will receive additional funding for broadband will likely be determined by the ongoing trajectory of overall federal funding for broadband.

---

[70] *Connecting America: The National Broadband Plan,* p. 152.

[71] See, for example, the proposal of the National Tribal Telecommunications Association for a "Tribal Broadband Factor" as part of USF reform. National Tribal Telecommunications Association, Ex Parte Communication to the FCC, *In the Matter of Connect America Fund, WC Docket No. 10-90; NTTA Proposal for a Tribal Broadband Factor*, June 19, 2015, available at https://prodnet.www.neca.org/publicationsdocs/wwpdf/62215ntta.pdf. There is also concern that parts of the CAF transition could reduce tribal broadband funding; see NTTA Ex Parte comments filed on February 23, 2016, available at http://apps.fcc.gov/ecfs/document/view?id=60001516284.

INDEX

C

D

E

F

G

H

I

J

L

M

N

O

P

Q

R

S

T

U

V

W